Heavy Gas
and Risk Assessment - II

*Proceedings of the Second Symposium on
Heavy Gases and Risk Assessment,
Frankfurt am Main, May 25–26, 1982*

edited by

SYLVIUS HARTWIG
Battelle-Institut e. V., Frankfurt am Main

D. Reidel Publishing Company

A MEMBER OF THE KLUWER ACADEMIC PUBLISHERS GROUP

Dordrecht / Boston / Lancaster

The Symposium was sponsored by Battelle-Institut e.V.

Library of Congress Cataloging in Publication Data

Symposium on Heavy Gas and Risk Assessment (2nd : 1982 : Frankfurt
    am Main, Germany)
    Heavy Gas and Risk Assessment II.

    "The Symposium was sponsored by Battelle-Institut e.V."
    Includes index.
    1. Gases, Asphyxiating and poisonous–Environmental aspects–
Congresses. 2. Air–Pollution–Congresses. 3. Atmospheric
diffusion–Congresses. I. Hartwig, Sylvius, 1938–   . II. Battelle-
Institut. III. Title.
TD885.S953   1982   604.7   83–4563
ISBN-13: 978-94-009-7153-0   e-ISBN-13: 978-94-009-7151-6
DOI: 10.1007/ 978-94-009-7151-6

Published by D. Reidel Publishing Company
P.O. Box 17, 3300 AA Dordrecht, Holland

Sold and distributed in the U.S.A. and Canada
by Kluwer Boston Inc.,
190 Old Derby Street, Hingham, MA 02043, U.S.A.

In all other countries, sold and distributed
by Kluwer Academic Publishers Group,
P.O. Box 322, 3300 AH Dordrecht, Holland

Table of Contents

TABLE OF CONTENTS

PREFACE

This book comprises the proceedings of the second symposium on heavy
gases and risk assessment held in May 1982 at the Battelle-Institut
e.V., Frankfurt am Main. The symposium was a sequel to a meeting on the
same subject held in September 1979. The second   symposium consisted
of four sessions. It was not always possible to definitively separate
the different sessions, as some of the papers discussed topics that
overlapped other sessions. The first section was concerned with
heavy gas dispersion and the associated aspects of modelling. The
Chairman for this session was Prof. M. Riethmüller from the Karman
Institute, Brüssels.

The second session was devoted to experimental aspects and topics related to heavy gas dispersion. The chairman for this session was Prof.
J. Havens.

The third session dealt with the field of explosion and fire.
The Chairman of this session was Dr. J. Pankrath of the Umweltbundesamt.

In the fourth session, some aspects of risk assessment in relation
to heavy gases was discussed. In every session two periods were set
aside for discussion, so that we had the chance, to settle some of
the controversies that arose.

I wish to express my appreciation to all those scientists who respon-
ded to our call for papers. Unfortunately, we were not able to in-

clude all the papers because of the limited period of time that we

had for the two-day symposium.

Organizing a symposium and publishing the proceedings carries with

it a considerable amount of hard work. My colleagues at the Battelle-

Institut e.V., W. Heudorfer, J. Kirsch and G. Schnatz were a steady

help during the organization of this symposium. I would like to ex-

press my gratitude to them.

I am also very much obliged to Mrs. R. Kra (Yale University) who

assisted me in the editorial work of this publication.

                                        Sylvius Hartwig

OPEN AND CONTROVERSIAL TOPICS IN HEAVY GAS DISPERSION AND RELATED

RISK ASSESSMENT PROBLEMS

S. Hartwig, Battelle-Institut e.V., Frankfurt am Main and

Universität Wuppertal

I would like to touch briefly on four areas, namely, connection of
dispersion to risk assessment, a very brief comment on the HSE trials,
improvements in the field since the last symposium and finally I
would like to emphasize three special topics on heavy gas modelling.
As mentioned previously, compared with the last symposium the additional session on risk assessment and consequence analysis testifies
to its growing importance. There are two reasons for this: On the one
hand, the general impetus for the examination of the reaction of
heavy gases is caused by their increasing industrial and general use
and the risks therein involved. The increase of the consumption is
in turn related to the increasing use of natural gas (LNG) and growing
industrial production as well as to our own consumption, as can be
seen in fig. 1.

On the other hand, the possible accidental releases and the risks
involved determine the special methods and models required for reliable assessments and predictions of possible consequences. A good
example is jet release.

When considering the hazard potential of the different heavy gases,

1

S. Hartwig (ed.), Heavy Gas and Risk Assessment - II, 1–25.
Copyright © 1983 by Battelle-Institut e.V., Frankfurt am Main, Germany.

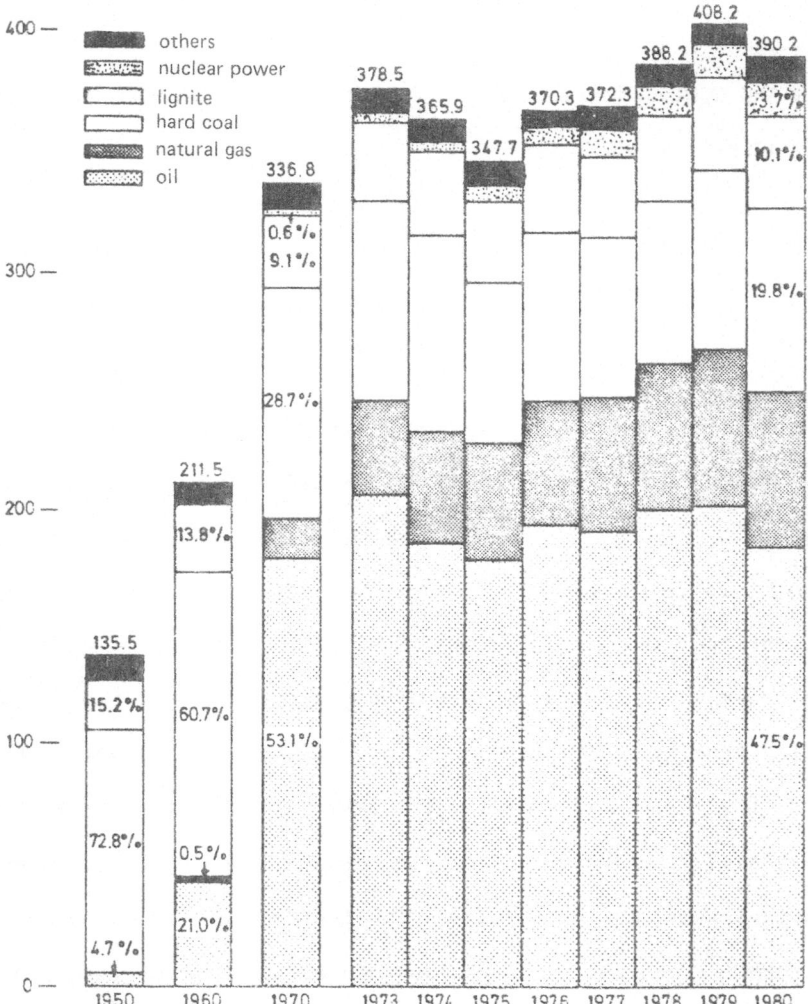

nach Bundesverband der deutschen Gas- und Wasserwirtschaft e.V. 1981

FIG.1: WEST GERMAN PRIMARY ENERGY DEMAND BY ENERGIES (MILLION TCE)

	natural disaster		aircraft		ship		rail		other		fire		all	
	I	II	I	II	I	II	I	II	I	II	I	II	I	II
Consequences	13781	479	650	161	1575	342	3082		1023	113	171	743	20282	1838
Number of Events	39	7	15	12	10	11	14		20	3	3	20	101	53

I = fatalities
II = material damage in 10^6 US $

statistic is based on reports of "Sigma der Schweizer Rück, 1982"

FIG. 2: INCOMPLETE STATISTIC OF REPORTED MAJOR DISASTERS FOR 1981

their importance and status should be seen in comparison with other

hazards. In explanation fig. 2 gives an overall view (but probably in-

complete) of the important major disaster that occurred in 1981 which

have caught the attention of the public. Figure 2 shows that, of the

more than 150 catastrophies that became public in 1981, none could be

related to the toxic effect of heavy gases. One or two perhaps, or

or none were related to the inflammability of heavy gases. In other

words catastrophies do seem in our field of interest, but not as often

as we may be led to believe. The reason for the uncertainty of evi-

dence in fig. 2 for inflammable gases is that the information on hand

may be incomplete. Under the category "fire", there are two fires on

tanks, 2 fires in refineries, one fire in an ammonia factory (Bruns-

büttel, West Germany, 21. Nov. 1981 - 10 mil. damage) and a gas ex-

plosion in a lignite mine. In these accidents, the participation of

heavy gases cannot be totally excluded.

Nevertheless, it is important to note that such an outline as is given

in fig. 2 does not represent the risk of one technique compared to others.

First, the data for one year are naturally an unreliable representa-

tion, second, the basis of comparison for the number of operations

without accidents is not given, and third, the risk of a technique is

the much larger spectrum of small accidents. Nevertheless such a sum-

mary gives some indication for the importance of a problem and its ranking.

For example, figure 3 shows the place of value of appreciated risks

in comparison with the LNG-Trucking operation for the United States.

I do wish to mention that the value and the uncertainty limit in this

figure should not be taken too literally. What are the current priori-

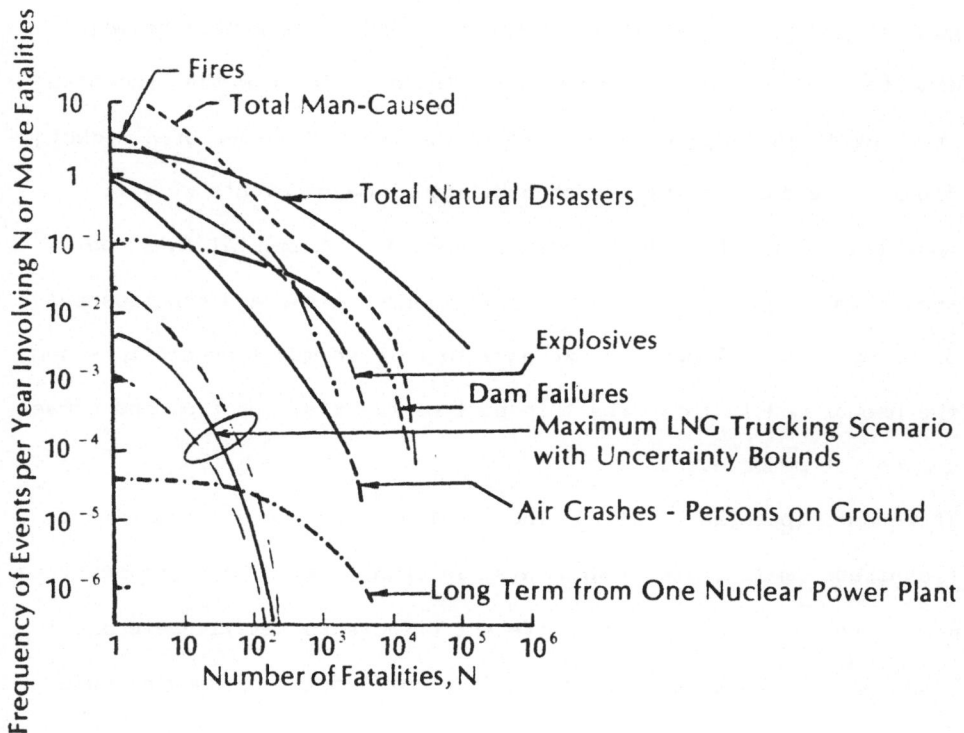

SOURCES: REACTOR SAFETY STUDY, WASH-1400, ADL RISK ANALYSIS OF LNG TRUCKING

FIG. 3: COMPARISON OF MULTI-FATALITY ACCIDENT PROBABILITIES

ties in research between heavy gas dispersion and risk assessment?
First of all, we know that, in most techniques, the risk is determined by a large number of small events. Secondly, we know that the
general public mainly perceives the risks and consequences of major
disasters in most of the techniques. Thirdly, we know that acceptability again is largely influenced by the major disaster of a technique.
Finally, because of the low probability of major events we will,
most likely, not be able to collect enough experimental data and
knowledge to quantify, in a reliable manner, risks and consequences
of major events. Therefore, we have to rely on model predictions and
the use of models to assess this part of the risk spectrum and consequences of major events.

If we anticipate and scale up the results of our models from small
events and small-scale experiments, we cannot use highly parametrized
models because the value of different effects for small-scale and
large-scale events may not only differ completely, but may produce
entirely unexpected results.

Therefore I suggest that we understand the physics as comprehensively
and conclusevely as is feasible. I consider this to be the absolute
requirement for the concept of risk analysis for the modelling of
heavy-gas dispersion which is in direct opposition to that which part of
the Petro-chemical Industry is doing.

Now I would like to point out how far our knowledge has come since the
last symposium in 1979: In the last few years both wind tunnel and
large-scale heavy gas experiments have been made. You will hear reports on most of them at this symposium, except, for the HSE experi-

ments which have been planned for this year. Therefore I will talk

briefly about those experiments.

The HSE experiments are the only large-scale experiments on an

isothermal heavy gas, not a cold gas, planned or undertaken until now.

The test site will be Thorney Island and the trials are called ac-

cordingly. In a first phase, 5 releases are planned with a fixed

density ratio of 2.0 with refrigerant 12, a fixed roughness corres-

ponding to rough pasture (i.e., 10 mm to 20 mm), a fixed quantity of

dense gas of 2000 m^3, a fixed source configuration, and a fixed ground

slope. The five releases are planned under the following atmospheric

conditions:

1. a release at low wind speed under conditions of high stability;

2. a release at moderate wind speed under neutrally stable conditions;

3. a release at high wind speed under neutrally stable conditions;

4. a release at moderate wind speed under moderately unstable con-

    ditions and

5. a passive, or neutrally-buoyant, release under the same conditions

    as in 4.

The estimated cost of the Thorney Island trials is just over

£ 1,000,000. A total of approximatelly £ 1,125,000 has been con-

tributed by 31 organisations (including HSE) distributed over the

USA, France, Germany, Italy, Sweden, Norway, Holland, the UK and the

European Community. In an extended programme, up to 15 additional

trials are planned to provide a better definition of the effects to

be investigated in a basic programme.

What are the developments since the last symposium? At the previous

heavy gas symposium, I discussed a number of problems which are not
yet clarified or still involve controversy. Some questions have not
been solved, whereas some progress has been made in others. I count
among the questions in which I have not seen convincing progress in
the last few years, the following: 1. 2-phase flow models, 2. self-
generated turbulence by release and simultaneous entrainment,
3. convection in seawater and time dependencies of heat sources,
4. coupling of different processes. However, improvements have been
made in the comprehension of 1. the change of the boundary layer
turbulence by the heavy gas cloud, and 2. the stratification by the
cloud. But these advances in our comprehension of the course of events
have not as yet - as far as I know- found expression in improved
models. I would like to enlarge on this here. Two major directions in
heavy-gas dispersion modelling have become apparent in the past few
years. The first of these attempts to improve box-models; the second
constructs and improves numeric models. The supporters of box-models
express the opinion that it is often not justified to use numeric
models for applications and calculations because they need much computer time and are very expensive. (J.Mc Quaid, 1982). Besides, considering current knowledge, it would not be justified to use numeric
models because the demands on the quality of data is high and the
data are often not available. On the other hand it must be recognized
that box-models are very highly parameterized and they cannot describe facts properly. Thus, the development of numeric models is
encouraged in many of institutions, both in Europe and in the USA.
In the USA there seems to be a whole family of daughter models of

the original SAI - Sigmet-Model. In a previous paper (Hartwig, 1979)
I showed, in an overview, several possibilities of release and the
following dispersion phases. Here I prefer to emphasize the demands
on the models that result from the discussions of the last two years.
Three points seem to be of importance:

1. As the heavy gas cloud most probably will alter the wind and
   diffusion profile, the exchange between and above the vapour cloud
   will be altered in a type of feed-back as well. Therefore, a better
   model, if any, of the perturbated profile is required.
2. Release of a heavy gas cloud normally does not occur in flat and
   unstructured terrain. It is more likely that orography, areas
   of factory sites or larger, roughness elements will play a more
   important role. There is growing need to model these factors.
3. Finally,there is the question of differences in the dispersion
   of cold and heavy gas clouds because cold gas clouds may induce
   further turbulence by ground heat which leads to a quicker dispersion and dilution.

Apart from the principal scientific interest, the main goal of all
efforts is to come to a more realistic conclusion about consequences
for example, about necessary safety distances to other industrial sites
in accidental release of heavy gas clouds.

The Turbulence Spectrum

The turbulence spectrum of the lower atmospheric boundary layer consists of the thermal part of turbulence and the mechanical part of the
turbulence which is produced by roughness elements. This can be seen,

S. HARTWIG

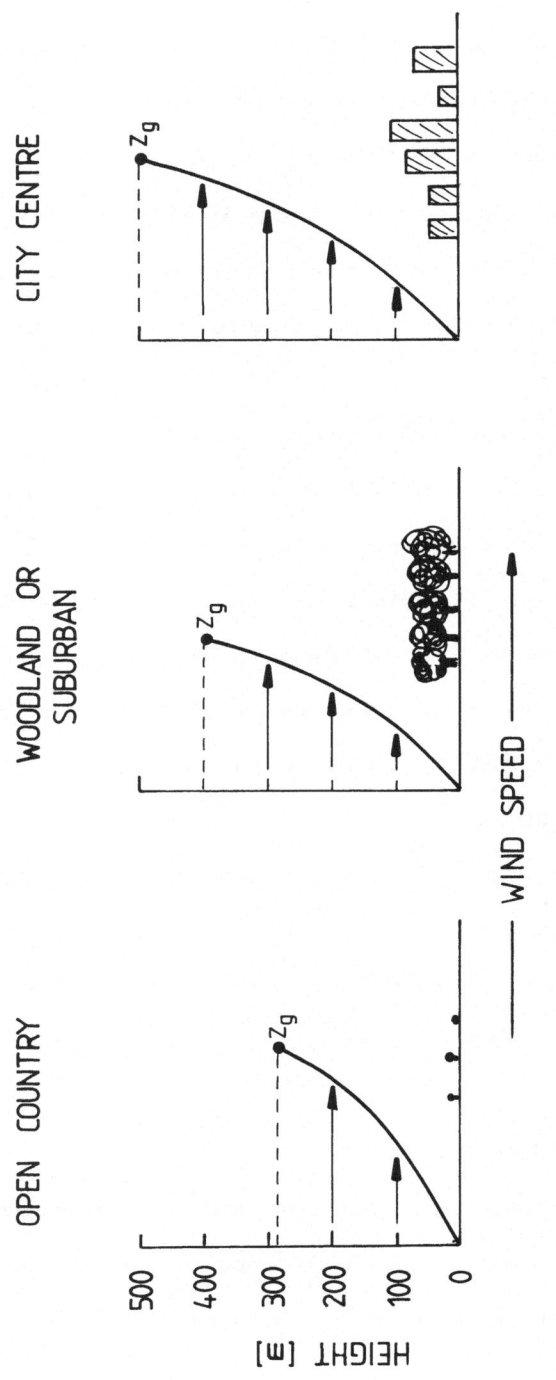

FIG. 4: CHANGE OF WINDPROFILE DUE TO ROUGHNESS ELEMENTS

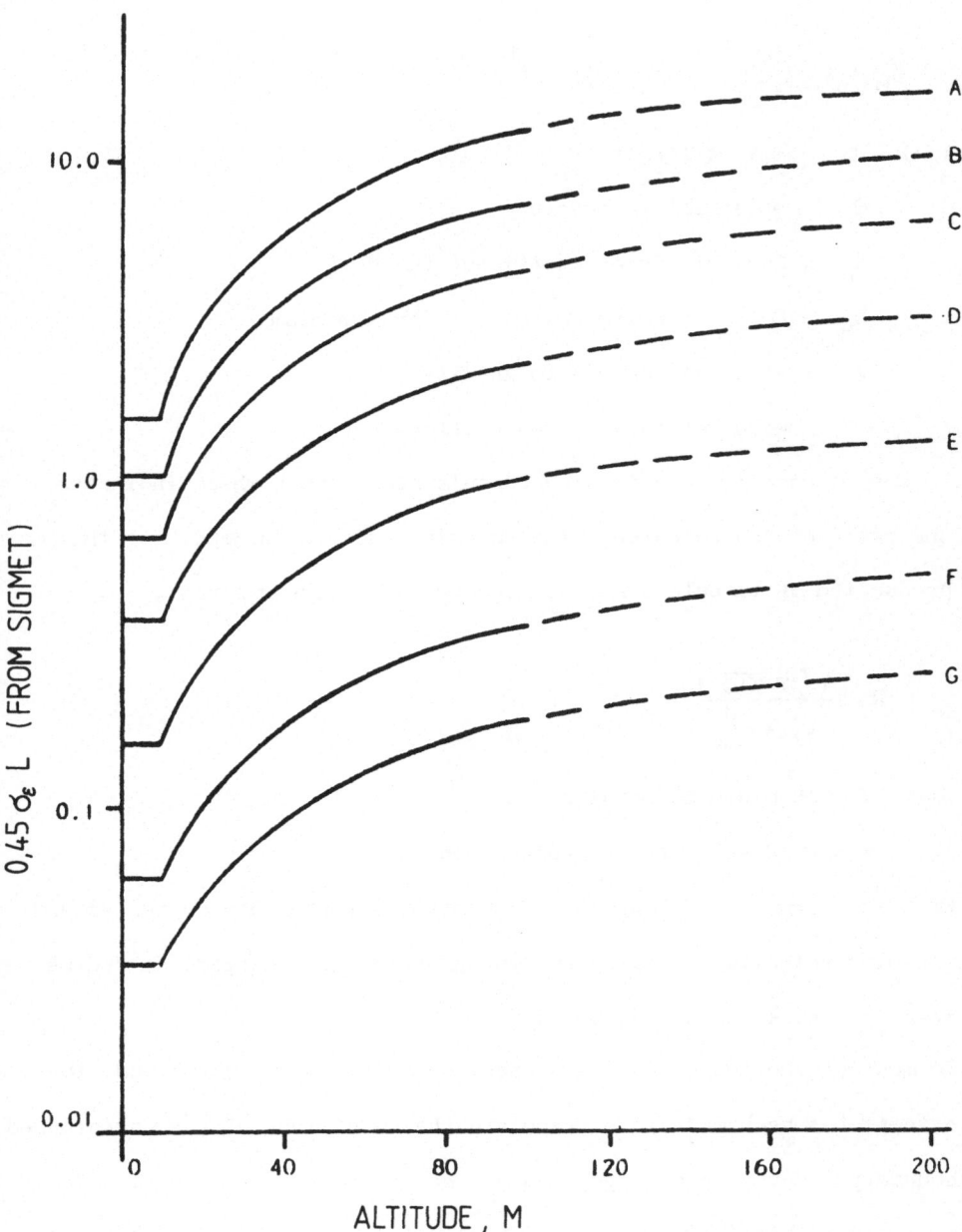

FIG.5: DIFFUSION COEFFIENTS USED IN THE
SIGMET - MODEL (AFTER HAVENS)

for instance, in Wu's formula of the diffusion coefficient (Wu, 1965):

$$K_M = (\quad (\frac{dv}{dz})^2 \quad - \frac{K_H}{K_M} \quad \frac{g}{\Theta} \quad \frac{d\Theta}{dz} \quad)^{1/2} \qquad L^2$$

with l   = mixing length

   Θ = potential temperature

   K_H = Eddy diffusion coefficent for heat

   K_M = Eddy diffusion coefficent for momentum

   g   = acceleration due to gravity

   v   = wind velocity in x-direction.

It can be equally seen in Richardson's flux number which indicates

the ratio of the turbulent kinetic energy due to buoyancy, to the

production of turbulent kinetic energy due to sheer stress.

$$R = \frac{(g/T \ \overline{w'\Theta} \ )}{v', \ w' \frac{dv}{dz}}$$

with T = absolute temperature

   w = wind velocity in z-direction.

With the number of roughness elements the turbulence increases and

the wind velocity decreases and the profiles are changed, as can be

seen in fig. 4 (Seinfeld, 1975).

In most of the models of the dispersion of the heavy gas cloud, the

attempt has been made to extrapolate the conditions of the undisturbed

boundary layer to the dispersion of the heavy gas cloud. Fig. 5 de-

monstrates this (Havens, 1979): The usual classifications of stability

according to Pasquill are used depending on the temperature- or

density gradient.

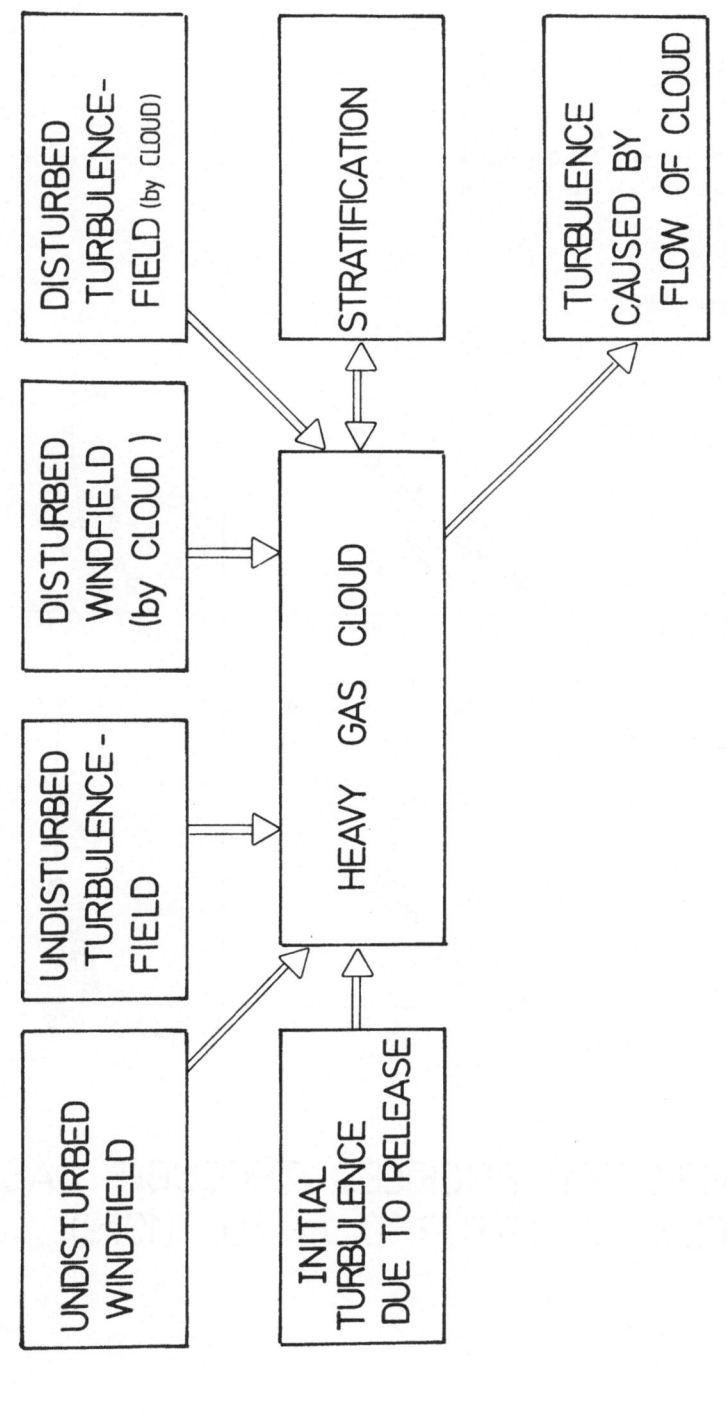

FIG. 6:   DYNAMIC VARIABLES INFLUENCING THE
VAPOR CLOUD

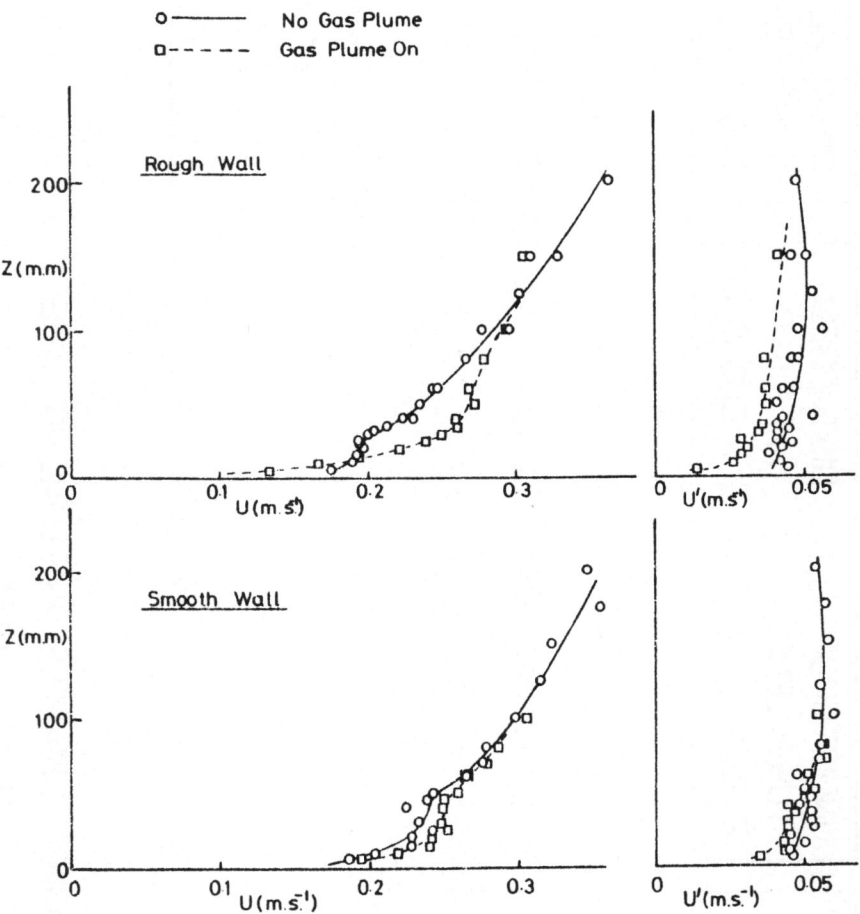

FIG. 7: VELOCITY PROFILES THROUGH GAS
PLUMES AFTER D. J. HALL (1979)

But the conditions of the reality of the heavy gas dispersion are more
complicated. Up to now, they could not be represented reasonable by
models because of our inadequate knowledge of reality.

Fortunately enough, in the last few years, we have gained some knowledge to   better understand the physics involved. The following
points should be considered when modelling (Fig. 6):

- the undisturbed wind-field
- the undisturbed field of turbulence
- the wind field changed by the heavy gas cloud
- the turbulence field changed by the heavy gas cloud
- the initial turbulence of the heavy gas cloud caused by the release
- the burbulence caused by the flow of the vapour cloud
- the turbulence or stratification caused by the temperature of the
  vapour cloud.

Wind tunnel experiments of recent years have shown that most of these
effects can play a role. D.J. Hall published the following results
when he made the wind tunnel experiments in the Warren Spring Laboratories in 1979. Fig. 7 (Hall 1979). The values shown in the figure
are average. It is not known to which period of average time these
values refer. The data were taken 250 mm downwind of the source. The
heavy gas cloud had density ratios of $\frac{\Delta\varrho}{\varrho}$ = 2,37 and 4,74 and was
10 to 15 mm high. The measurements were taken with and without rough-
ness elements.

These roughness elements consisted of screws 20 mm high, that is of
obstacles approximately as high as the heavy gas cloud, itself. The
screws could be removed easily for"smooth-wall" measurements.

The heavy gas was released quasi-continually over a period of 2 to
20 sec. The Froud number fluctuated in the three experiments between 0.09 to 0.143. (Froud number = $\frac{u}{\sqrt{g \cdot L}}$).

The Richardson number had values between 233 and 291. (Ri= $g \frac{\Delta g}{g} \frac{L}{u^2}$).

The upper part of the figure shows measurements with roughness elements
the lower part without. In the left part of the figure, the
wind velocity is shown dependent on the height; in the right part,
the turbulence is shown dependent on the height. The following can
be inferred if there is a rough ground:

- the wind velocity in the heavy gas cloud falls considerably
- the wind velocity above the heavy gas cloud increases (because
  roughness elements are covered)
- the turbulence within and above the heavy gas cloud falls
  remarkably;

if there is smooth ground:

- the wind velocity increases to a lesser extent above the cloud
- the turbulence falls to a small degree.

In other words, the heavy gas cloud alters the wind profile which,
itself, influences considerably the size of the gravity- heavy-gas
cloud. The strength of turbulence above the gas cloud which covers the
ground roughness decreases. In this way, the mechanical part of
the turbulence is suppressed. If I referred to Wu's formula (which
certainly is not within the area of definition here) this would mean
that the term with $(\frac{dv}{dz})^2$ decreases and, consequently the diffusion
coefficient becomes smaller. Accordingly, the Richardson gradient
number should increase, which means that higher stability values

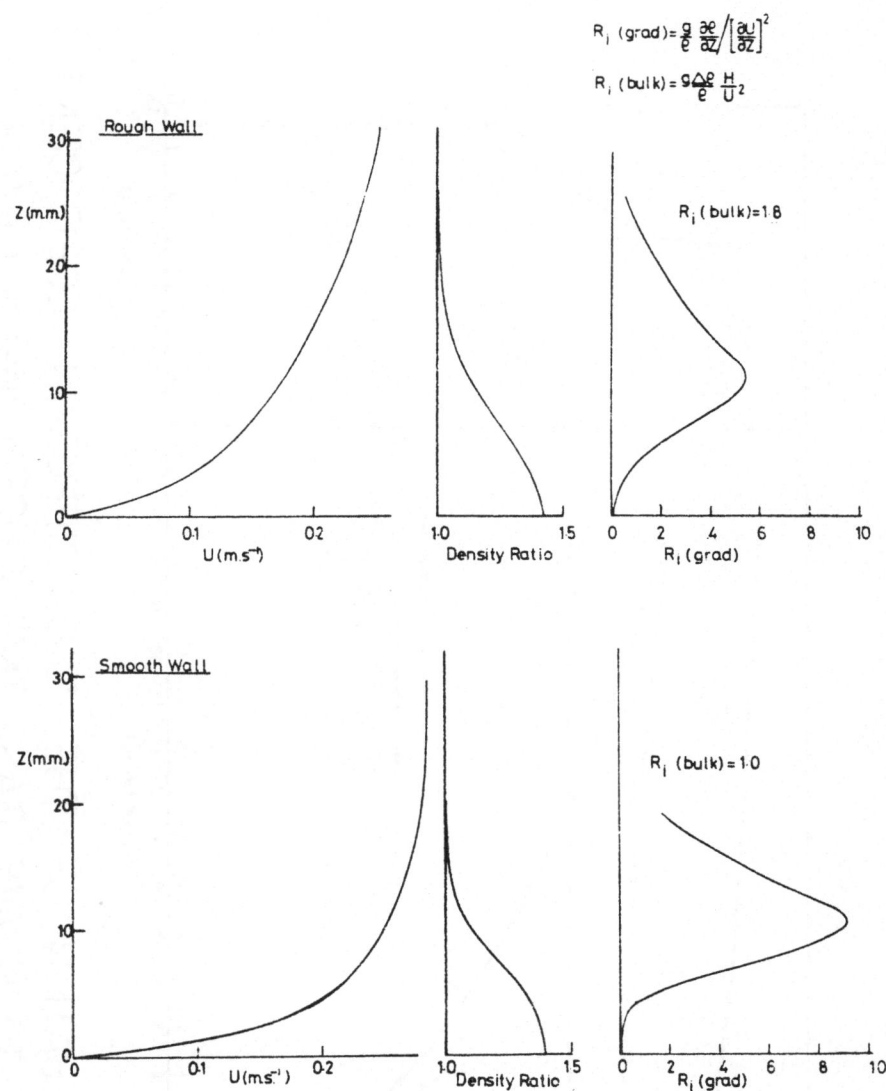

FIG.8: GRADIENT RICHARDSON No.'s FOR PLUMES ON ROUGH AND SMOOTH WALLS (AFTER D. J. HALL, 1979)

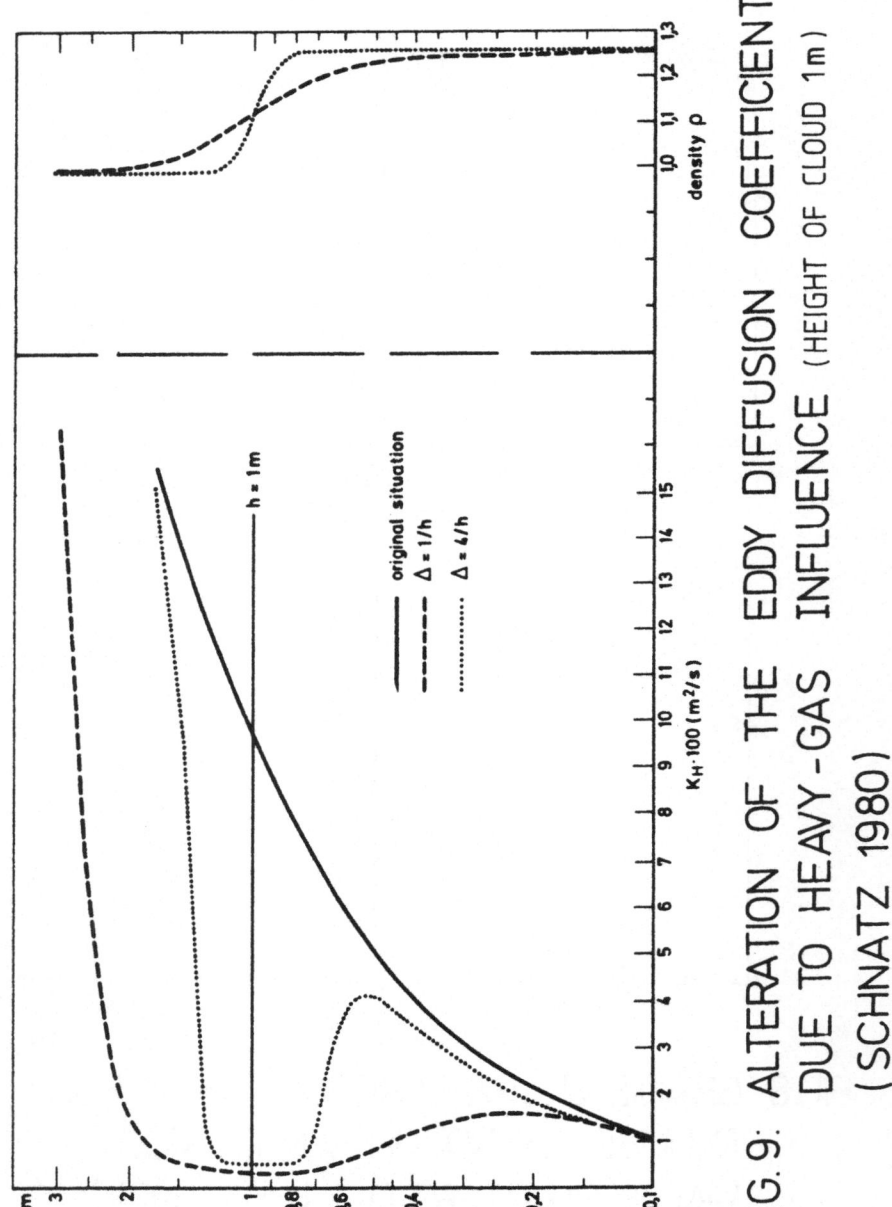

FIG. 9: ALTERATION OF THE EDDY DIFFUSION COEFFICIENT DUE TO HEAVY-GAS INFLUENCE (HEIGHT OF CLOUD 1m) (SCHNATZ 1980)

should occur. This can also be seen in figure 8, where the Richardson

gradient number, the density, and the wind velocity, dependent on

altitude, as shown. The most remarkable result is that Ri (grad) does

not have its highest values at the bottom of the cloud where the

highest density exists, but in the boundary layer between the cloud

and the atmosphere. This is understandable likewise qualitatively

because of the strong "driving back forces" for a vertical displacement

of a heavy gas cloud at the boundary layer because of bouyancy. Within

the gas cloud near the ground, mechanical turbulence produced by the

flow pattern of the cloud seems to occur. However, if the last as-

sumption is correct, there will be differences in the vertical pro-

file of turbulence between stationary and moving gas clouds. I be-

lieve that the results shown here are not only of academic nature,

but that they are important at least for the dispersion of flammable

gases where LFL-values of ca. 5 % occur. These LFL-values are pro-

pably reached or kept near the source of the gas cloud. The situation

may be different if toxic gases are considered when diffusion is im-

portant until the ppm-values are reached, and at greater distances

the near-cloud conditions may not be so important.

The following requirements result from these findings:

- modelling of vertical diffusivity has to be more sophisticated

  in the future.

- Time-dependancy of diffusivity has to be considered because of a

  possible difference for stationary and drifting vapour clouds.

- I do not foresee that it will be possible to model orographic

  effects in a general way in the near future.

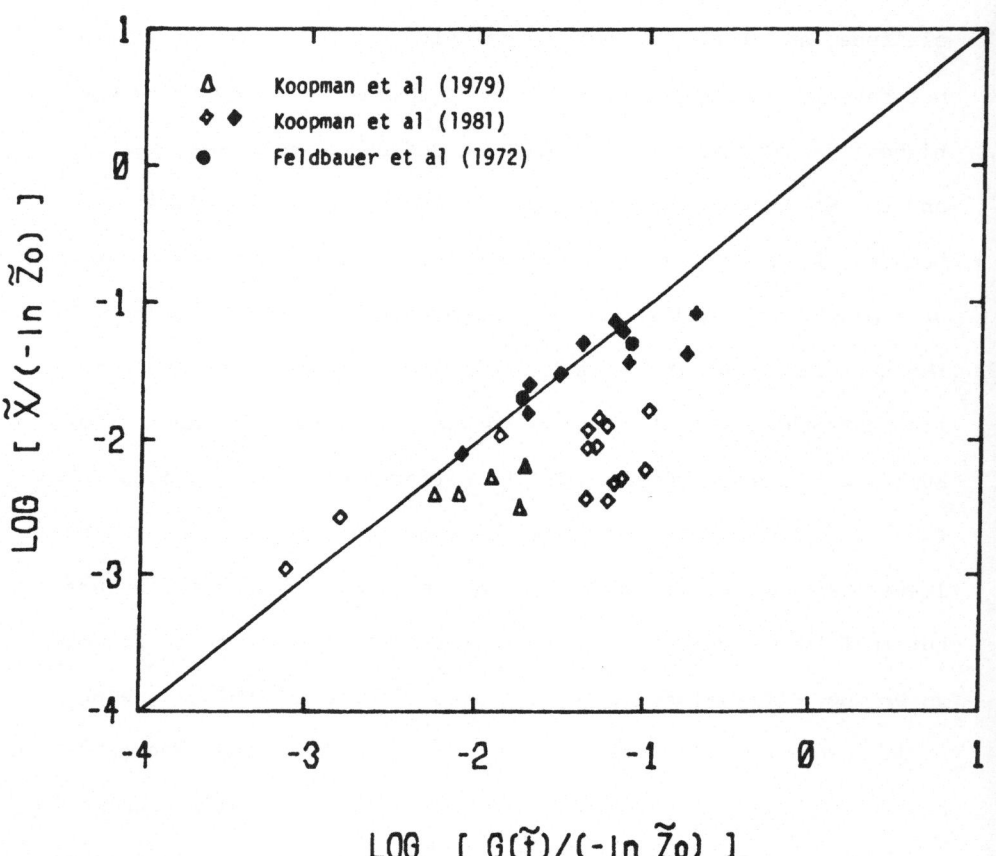

FIG.10: DISTANCE MEASURED IN LNG FIELD TESTS OVER
LEVEL (CLOSED SYMBOL) OR SLOPING (OPEN
SYMBOL) SURFACE, AS A FUNCTION OF TIME,
COMPARED WITH CALCULATED VALUES OF A
PARAMETRIC MODEL. (AFTER FAY)

I am afraid we will have to confine ourselves to case-by-case
modelling.

Smaller obstacles should be parameterized by the roughness-lenght
approach; larger ones by their fluid dynamic behaviour.

Small and large has to be seen here in relation to the height of
the heavy gas cloud.

Finally figure 9 shows the vertical diffusion coefficient, as evaluated
in the numerical model Hegas (Schnatz 1980). I do not believe that this
model is an ideal solution, but I do believe that it is headed in the
right direction for solving the problem.

Difference in Dispersion of a Cold Gas to that of an Isothermal Gas

One of the questions which came up recently (Fay 1981) was whether
a difference exists between the dispersion of a cold gas like LNG
to the dispersion of an isothermal gas like Freon 12
at ambient temperature. In reviewing all available experimental data
and comparing these data with fitted model calculations, Fay found a
difference in the LNG experiments compared to the isothermal experiments.

Figure 10 and 11 show a comparison of the China Lake (Koopman 1979,
Koopman 1981) and the Matagarda Bay experiments (Feldmann 1972)
through model calculations based on isothermal reasoning.

The Figures   show the calculated scale, the ordinate and measured
scale in the abscissa. Figure 10 gives distance values in dimensionless variables in a double logarithmic scale as a function of time,

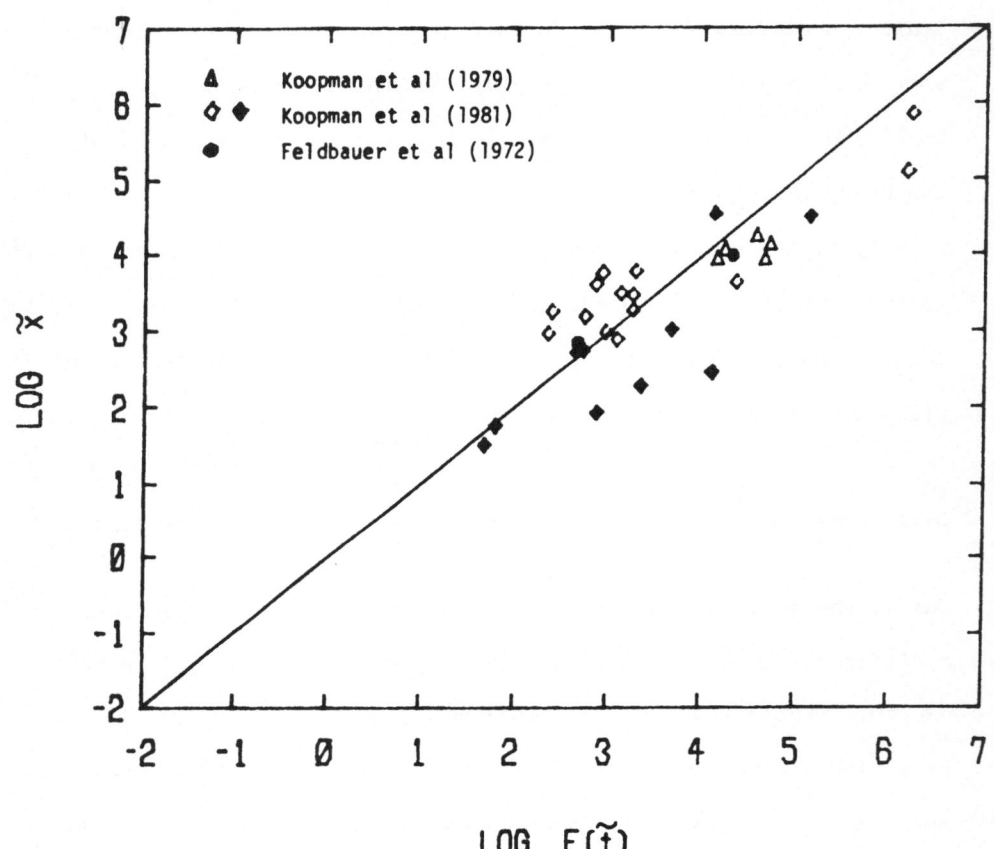

FIG.11: CONCENTRATION MEASURED IN LNG FIELD TESTS
OVER LEVEL (CLOSED SYMBOL) OR SLOPING
(OPEN SYMBOL) SURFACE, AS A FUNCTION OF
TIME, COMPARED WITH VALUES OF A
PARAMETRIZED MODEL. (AFTER FAY)

figure 11 gives distance values as a function of concentration. In both, the values are too low compared to expected calculated values, suggesting additional entrainment or higher mixing as expected. Fay proposes two effects which might give rise to increased vertical mixing rates for LNG vapour clouds. The bottom layer of such clouds, being warmed by the ground, may be unstably stratified. Convection currents originating in this layer may penetrate to the top of the cloud and generate internal turbulence. The other suggestion is that the upper surface of the cloud where the air/vapour mixing ratio is large will become unstably stratified because of vapour condensation as proposed by Haselman (1980). Fay argues that both of these mechanisms could increase the entrainment rate of the cloud without significantly altering the bulk Richardson number from its adiabatic value. Without a detailed discussion of the experimental data, it is difficult to judge these proposed effects. Nevertheless, I do not think I could agree with Fay's first effect.

First, there seems to be some ambiguity about which concentration peak (page 40, Fay's paper) is used for the comparison; he sometimes uses a smaller one. Second, if heat transfer from the ground to the bottom layer of the cloud plays an important part, then I would expect that the Matagarda Bay experiments on water would show the effect more clearly. But the opposite is the case. These data show no unusual effect at all.

If it were possible to discuss single concentration data as a function of time, I would feel that a clearer understanding could be gained.

But it is my general view that time and heat source are not sufficient

to account for the additional dilution effects.

References

Fay (1981):
    James A. Fay and Dale Ranck: Scale Effects in Liquefied Fuel
    Vapor Dispersion; Final Technical Report for USDOE Contract No.:
    DE-A CO2-77 EV 04204

Hall (1979):
    D.J. Hall: Further Experiments on the Model of an Escape of
    Heavy Gases; Warren Spring Laboratory; LR 382 (AP)

Hartwig (1980):
    Sylvius Hartwig and Ditmar Flothmann: Open and Controversial
    Problems in the Development of Models for the Dispersion of
    Heavy Gases in S. Hartwig: Heavy Gases and Risk Assessment,
    D. Reidel, Dordrecht, Holland 1980

Hartwig (1982):
    Sylvius Hartwig: Identification of Problem Areas Related to the
    Dispersion of Heavy Gases; von Karman Institute Lectures Series
    1982-03, Brüssel 1982

Haselman (1980):
    L.C. Haselman: Effect of Humidity on the Energy Budget of a LNG
    vapor cloud. Liquified Gaseous Fuels in Safety and Environmental
    Control Assessment Program: Second Status Report. DOE-IEV-0085

Havens (1977):
    Jerry A. Havens: Predictability of LNG Vapor Dispersion from
    Catastrophic Spills onto Water; an Assessment Report of the Coast
    Guard; 1977

Havens (1979):
    Jerry A. Havens: A Description and Assessment of the Sigmet LNG
    Vapor Dispersion Model Report to the Coast Guard; Contract No.:
    DOT-CG-74676-A

Koopman (1979):
    R.P. Koopman: Data and Calculations of Dispersion on 5 m^3 LNG
    Spill Tests UCRL-52876; Lawrence Livermore Laboratories

Koopman (1981):
    R.P. Koopman et al: Burro Series data report, LLNL/NWC 1980
    LNG spill tests

Lederwall (1981):
    R.T. Lederwall et al: Burro Series 40 m³ LNG Spill Experiments;
    Preprint UCRL-86704 Lawrence Livermore Laboratories

Mc Quaid (1982):
    Jim Mc Quaid: The HSE Large Scale Experiments; von Karman
    Institute Lecture Series 1982-03, Brüssel 1982

Schnatz (1980):
    Gottfried Schnatz and Ditmar Flothmann: A K-Model and its
    Modification for the Dispersion of Heavy Gases in S. Hartwig:
    Heavy Gases and Risk Assessment, D. Reidel, Dordrecht, Holland,
    1980

Seinfeld (1975):
    John H. Seinfeld: Air Pollution; Mc Graw Hill; New York, 1975

Wu (1965):
    A Study of Heat Transfer Coefficient in the Lowest 400 Meters
    of the Atmosphere, Journal of Geoph. Res. 70, pp 1801-1807, 1965

FORMULATIONS OF THE DENSE GAS DISPERSION PROBLEM

S.F. Jagger, United Kingdom Atomic Energy Authority
Warrington, GB

1. INTRODUCTION

The recent increase in use, in the chemical and process industries, of
materials which, when released to the atmosphere, form denser-than-air-
mixtures, has highlighted the distinct lack of knowledge on the way
in which such substances are dispersed. The fact that many of these
materials are highly flammable or toxic means that this topic is a
vital safety issue demanding attention.

The particular situation to be considered here is the sudden release of
a "puff" of dense gas resulting from a catastrophic failure of a containment tank. Some experiments and simple models indicate that such
a released cloud would drift at a speed related to the ambient wind
speed, while initially, it slumps, as would a liquid column. The cloud
is diluted by a variety of mechanisms; turbulence generated by cloud
slumping, ambient atmospheric turbulence, or convective turbulence
generated by temperature gradients within the cloud. Eventually, the
cloud height begins to grow and, when its density approaches that of
air, it disperses as would a passive contaminent.

To describe these processes, an approach much more sophisticated than
for passive pollutants is needed. This is because the contaminant is
said to be active, which means that it affects the properties of its
host gas, in this case air, by modifying specific heats, conductivities,
and turbulence velocity profiles, etc., due to its high concentration.
Buoyancy forces must also be considered due to the action of gravity
and their stabilising action.

A variety of models are now available to describe these processes.
Each model attacks the problem on a different level. However, before
we consider a rigorous deviation from a dense-gas model, the advantages
of such a formulation will be discussed and some comparison with simpler approaches will be made.

27

S. Hartwig (ed.), Heavy Gas and Risk Assessment - II, 27–52.
Copyright © 1983 by Battelle-Institut e.V., Frankfurt am Main, Germany.

2. BACKGROUND TO DEVELOPMENT

2.1 The Need for the Model

At present,three main types of dense gas models are regularly used
for hazard evaluation of dense gas clouds. These are closely examined
in the recent issue of the Journal of Hazardous Materials / May 1982/
which contains many reviews on the subject. The three modelling areas
are

. "Box" models
. Similarity models
. Numerical formulations.

The simplest of these are "box" or bulk property models in which the
cloud is assumed of uniform composition, temperature, etc. /Cox and
Carpenter, 1979; Fryer and Kaiser, 1978;Eidsvik,1980/. In such models,
the main processes   responsible for dispersion are identified.

Cloud development is then followed by solving equations consisting of
empirical expressions for these processes. Thus, relationships are collected for entrainment, slumping, advection, ground heating, etc.
Such an approach is useful in that it is very simple to compute, it in-
corporates many processes, and produces reasonably satisfactory correlations with experimental data. However, this model requires much
empirical input and cannot account for several important factors in
the dispersion. Thus, it cannot explain the complicated interaction
of the cloud and the mean flow. In addition, it is unable to describe
a time-varying source of vapour or dispersion in regions with buildings
or significant topography.

The advanced similarity profile model is charcterised by the formulations of Colenbrander /1980 /and Fannelop /1982/. In such models,
specified profiles of temperature, concentration, and velocity are
adopted both through and across the cloud. Clearly, such representations of the cloud are more realistic than those of "box" models and
produce a clearer picture of hazards represented by the cloud. In addition, such models should provide a better basis for computation of
entrainment from simplified forms of the diffusion equation. In general,
however, these models basically offer the same advantages and suffer
the same drawbacks as "box" models.

Currently, much effort is directed to fully three-dimensional solutions
of the Navier-Stokes equations - the coupled time-dependent conservation equations of mass, momentum, energy, and species, which suffer
from none of the problems noted above /Havens 1979; Chan et al; 1981/.
By combining these equations with a suitable turbulence closure scheme,
it should be possible to describe all processes important in cloud
development. But this approach suffers from two serious drawbacks.
Solutions to the resulting equations must be obtained by complex finite
difference or finite element schemes and these are very expensive to
compute; existing codes for 2-dimensional dense gas dispersion take
10^2 -10^3 s (Deaves, private communication) central processer time, to
follow clouds to relativly high concentrations which are the lower

flammable limits of flammable materials. For toxic clouds, it is often
necessary to follow the dispersion to even lower concentrations. At
these low concentrations there are other difficulties - for example,
errors in model predictions can become large due to numerical in-
stabilities and swamp real physical changes. In addition, the extreme
conditions of low temperature and high density often encountered in
such clouds may invalidate conventional approximations used to simplify
the equations for which solutions are required.

However, these models do ffer considerable advantages. They can model
turbulence-induced slumping and the interaction of ambient flow and
cloud regime. In theory, it is also possible to model the transport of
turbulent energy throughout the cloud, providing a suitable turbulence
model is chosen. It is also practical, providing computational time
problems can be overcome, to model the effect of buildings or topography, which other model typescan do. Any source can also be considered,
so that by modification of boundary conditions, a time varying source
or a source with some initial momentum can be modelled. Again, this is
extremely difficult with any other formulation.

Thus, there are many reasons for developing such a model.

2.2 Conditions in Released Clouds

As a preliminary to the formulation of a dense gas model, a study of
materials and situations which the model would be expected to handle
should be made. This would permit a choice of the most appropriate
set of equations and a turbulence model of the correct order, so that
solutions of the necessary accuracy and optimum computational requirements can be obtained. It would also allow for an assessment of the
various approximations which can be applied to the basic equations.
Of course, similar considerations must also govern the selection of a
numerical solution scheme. This will be the subject of an other report
and is not treated here. A prescription is given here for a choice of
model formulation for either "puff" or continuous releases of a dense
gas.

The release conditions can appreciably affect the dispersion of
a dense cloud. In fact, for ammonia, the release mechanism can determine
whether the cloud is or is not denser-than-air. Thus, if a release results from failure of pressurized storage, the cloud can have considerable initial momentum and high initial velocities are observed with
highly turbulent and large mixing. Usually, however, such initial phases are often considered in an ad hoc way with initial dilutions and
expansions allowed for by empirical methods. There is no need to model
these processes. Such releases also often give rise to two phase
flashing flow. The modelling of these flows is extremely difficult and
no attempt to calculate their behaviour has been attempted. If the
initial processes are merely regarded as producing initial conditions
for a dispersion model,then many difficulties are by-passed. The
release time scale is relatively large and the large initial dilutions

produce slow contaminant concentrations and near ambient cloud temperature at commencement of the modelled dispersion process. In these cases
it may even be possible to apply a constant density dispersion model.

Similar characteristic release time scales result from other commonly
envisaged releases. Evaporation of a volatile pool or a long period release from low pressure storage would result in small efflux velocities
with variation of source over long time periods. The releases can
then be regarded as steady but there is little dilution at the source
and clouds or plumes of high concentration must be considered.

The situation of the release must also be regarded as a factor in
modelling the dispersion. A dense cloud or plume remains near ground
for a long period and topography and building effects must be of relatively greater importance for a dense release than a passive one.
Simple models cannot treat such effects and numerical codes remain the
only possible way of accounting for their presence. However, the choice
of a turbulence model is very critical to correctly describe the behaviour of such an event. K-theory is known to have weaknesses in
these areas; a higher order closure is necessary.

3. DERIVATION OF EQUATIONS FOR A DENSE GAS MODEL

3.1 The Basic Equations

All existing and proposed dense gas models based on hydrodynamic formulations are based on the same equations which express the balance of
mass, momentum, energy, and species for a fluid element.

$$\frac{\partial \rho}{\partial t} + U_j \frac{\partial \rho}{\partial x_j} + \rho \frac{\partial U_j}{\partial x_j} = 0 \qquad\qquad a$$

$$\rho \frac{\partial U_i}{\partial t} + \rho U_j \frac{\partial U_i}{\partial x_j} = -\frac{\partial p}{\partial x_i} + \rho g_i + \frac{\partial \tau_{ij}}{\partial x_j} \qquad\qquad b$$

$$\qquad\qquad 3.1$$

$$\rho \frac{\partial H}{\partial t} + \rho U_j \frac{\partial H}{\partial x_j} = \frac{\partial p}{\partial t} + U_j \frac{\partial p}{\partial x_j} + \frac{\partial}{\partial x_j}\left(k \frac{\partial T}{\partial x_j}\right) + \tau_{ij}\frac{\partial U_i}{\partial x_j} + S_0 \qquad\qquad c$$

$$\frac{\partial c}{\partial t} + U_j \frac{\partial c}{\partial x_j} + c \frac{\partial U_j}{\partial x_j} = \frac{\partial}{\partial x_j} D \frac{\partial c}{\partial x_j} \qquad\qquad d$$

With 3.1a the above can be placed in the more convenient flux divergence form

$$\frac{\partial \rho}{\partial t} + \frac{\partial \rho u_j}{\partial x_j} = 0 \qquad\qquad a$$

$$\frac{\partial \rho u_i}{\partial t} + \frac{\partial \rho u_i u_j}{\partial x_j} = - \frac{\partial p}{\partial x_i} + \rho g_i + \frac{\partial \tau_{ij}}{\partial x_j} \qquad\qquad b$$

$$\frac{\partial \rho H}{\partial t} + \frac{\partial \rho u_j H}{\partial x_j} = \frac{\partial p}{\partial t} + u_j \frac{\partial p}{\partial x_j} + \frac{\partial}{\partial x_j}\left(k \frac{\partial T}{\partial x_j}\right) + \tau_{ij} \frac{\partial u_i}{\partial x_j} + S_o \qquad 3.2\ c$$

$$\frac{\partial \rho C_\alpha}{\partial t} + \frac{\partial \rho u_j C_\alpha}{\partial x_j} = \frac{\partial \rho D}{\partial x_j} \frac{\partial C_\alpha}{\partial x_j} \qquad\qquad d$$

Several auxiliary relations are also required which include
a. an expression for the viscous stress tensor, τ_{ij}

$$\tau_{ij} = \mu \left(\frac{\partial u_i}{\partial x_j} + \frac{\partial u_j}{\partial x_i} \right) + \delta_{ij} \mu^* \frac{\partial u_m}{\partial x_m}$$

b. a relation for the specific enthalpy of the mixture

$$H = \left[(1 - C_\alpha) C_{pa} + C_\alpha C_{pg} \right] T + rL (1 - C_\alpha) f(T)$$

where

$$f(T) = \begin{cases} 0 & T \leqslant 263\ K \\ (T-263)/20 & \text{for} \quad 263 < T \leqslant 283\ K \\ 1 & T > 283\ K \end{cases} \quad \text{(Havens 1979)}$$

and C_α is the mass fraction of contaminant $= \frac{c}{\varsigma}$

c. an equation of state for the gaseous mixture

$$P = \varsigma \frac{RT}{M_a} (1 + \mu C_\alpha)$$

Equations 3.1c and 3.2c expressing conservation of enthalpy, are used here in place of a temperature equation, to account for the effect of the contaminant on the specific heat of the mixture. It is derived from the first law of thermodynamics expressing the change of internal energy of a fluid particle as a result of changes in state. If we select S and V as thermodynamic variables and refer to a component mixture of unit mass

$$du = \left(\frac{\partial u}{\partial S}\right)_{V\,C_{\alpha_1},C_{\alpha_2}} dS + \left(\frac{\partial u}{\partial V}\right)_{S\,C_{\alpha_1},C_{\alpha_2}} dV + \left(\frac{\partial u}{\partial C_{\alpha_1}}\right)_{S\,V\,C_{\alpha_2}} dc_{\alpha_1} + \left(\frac{\partial u}{\partial C_{\alpha_2}}\right)_{S\,V\,C_{\alpha_1}} dc_{\alpha_2} \quad 3.3$$

Substituting $\left(\frac{\partial u}{\partial C_{\alpha_1}}\right)_{S\,V\,C_{\alpha_2}} = \mu_1$,   the chemical potential permit mass of species 1

$$\left(\frac{\partial u}{\partial C_{\alpha_2}}\right)_{S\,V\,C_{\alpha_1}} = \mu_2 , \quad \text{similarly defined}$$

$$\left(\frac{\partial u}{\partial S}\right)_{V\,C_{\alpha_1}\,C_{\alpha_2}} = T$$

$$\left(\frac{\partial u}{\partial V}\right)_{S\,C_{\alpha_1}\,C_{\alpha_2}} = -P$$

To maintain constancy of mass of the fluid element $dc_{\alpha_1} = -dc_{\alpha_2}$

$$du = TdS - pdV + (\mu_1 - \mu_2) dC_\alpha \qquad\qquad 3.4$$

Following the motion of the fluid particle of unit mass

$$\frac{Du}{Dt} = T\frac{DS}{Dt} - p\frac{DV}{Dt} + (\mu_1 - \mu_2)\frac{Dc_\alpha}{Dt} \qquad\qquad 3.5$$

Hence, using H = U + pV, and 1a and 1d which show $\frac{Dc_\alpha}{Dt} = 0$ when diffusion is neglected.

$$\varsigma\frac{DH}{Dt} = \varsigma T\frac{DS}{Dt} + \frac{Dp}{Dt} \qquad\qquad 3.6$$

Thus, if the fluid particle is isolated from its surroundings all changes are adiabatic and dS = 0. Hence, 3.1c becomes

$$\varsigma\frac{DH}{Dt} = \frac{Dp}{Dt} \qquad\qquad 3.7$$

However, if heat is supplied to the fluid particle, then

$$T \frac{DS}{Dt} = \frac{DQ}{Dt}$$ 3.8

Bachelor /1967/ pointed out that this can be accomplished in a number
of ways through conduction, diffusion of contaminant, or by viscous
dissipation of internal turbulent energy. Internal sources of heat or
any radiant flux can also be incorporated in the relation above. The
rate of heat supplied per unit mass is then given by

$$\frac{DQ}{Dt} = \frac{1}{\rho} \left(\frac{\partial}{\partial x_j} k \left(\frac{\partial T}{\partial x_j} \right) + \tau_{ij} \frac{\partial u_i}{\partial x_j} + S_o \right)$$ 3.9

Here, S_o represents internal heat generation per unit volume due to
external causes such as radiation and the second term on the RHS of 9
is heating due to mechanical dissipation of energy. Any diffusive flux
has been neglected. Substitution of 3.9 into 3.6 then becomes the
equation for enthalpy, 3.1c.

3.2 Approximations to the Equations

The equations,3.2a-d represent a very complex set which, especially when
effects of turbulence are incorporated, would present a very formidable
solution problem. Many researchers have consequently sought approximations to these equations to simplify the problem. There are a number
of available approaches. Foussant /1981/ adopted a rigorous set of
Boussinesq equations; Ermak et al /1981/ used a generalised set of
anelastic equations, and Farmer /1982/ produced a less demanding set
of Boussinesq equations. In this section I examine the available approximations and their validity to the dense gas dispersion problem. An
analysis similar to those performed in Farmer /1982/ and Donaldson
/1973/ was made to examine the usefulness of several approximations
to the various materials.   Omitting the details of `
this analysis, the following conclusions were drawn.
a. The boussinesq equations used by Foussant and the generalised
Boussinesq equations proposed by Farmer are not valid for all materials
close to the source. This is due both to the large temperature differences and also to the large turbulent fluctuations observed in experiments. For example, Puttock et al /1982/ observed peak to mean
concentration ratios of 1.4 in the Maplin Sands dense gas experiments.

b. The anelastic equations of Ermak et al /1981/ was also invalid for
the same reasons. Lack of validity applies for all materials near the
source and for such cold materials as LNG even when large amounts of
air have been entrained. This may occur even at appreciable distances
from the source.

The only approximations that seem to have validity are set out below.
a. Neglect of molecular diffusion terms away from solid walls. The
flows involved in this dense gas are extremely turbulent, and at the
high Reynolds numbers, the turbulent transport fluxes far exceed those
due to molecular transfer.

b. The flow is essentially incompressible. Thus, at low flow speeds
terms concerned with viscous dissipation and heating can be neglected
as they are of the order u^2/c^2 compared to the other terms (u being a
typical flow velocity).
This largely involves terms in the energy equation. The final approximation equations for the dense gas model become

$$\frac{\partial \rho}{\partial t} + \frac{\partial \rho u_j}{\partial x_j} = 0$$

$$3.10$$

$$\frac{\partial \rho u_i}{\partial t} + \frac{\partial \rho u_i u_j}{\partial x_j} = - \frac{\partial p}{\partial x_i} + \rho g_i$$

$$\frac{\partial \rho H}{\partial t} + \frac{\partial \rho u_j H}{\partial x_j} = 0$$

$$\frac{\partial \rho C_\alpha}{\partial t} + \frac{\partial \rho u_j C_\alpha}{\partial x_j} = 0$$

4. INCORPORATION OF TURBULENCE

4.1 Decomposition and Averaging

Equations 3.10, if solved, will give instantaneous values of the variables throughout the spatial domain considered. Problems, however,
include imposition of boundary and initial conditions due to the
randomness of the turbulence and the fact that the important features
of the turbulence have very small temporal and spatial size, often
10^{-1}s and 10^{-3}m, respectively. In order to overcome these problems
it is first assumed that all flow variables can be decomposed into
a slowly varying mean flow and a rapidly varying turbulence component.
The mean flow is then determined by taking an ensemble average over
a number of different flow realisations. Such averages are rarely
available from data and are difficult to relate to experimental quantities Chatwin, /1981/. For a non-steady problem, it is difficult to
relate ensemble averages to values occurring at a point. For a steady
problem, the ensemble averages do not change with time and may be
replaced by time averages at one point in a fluid.

Following the convention of Reynolds, the variables are expressed as
a mean part, denoted by an overbar, and a primed quantity - the fluctuating part. Hence, for velocity and density

$$u_i = \bar{u}_i + u_i' \qquad\qquad \rho = \bar{\rho} + \rho' \qquad\qquad 4.1$$

The ensemble averaging, denoted by the overbar, represents:

$$\bar{u}_i = \lim_{N \to \infty} \frac{1}{N} \sum_{p=1}^{N} u_i^{(p)} \qquad\qquad 4.2$$

This averaging process obeys the following simple rules

$$\overline{U_i'} = \overline{\varsigma'} = 0$$

$$\overline{\varsigma U_i} = \overline{\varsigma}\,\overline{U_i} + \overline{\varsigma' U_i'} \qquad\qquad 4.3$$

$$\overline{U_i + V_i} = \overline{U_i} + \overline{V_i}$$

$$\overline{a\,U_i} = a\,\overline{U_i}$$

The relations are then substituted into the original full equations which are then averaged. Recording the continuity and momentum equations only, the equations for the average field variables are:

$$\frac{\partial \overline{\varsigma}}{\partial t} + \frac{\partial}{\partial x_j}\left(\overline{\varsigma}\,\overline{U_j}\right) = -\frac{\partial \overline{\varsigma' U'}}{\partial x_j} \qquad\qquad 4.5$$

$$\frac{\partial(\overline{\varsigma}\,\overline{U_c})}{\partial t} + \frac{\partial(\overline{\varsigma}\,\overline{U_i}\,\overline{U_j})}{\partial x_j} = -\frac{\partial \overline{\varsigma}}{\partial x_i} + \overline{\varsigma} g_i - \frac{\partial}{\partial x_j}\left(\overline{\varsigma}\,\overline{U_i' U_j'} + \overline{U_i}\,\overline{\varsigma' U_j'}\right.$$
$$\left. + \overline{U_j}\,\overline{\varsigma' U_i'} + \overline{\varsigma' U_i' U_j'}\right) \qquad 4.6$$

Averaging the equations results in loss of information and the introduction of new unknowns - in this case, the Reynolds stress tensor $\overline{U_i' U_j'}$, the turbulent mass fluxes $\overline{\varsigma' U_j'}$ and the triple density-velocity fluctuation correlation $\overline{\varsigma' U_i' U_j'}$. Averaging the enthalpy and species conservaticn equations produces other unknown correlations for quantities such as the turbulent enthalpy and species fluxes $\overline{U_i' H'}$ and $\overline{U_i' C_\alpha'}$. It is not possible, through algebraic manipulation or averaging, to produce a closed finite set of equations for the averages shown above with the same number of equations as averages. Equations for these quantities can be derived from the original conservation equations, but equations for averages of products of influctuations contain averages of products of n + 1 fluctuations. This is called the 'closure' problem and a 'turbulence model'is a method or series of equations by which the unknown quantities, such as those in the equations, above, are expressed in terms of known quantities.

The decomposition and averaging employed above was developed for constant density flows or those for which all correlations involving the density fluctuations disappear. Hence,

$$U_j'\,\varsigma' = 0$$

This assumption is in fact used in the SIGMET code and its derivatives Havens, /1979 / and the CAFE suite of programmes / Deaves, Private Communication, 1982 /. Thus, all turbulent motions in such a model are assumed Boussinesq and the interaction of density fluctuations with the velocity and other fields are neglected. This procedure is adopted largely for ease of solution of equations of the type represented by

4.4 but possible important aspects of the problem may be lost. Another possible important approach for variable density flow was suggested by Favre / 1969 / and adopted by Jones / 1980 / for flows in furnaces and Jones / Private Communication / for flows in liquid sodium. This is the technique of density-weighted averaging and can be compared with the conventional unweighted averaging outlined above.

Density weighted averaging is defined by

$$U_i = \tilde{U}_i + U_i'' \qquad\qquad C_\alpha = \tilde{C}_\alpha + C_\alpha'' \qquad\qquad 4.7$$

Where

$$\overline{\varrho\, U_i''} = 0 \qquad\qquad\qquad \overline{\varrho\, C_\alpha'} = 0$$

$$\overline{U_i''} \neq 0 \qquad\qquad\qquad \overline{C_\alpha''} \neq 0 \qquad\qquad 4.8$$

From the definitions of weighted and unweighted density averaging, the relationships between the quantities can easily be obtained. Thus,

$$\overline{U_i} = \tilde{U}_i + \overline{U_i''} \qquad\qquad\qquad\qquad a$$

$$\overline{U_i''} = - \overline{\varrho' U_i'}/\bar{\varrho} \qquad\qquad\qquad b$$

$$\qquad\qquad\qquad\qquad\qquad\qquad\qquad 4.9$$

$$\overline{\varrho' U_i''} = \overline{\varrho' U_i'} \qquad\qquad\qquad\qquad c$$

$$\overline{U_i' U_j'} = \overline{U_i'' U_j''} - \overline{\varrho' U_i'' U_j''}/\bar{\varrho} + \overline{U_i''}\,\overline{U_j''} \qquad d$$

It should be noted that the pressure and density fields do not take part in this density weighting process. In the above, tilted quantities refer to a density weighted average, the first term on the RHS of 4.9 d representing the density weighted Reynolds stress. Using these procedures the averaged equations for the mean quantities of mass, velocity, enthalpy, and species can be written as

$$\frac{\partial \bar{\varrho}}{\partial t} + \frac{\partial \bar{\varrho}\tilde{U}_i}{\partial x_j} = 0 \qquad\qquad\qquad a$$

$$\frac{\partial \bar{\varrho}\tilde{U}_i}{\partial t} + \frac{\partial \bar{\varrho}\tilde{U}_i\tilde{U}_j}{\partial x_j} = - \frac{\partial \bar{\varrho}}{\partial x_i} + \bar{\varrho} g_i - \frac{\partial \overline{\bar{\varrho} U_i'' U_j''}}{\partial x_j} \qquad b$$

$$\qquad\qquad\qquad\qquad\qquad\qquad\qquad 4.10$$

$$\frac{\partial \bar{\varrho}\tilde{H}}{\partial t} + \frac{\partial \bar{\varrho}\tilde{U}_j\tilde{H}}{\partial x_j} = - \frac{\partial (\overline{\bar{\varrho} U_i'' H''})}{\partial x_j} \qquad c$$

$$\frac{\partial \bar{\varrho}\tilde{C}_\alpha}{\partial t} + \frac{\partial \bar{\varrho}\tilde{U}_j\tilde{C}_\alpha}{\partial x_j} = - \frac{\partial (\overline{\bar{\varrho} U_i'' C_\alpha''})}{\partial x_j} \qquad d$$

There seems considerable justification for the use of density weighted averaging as noted by Jones / 1980/. It is true that turbulence closure used for flows with large density draw heavily on assumptions used in constant density flows. Questions as to whether such assumptions require modification before transfer remain to be answered but the equations are simpler in density weighted form and many experimental instruments such as pitot tubes and possibly laser doppler anemometers measure a quantity closely related to a density weighted quantity. However, such differences should be immaterial since it should be possible to calculate both weighted and unweighted quantities.

4.2 Methods of Closing the Equations

The literature is filled with papers on methods to close equations of the type shown above, e.g. Reynolds and Cebeci (1976). In general, the available methods can be classified into groups

- Zero-equation models
- One-equation models
- Two-equation models
- Stress equation modelling
- Large eddy simulations

All models up to and including the level of two-equation formulations are based on the eddy viscosity idea first attributed to Boussinesq. This relates the Reynolds stress to the gradient of mean velocity by analogy with molecular effects. Thus,

$$-\overline{u_i' u_j'} = \nu_t \left(\frac{\partial \overline{U_i}}{\partial x_j} + \frac{\partial \overline{U_j}}{\partial x_i} \right) - \frac{2}{3} k \, \delta_{ij} \qquad 4.11$$

The last term was recently introduced so that the sum of the three normal stresses is twice the kinetic energy of the turbulence. The proportionality constant is the turbulent eddy viscosity. It can be linked to the velocity and length scales of turbulence by

$$\nu_t \propto U_t \, L_t \qquad 4.12$$

It is extended to the transport of other quantities such as heat or species through an analogous relation of flux and gradient of mean transported quantity

$$-\overline{u_i' \phi'} = \Gamma_t \, \frac{\partial \overline{\phi}}{\partial x_i} \qquad 4.13$$

where Γ_t is the turbulent diffusivity of heat or species. Γ_t can be related to ν_t through a turbulent Schmidt or Prandtl number.

$$\Gamma_t = \frac{\nu_t}{\sigma_t} \qquad 4.14$$

Although many researchers use $\widetilde{\sigma_t}$ = constant, indications are that it is strongly affected by buoyancy and streamlined curving.

Such a framework for a turbulence model breaks down in certain flow regions and the use of a scalar quantity for ν_t and Γ_t is a simplification of limited realism, especially in flows exposed to strong directional influence such as gravity or topography. This can only be overcome by use of a higher order model such as a stress-equation model.

a. Zero equation models. These models require no additional solution for ν_t . They simply prescribe a form for ν_t in terms of parameters of the flow. These are widely used for flows with passive contaminants but have not been developed for buoyancy dominated flows. SIGMET uses such a prescription extending work originally intended for thermally stratified flows. In such theories, the eddy coefficients are defined by

$$\overline{u_i' u_j'} = K_m \frac{\partial \bar{u}_i}{\partial x_j}$$

There are many extensions of this method to thermally stratified media. That used in SIGMET is a typical example. K is dependent on some stability parameters which in turn vary with the vertical temperature gradient. Thus, the vertical turbulent momentum diffusivity K_v , is given by a formula by Hanna (1968)

$$K_v = 0.45\, \sigma_t\, \bar{u}\, L \qquad\qquad 4.15$$

The Reynolds analogy for equality of diffusivities of heat, momentum, and species is assumed, and the vertical and horizontal diffusivities are related by an empirically fixed constant. In the above, L is a turbulent length scale tabulated against stability by Havens (1979) and σ_t is the standard deviation of the horizontal wind direction, again dependent on stability.

It is not clear how successful the transfer of such theory to buoyancy dominated flows can be. The prescription used above seems as applicable as any other available but it is bound to suffer from the same disability as all other prescribed eddy viscosity closures, in that it assumes local equilibrium of turbulence, not allowing for transport of turbulent energy from other regions.

Another method used for spatial variation of ν_t is the mixing length hypothesis. Providing the flow has only one significant velocity gradient ($\partial \bar{u}/\partial x_i$), it is supposed that ν_t is given by

$$\nu_t = \ell_m^2 \left| \frac{\partial u}{\partial x_i} \right| \qquad\qquad 4.16$$

where L_m is the mixing length and must be prescribed empirically. For simple flows prescription of L_m may be possible but for complex re-circulating flows the L_m variation may be too complex for practical application. Empiricism can also be invoked to account for effects of buoyancy. The DEPYS formula giving L_m as a function of Richardson number is a typical example (Panofsky, 1963). Leslie (1979) has com-prehensively reviewed the extension of the mixing length formulation for buoyant flows.

However, this suffers from the same problem as the prescribed eddy viscosity formulation in that it assumes local equilibrium of turbulence and should not be applied   when convective and diffusive transport of turbulence is important.

b. One equation models. In order to overcome the above-mentioned limitations of zero equation models, turbulence specifications that account for transport of turbulence quantities by solution of differ-ential transport equations were introduced. Instead of a relationship between the fluctuating velocity scale and the mean velocity gradients, the velocity scale is determined from a transport equation. The scale chosen is $k^{\frac{1}{2}}$, where $k = \frac{1}{2}(u_i' u_i')$. The eddy viscosity can then be given by

$$ \nu_t = C_\mu' \, k^{\frac{1}{2}} \, L_t \qquad\qquad 4.17 $$

where C_μ' is an empirical constant. An exact transport equation for k can be obtained from the Navier Stokes equations at high Reynolds numbers, accounting for convective and diffusive transport, production and dissipation.

In such models L_t now needs to be specified. Leslie (1980) has re-commended this approach rather than extending the model to include a further transport equation for L_t, believing that, for buoyancy-dominated flows, the results produced would be as reliable. Various prescriptions are available and again the KEPYS-type formulae used for mixing length can be introduced to account for buoyancy effects. An alternative is to relate L_t to local derivatives of the mean velocity, ie,

$$ L_t \propto \left(\left| \partial \bar{u}/\partial x_i \; \Big/ \; \partial^2 \bar{u}/\partial x_i^2 \right| \right) \qquad\qquad 4.18 $$

Recently, several workers used this formulation but replaced $\partial \bar{u}/\partial x_i$ by ψ (given by $k^{\frac{1}{2}}/L_t$). This gives the length scale

$$ L_t = K \, \frac{\psi}{\partial \psi/\partial x_i} \qquad\qquad 4.19 $$

where K is the Von Karman constant. This formula has the advantage that buoyancy effects on L_t are automatically accounted for through the buoyancy term in the k equation.

Such models have certain advantages including accounting for transport of turbulence energy. However, because of the difficulty of
specifying the length-scale distribution in complex flows, it is
possibly more applicable to simpler flow situations. Thus, there is
a recent trend to move to a model in which L_t is determined from a
transport equation also.

c. Two-equation models. By analogy with one-equation turbulence models
two-equation models involve solution of two transport equations for
quantities from which the eddy viscosity can be compiled at any point.
From 4.11, the two quantities involved are the turbulence velocity
and length scales. As in one-equation models, 4.16 is used and the
velocity scale is replaced by turbulent kinetic energy. Also, direct
use of transport equations for L_t have had little success and attention has been shifted to various related parameters. The most common
of these is the dissipation of turbulent energy ε , which, at a high
Reynolds number, is virtually identical with the isotropic dissipation.
The model so formed is the well-known formulation which will be discussed in greater detail in Section 4.3.

The main disadvantage of two-equation models is that the still retain
the eddy viscosity concept in which ν_t is assumed to be a scalar and
the turbulent Schmidt and Prandtl numbers, σ_t must be input to the
model. Thus, when ν_t is highly anisotropic and when flows have significant body force effects, two-equation models have limited application. Further information must be obtained from either stress-equation modelling or the simpler algebraic stress-equation modelling.

d. Stress-equation modelling. Models discussed so far assume the local
state of turbulence can be characterised by the one-velocity scale and
that the individual Reynolds stresses can be related to this scale.
This sometimes implies that the transport of individual stresses is
not adequately considered, even if the transport of the characterising
velocity scale is. To allow for the differential development of the
Reynolds stresses (representing the various velocity scales in complex flows) and to account properly for their transport, models that
employ transport equations for individual stresses $\overline{u_i' u_j'}$ (or $\overline{u_i'' u_j''}$)
have been developed. Analogous transport equations have been introduced
for the turbulent heat and mass fluxes $\overline{u_i' \phi'}$ (or $\overline{u_i'' \phi''}$. These equations
are derived in their exact form from the initial set of conservation
equations but they, of course, contain averages of products of three
fluctuating quantities. To obtain a closed system, these and other
terms must be modelled in some way. The derivation of exact equations
has the advantage of automatically including buoyancy and other special
effects.
Recently, such models have been extended and exact transport equations
for the triple correlation terms obtained,(Lumley et al 1978; Hanajalic and Launder, 1972) these equations being closed by modelling of
the higher order terms appearing here. However, even at the lower
level of closure, stress equation modelling is a very complex formulation and consequently, is little used for practical application.

They involve solution of equations of the type shown below for the density weighted Reynolds stress.

$$\bar{\rho}\,\frac{\partial \overline{u_i'' u_j''}}{\partial t} + \bar{\rho}\,\tilde{u}_j\,\frac{\partial \overline{u_i'' u_j''}}{\partial x_j} = -\left\{\bar{\rho}\,\overline{u_i'' u_j''}\,\frac{\partial \tilde{u}_i}{\partial x_j} + \bar{\rho}\,\overline{u_i'' u_j''}\,\frac{\partial \tilde{u}_j}{\partial x_i}\right\}$$

$$-\frac{\partial}{\partial x_i}\,\bar{\rho}\,\overline{u_i'' u_j'' u_i''} - \left\{\tilde{u}_i''\frac{\partial p}{\partial x_j} + \tilde{u}_j''\frac{\partial \bar{\rho}}{\partial x_i}\right\} \qquad 4.20$$

$$-\left\{\overline{u_i''\frac{\partial p}{\partial x_j}} + \overline{u_j''\frac{\partial p}{\partial x_i}}\right\} - \left\{\overline{\tau_{ii}\frac{\partial u_j''}{\partial x_i}} + \overline{\tau_i\frac{\partial u_i''}{\partial x_i}}\right\}$$

A simple three-dimensional pollution dispersion problem would involve solution of nine equations of the type represented by 4.19, in addition to the equations for the averaged flow variables - a total of 15 equations. Thus, despite their great potential, their practical use is still small. In addition, Jones (1979) suggested that for flows with significant density fluctuations results obtained from a two-equation model are equally reliable at this stage since modelling of many of the terms of 4.11 is still very uncertain.

Some of the difficulties mentioned above have been overcome by a simplified application of the stress equation. Launder et al (1981) examined dispersial of a passive pollutant using a two-equation model, yet have obtained an idea of the anistorpy of the turbulence by drastically simplifying exact transport equations for turbulent stresses. The resulting equations are algebraic and can be easily solved. Though transport of $\overline{u_i' u_j'}$ and $\overline{u_i' \phi'}$ components is not always realistically accounted for, any effect on the turbulence entering through the source/sink terms in the $\overline{u_i' u_j'}$ and $\overline{u_i' \phi'}$ is, in particular, that of streamline curvature and buoyancy. The extension of the k-ε model to examine anisotropy of the turbulence in some way is a practical possibility and a feasible development of such a model but a full solution of the transport equations for the turbulent fluxes remains a long-term computational problem.

e. Large eddy simulations. Perhaps the most promising area for long-term modelling of turbulent flows is the field of large eddy simulations, also called subgrid scale modelling. Leslie has reviewed much at the literature in this field.

The idea is to attempt a three-dimensional time-dependent simulation of large-acale turbulence and model the effects of smaller scale motions, the dimensions of which are determined by the fact that they must be less than the dimensions of the finite difference grid. They differ from the models described above in that only the smaller scale motions are modelled rather than all scales. From the point of view of pollution dispersal in the atmosphere, however, such models, though conceptually attractive, will clearly be extremely impractical. Their computational requirements are immense(Jones, 1982 Private Communication), even for the simplest application, no work has yet been attempted in regions of significant fluctuations in density. In addition to these computational difficulties, there are problems associated with imposition of boundary conditions.

4.3 The k-ϵ Model

As discussed in 4.2c the k-ϵ model is a particular form of the two-equation model for turbulent flows. The turbulent eddy viscosity is related to the turbulent velocity and length scales by 4.11, inferred from dimensional analysis. The principal of the k-ϵ model is that transport equations should be solved for these two or closely related quantities so that transport effects for turbulence as well as production and dissipation of turbulent energy can be considered.

The physically most meaningful velocity scale chosen is $k^{\frac{1}{2}}$, k being the turbulent kinetic energy per unit mass, which is related to the turbulent velocity scale by:

$$k = \tfrac{1}{2} \, \overline{u_i' u_i'} \qquad\qquad 4.21$$

When this is used, the turbulent eddy viscosity can be calculated from 4.16, here repeated:

$$\nu_t = c_\mu' \, k^{\frac{1}{2}} \, L_t$$

where c_μ is an empirical constant.

The length scale, L_t, characterising the size of the large energy-containing eddies is subject to the same transport processes as the energy k, for example, dissipation which destroys the smallest eddies and effectively increases the eddy size. Also, vortex stretching reduces eddy size. The balance of these processes can be expressed as a transport equation for L_t. However, the use of such transport equations has proved disappointing, so that other variables connected to the turbulence length scale have been tried. The most widely adopted parameter is the turbulence energy dissipation rate, which at high Reynolds number is virtually identical with the isotropic dissipation. ϵ is given by

$$\epsilon = c_\epsilon' \, k^{\frac{3}{2}} / L_t \qquad\qquad 4.22$$

Thus combining 4.16 and 4.21

$$\nu_t = c_\mu \, k^2 / \epsilon \qquad\qquad 4.23$$

Exact transport equations for both these quantities can be easily obtained. That for k can be obtained by concentration of the transport equation for Reynolds stress $\overline{u_i' u_j'}$, this being derived in turn from the Navier Stokes equations. This simply involves taking the i and j component equations and multiplying by the fluctuating velocity components u_j' and u_i' respectively. The resulting equations are then added and averaged. Similarly, a transport equation for ε can be obtained. At high Reynolds number ε is identified with the isotropic dissipation and is identical to the averaged fluctuating vorticity multiplied by the molecular kinematic viscosity.

$$\varepsilon = \nu \overline{\left(\frac{\partial u_i'}{\partial x_j} \frac{\partial u_i'}{\partial x_j} \right)} \qquad 4.24$$

Thus, a transport equation can be formed by differentiating the 1 component Navier Stokes equation with respect to x_j, multiplying by $\partial u_i/\partial x_j$ and averaging. These transport equations in their exact form, derived from the generalised Boussinesq equations of Farmer (1982), are at high Reynolds number

$$\frac{\partial k}{\partial t} + \overline{U}_i \frac{\partial k}{\partial x_i} = \underbrace{\frac{\partial}{\partial x_i} \left\{ \overline{u_i' \left(\frac{u_i' u_j'}{2} + \frac{g'}{f} \right)} \right\}}_{a} - \underbrace{\overline{u_i' u_j'} \frac{\partial \overline{u_i}}{\partial x_j}}_{b}$$
$$\underbrace{- \frac{g_i}{T_0} \overline{u_i' T'} - \mu g_i \overline{u_i' C_\alpha'}}_{c} \underbrace{- \nu \overline{\frac{\partial u_i'}{\partial x_j} \frac{\partial u_i'}{\partial x_j}}}_{d} \qquad 4.25$$

$$\frac{\partial \varepsilon}{\partial t} + \overline{U}_i \frac{\partial \varepsilon}{\partial x_i} = \underbrace{- \frac{\partial}{\partial x} \left(\overline{u' \varepsilon'} \right)}_{a} \underbrace{- 2\nu \overline{\frac{\partial u_i'}{\partial x_k} \frac{\partial u_i'}{\partial x_j} \frac{\partial u_k'}{\partial x_j}}}_{b} \underbrace{- 2 \left(\nu \overline{\frac{\partial u_i'}{\partial x_j} \frac{\partial u_i'}{\partial x_j}} \right)^2}_{c} \quad 4.26$$

In order to achieve closure without recourse to higher-order equations, the terms of these equations must be modelled in some way. This involves expressing the unknowns in terms of known quantities. For the k equation, there is little controversy over this and the terms with their model assumptions and types are

   a. Diffusive transport, $\dfrac{\nu_t}{\sigma_k} \dfrac{\partial k}{\partial x_i}$

   b. Production by shear

   c. Production/dissipation by gravitational buoyancy flows

$$g_i \left(\frac{\nu_t}{\sigma_{kt} T_0} \frac{\partial \overline{T}}{\partial x_i} + \frac{\mu \nu_t}{\sigma_{ct}} \frac{\partial \overline{C}_\alpha}{\partial x_i} \right)$$

   d. Viscous dissipation, ε

Here (a) involves a gradient diffusion approximation and σ_k is an empirical constant. σ_{Tt} and σ_{ct} are turbulent Prandtl and Schmidt numbers and must be input to the model.

However, there is great uncertainty attached to modelling the terms of the ε equation and many workers (e.g., Leslie, 1979) have suggested that these uncertainties negate any advantage gained in using the model. However, for complex flows of a dense gas cloud, any prescription of a length scale would be extremely difficult; hence, use of a modelled ε equation is continued.

The terms (a) (b) and (c) of 4.25 are due to diffusive transport, vortex stretching, and viscous destruction, respectively. Although there are largely differing views on how these terms should be modelled, the most widely accepted approximation are adopted here. The diffusive transport term is usually modelled using a gradient diffusion approximation. Hence, with σ_ε an empirical constant:

$$-\overline{u_i'\varepsilon'} = \frac{\nu_\varepsilon}{\sigma_\varepsilon} \frac{\partial \varepsilon}{\partial x_i} \qquad\qquad 4.27$$

Rodi (1971) has argued that the generation term due to vortex stretching and the dissipation term cannot be modelled separately since both are infinite Reynolds number. The following expression is adopted to give the correct source term in shear turbulence

$$-2\nu \frac{\partial u_i'}{\partial x_k} \frac{\partial u_i'}{\partial x_i} \frac{\partial u_k'}{\partial x_i} - 2\left(\nu \overline{\frac{\partial^2 u_i'}{\partial x_i \partial x_i}}\right)^2 = \left(C_{1\varepsilon} \frac{P}{\varepsilon} - C_{2\varepsilon}\right)\frac{\varepsilon^2}{k} \qquad 4.28$$

In 4.28 $c_{1\varepsilon}$ and $c_{2\varepsilon}$ are empirical constants and P is the shear production term (b) from the energy equation 4.24. Lumley and Khajeh-Noun (1974) have discussed the modelling of these terms and suggest that P/ε in 4.27 must be replaced by a parameter expressing the anisotropy of turbulence. In buoyancy-controlled flows, many researchers, e.g., Zeman, suggest that a buoyancy contribution should be included and the RHS of 4.28 replaced by

$$\left(C_{1\varepsilon}\left\{\frac{P+G}{\varepsilon}\left(1 + C_{3\varepsilon} R_f\right)\right\} - C_{2\varepsilon}\right)\frac{\varepsilon^2}{k} \qquad\qquad 4.29$$

Here G is a term from the k equation expressing the buoyant production destruction of turbulent energy from all sources and $c_{3\varepsilon}$ is an additional empirical constant. R_f is the flux Richardson number which is normally defined as the negative ratio of production of turbulent kinetic energy by buoyancy to that by stress. Using this definition Rodi found that two different values of $c_{\varepsilon 3}$ are needed for different geometries in shear layers. He redefined R_f to include only production of lateral turbulent kinetic energy by buoyancy (G_L) in the numerator, and total production, due to buoyancy and shear in the denominator. Hence,

$$R_f = -\frac{1}{2} G_L/(P+G) \qquad\qquad 4.30$$

with this definition of R_f, only a single value of $c_{\epsilon 3}$ suffices for many flow situations.

Clearly the k-ϵ model, which uses the eddy viscosity concept, cannot imply anisotropies in the turbulence structure. However, the model can be extended by using the algebraic stress-equation modelling technique, to deal especially well with directional effects such as gravity or the presence of walls. Modelled transport equations for the stresses and turbulent fluxes are simplified to yield algebraic relations.

These equations contain all the relevant terms which closely approximate the effects of buoyancy, allowing for anisotropic eddy viscosities and computation of turbulent Prandtl and Schmidt numbers.

However, even in its simple form, the k-ϵ model has had wide success in predicting many turbulent flows including recirculating flows. The same model constants cannot be expected to apply in all situations but with judicious selection of values some satisfactory predictions can be obtained. Where agreement with the experiment has been poor it is generally the ϵ equation that can be identified as the cause. Recent investigations of other forms of this equation have had little success (Hanajalic and Launder, 1978).

For flows with significant density variations, a similar k-ϵ model can be derived by analogy with that for constant density flows. It is clear that buoyancy effects must have some influence on the closure assumption adopted, but these are neglected. It is assumed first that similar closure assumptions for constant density flows can be applied to equations for significant density variations derived from equations 4.9. Also, lack of experimental knowledge of these techniques implies that predictions from the k-ϵ formulation will be as reliable as those obtained by application of higher-order closures (Jones, 1979).

Similarly, a dissipation equation can be derived. However, this equation seems so uncertain that Jones (1979) believes there is little point in deriving an exact equation with density variations. He uses the form of the constant density equation of Hanajalic and Launder (1972) and adapts it to the density weighted form. The modelled form of this equation is written as

$$\bar{\rho}\,\frac{\partial \epsilon}{\partial t} + \bar{\rho}\tilde{u}_j\,\frac{\partial \epsilon}{\partial x_j} = \frac{\partial}{\partial x_j}\left(\frac{\mu^T}{\sigma_\epsilon}\,\frac{\partial \epsilon}{\partial x_j}\right) - c_{\epsilon 1}\,\bar{\rho}\,\epsilon\,\tilde{u}_j\,\frac{\partial \tilde{u}_j}{\partial x_j} - c_{\epsilon 2}\bar{\rho}\,\frac{\epsilon^2}{k} \qquad 4.31$$

This equations still do not form a closed set for the unknown correlation $\overline{\rho' u_i''}$ appears in 4.31 a closure assumption is required for this quantity. This can be done by forming a transport equation for $\overline{\rho' u_i''}$ and approximating the terms of this equation to form an algebraic expression. This can be considerably simplified as a first approximation regarding buoyancy effects as a secondary effect. In this situation, terms involving mean rates of strain and mean pressure gradients are zero. The closure approximation for $\overline{\rho' u_i''}$ then becomes:

$$\overline{\S' u_i''} = -\frac{1}{C_{\phi_1}}\frac{k}{\epsilon}\,\overline{u_i'' u_j''}\,\frac{\partial\bar{\S}}{\partial x_j} \qquad\qquad 4.32$$

Thus, the model is complete when values of the constants are provided, for a closed set of equations has been formed for the mean quantities $\tilde{U}$, $\tilde{H}$, $\tilde{C}_\alpha$, $\bar{P}$, $\bar{\S}$ and the density weighted fluxes $\overline{u_i'' u_j''}$ $\overline{u_i'' H''}$ $\overline{u_i'' C_\alpha''}$

4.4 The Final Proposed Model

Derivations of the previous sections are summarised and values for the model constants are given below. Equations for the mean flow are

$$\frac{\partial\bar{\S}}{\partial t} + \frac{\partial\bar{\S}\tilde{U}_j}{\partial x_j} = 0$$

$$\frac{\partial\bar{\S}\tilde{U}_i}{\partial t} + \frac{\partial\bar{\S}\tilde{U}_i\tilde{U}_j}{\partial x_j} = -\frac{\partial p}{\partial x_i} + \bar{\S}g_i - \frac{\partial}{\partial x_j}\bar{\S}\,\overline{u_i''\tilde{u}_j''}$$

$$\frac{\partial\bar{\S}\tilde{H}}{\partial t} + \frac{\partial}{\partial x_j}\bar{\S}\tilde{U}_j\tilde{H} = -\frac{\partial}{\partial x_j}\left(\bar{\S}\,\overline{u_j''\tilde{H}''}\right)$$

$$\frac{\partial\bar{\S}\tilde{C}_\alpha}{\partial t} + \frac{\partial}{\partial x_j}\bar{\S}\tilde{U}_j\tilde{C}_\alpha = -\frac{\partial}{\partial x_j}\left(\bar{\S}\,\overline{u_j''C_\alpha''}\right)$$

By comparison with equations for constant density flows, the following relations can be made for the density-weighted Reynolds stress, density-weighted enthalpy flux, and density-weighted concentration flux:

$$\bar{\S}\,\overline{u_i'' u_j''} = \frac{2}{3}\,\delta_{ij}\,\bar{\S}\,k\ -\mu_T\left(\frac{\partial\tilde{U}_i}{\partial x_j} + \frac{\partial\tilde{U}_j}{\partial x_i}\right) \qquad\qquad 4.33$$

$$\bar{\S}\,\overline{u_i''\tilde{H}''} = -\frac{\mu_T}{\sigma_{T_t}}\,\frac{\partial\tilde{H}}{\partial x_i} \qquad\qquad 4.34$$

$$\bar{\S}\,\overline{u_i''\tilde{C}_\alpha''} = -\frac{\mu_T}{\sigma_{\epsilon t}}\,\frac{\partial\tilde{C}_\alpha}{\partial x_i} \qquad\qquad 4.35$$

Where μ_T is the turbulent density-weighted eddy viscosity and

$$\mu_T = C_\mu \bar{\rho} \frac{k^2}{\varepsilon}$$

σ_{Tt} is some form of turbulent Prandtl number,
σ_{ct} is some turbulent Schmidt number,

and k is now defined by

$$k = \frac{1}{2} \widetilde{u_i'' u_i''} \qquad\qquad 4.36$$

Again, an energy equation can be derived by contraction of the Reynolds stress equation (density weighted), 4.20. The resulting relation takes the form

$$\bar{\rho} \frac{\partial k}{\partial t} + \bar{\rho} \tilde{U}_i \frac{\partial k}{\partial x_i} = \frac{\partial}{\partial x} \overline{\bar{\rho} \, u_i'' \left(u_i'' u_j'' + \frac{p'}{\bar{\rho}} \right)} - \bar{\rho} \widetilde{u_i'' u_j''} \frac{\partial \tilde{U}_i}{\partial x_j} + \overline{\frac{p' u_i''}{\bar{\rho}} \frac{\partial \bar{p}}{\partial x}} - \bar{\rho} \varepsilon \quad 4.37$$

There are several possible modelling approaches for the first term of the RHS of this equation. Usually, a gradient transport assumption suffices and the quantity in brackets is represented by

$$\frac{\mu_T}{\sigma_k} \frac{\partial k}{\partial x_j}$$

However, another form by Daly and Harlew is possible. This involves the neglect of pressure diffusion, which is considered small, and the triple correlation is given by

$$\overline{u_i'' \widetilde{u_i'' u_j''}} = C_s \frac{k}{\varepsilon} \widetilde{u_i'' u_j''} \frac{\partial k}{\partial x_j}$$

Here c_s is an empirical constant.

Also, the equation of state and a relation for enthalpy are required.

$$P = \bar{\rho} \frac{RT}{M_a} (1 + \mu C_\alpha)$$

$$H = \left[(1 - C_\alpha) C_{pa} + C_\alpha C_{pg} \right] T$$

Closures are required for the turbulent fluxes.

$$\bar{\rho} \widetilde{u_i'' u_j''} = \frac{2}{3} \delta_{ij} k \bar{\rho} - \mu_T \left(\frac{\partial \tilde{U}_i}{\partial x_j} + \frac{\partial \tilde{U}_j}{\partial x_i} \right)$$

$$\bar{\rho}\,\overline{u_i'' H''} = -\frac{\mu_T}{\sigma_{Tt}}\frac{\partial \tilde{H}}{\partial x_i}$$

$$\bar{\rho}\,\overline{u_i'' C_\alpha''} = -\frac{\mu_T}{\sigma_{\epsilon t}}\frac{\partial \tilde{c}_\alpha}{\partial x_i}$$

Thus, σ_{Tt} and σ_{ct}, the turbulent Prandtl and Schmidt numbers, are to be input to the model. μ_T is computet from the equations for turbulent kinetic energy and dissipation transport.

$$\mu_T = C_\mu \bar{\rho}\, k^2/\epsilon$$

Where k and ϵ are derived from

$$\bar{\rho}\frac{\partial k}{\partial t} + \bar{\rho}\tilde{u}_j\frac{\partial k}{\partial x_j} = \frac{\partial}{\partial x_j}\left(\frac{\mu_T}{\sigma_k}\frac{\partial k}{\partial x_j}\right) - \bar{\rho}\overline{u_i' u_j''} - \frac{1}{C_{\phi 1}}\bar{\rho}\,\frac{k}{\epsilon}\,\overline{u_i'' u_j''}\frac{\partial \bar{\rho}}{\partial x_j}\frac{\partial p}{\partial x_j}$$

$$\bar{\rho}\frac{\partial t}{\partial t} + \bar{\rho}\tilde{u}_j\frac{\partial \epsilon}{\partial x_j} = \frac{\partial}{\partial x_j}\left(\frac{\mu_T}{\sigma_\epsilon}\frac{\partial \epsilon}{\partial x_j}\right) - C_{\epsilon 1}\bar{\rho}\,\epsilon\,\tilde{u}_j\frac{\partial \tilde{u}_i}{\partial x_j} - C_{\epsilon 2}\bar{\rho}\,\frac{\epsilon^2}{k} - \bar{\rho}\epsilon$$

Further constants, σ_k, $C_{\phi 1}$, σ_ϵ $C_{\epsilon 1}$, $C_{\epsilon 2}$ and finally cμ must be supplied.
Values of these constants have been taken from Jones (1979) and Rodi (1979) and are summarised in Table 1

Table 1 : Model Constants

C_μ	σ_k	σ_ϵ	$C_{\phi 1}$	$C_{\epsilon 1}$	$C_{\epsilon 2}$	σ_{Tt}^*	σ_{ct}^*	Constant
0.9	1.0	1.3	4.3	1.45	1.90	1.0	1.0	Value

* The values of σ_{Tt} and σ_{ct} are not strictly constant and these parameters must obviously be subject to some model 'tuning'.

Boundary conditions must be specified, consisting of initial values, at all points in the computational domain of the programme variables, and subsequently, values at the boundaries of this domain of these same variables. The provision of such data presents no problem. At all points away from any rigid underlying surface and free-stream typical atmospheric values can be supplied. However the k-ϵ model does not hold near the underlying surface, were there are rapid gradients of mean velocities, turbulent quantities and great difficulty in matching the free space flow to the boundary values at the wall. This would require a very fine computational mesh near the wall. To avoid this, the velocity profile is matched to a logarithmic profile depending on the wall shear stress.

Thus, we identify a logarithmic layer in which local turbulent equilibrium is supposed to exist, that is, turbulence energy and dissipation rate are equal. This allows us to fix logarithmic wall functions for velocity and identification of matching functions for turbulence kinetic energy and dissipation rate. If we initially assume that there is no exchange of mass, concentration, or heat flux with the underlying surface, then a boundary condition for these quantities can be set as $\partial \phi / \partial z = 0$

4.5 Simplifications to the Proposed Model

It is possible that the model proposed above could present some computational difficulties since it requires the solution of two additional transport equations. A simplified model would be useful with which we could assess the utility of such a formulation to the dense gas problem.

There are several ways to accomplish this. A simple 2D model could be used. However, there seems little hope of validation of such a model and Deardorff has drawn attention to the essential difference between 2 and 3D turbulence, noting that 2D turbulence is not just a section of 3D turbulent flow. Thus, there seems little advantage in proceeding along this path.

Another approach would be to average the flow in one direction, thus removing dependence of the flow variables on one coordinate. Clearly, it is possible to average either across the flow or through the flow in a vertical direction. Either of these approaches seems equally applicable and feasible. However, first we derived a depth-averaged model, using a hydrostatic pressure variation. Such models have also been applied to shallow water flow and have been considerably successful in predicting turbulent flows in channels and estuaries. The imposition of a hydrostatic pressure variation, though perhaps not strictly applicable to a density current head, (Chan and Gresho 1981) is a necessary part of such a model. Zeman (1981) suggested that the distribution of flow variables through the depth of the flow is very uniform because of internal convection due to temperature differences across the cloud. Thus, a depth-averaged model seems most appropriate. For a cloud of height in the averaged mean flow, equations 4.10 become:

$$\frac{\partial h}{\partial t} + \frac{\partial h \bar{u}}{\partial x} + \frac{\partial h \bar{v}}{\partial y} = 0$$

$$\frac{\partial \bar{u}}{\partial t} + \bar{u} \frac{\partial \bar{v}}{\partial x} + \bar{v} \frac{\partial \bar{u}}{\partial y} = -g \frac{\partial h}{\partial x} + \frac{1}{\rho h} \frac{\partial}{\partial x} \left(h \overline{\widetilde{u_i'' u_j''}} \right) + \frac{1}{\rho h} \frac{\partial}{\partial y} \left(h \overline{u'' v''} \right)$$

$$+ \frac{1}{\rho h} \frac{\partial}{\partial x} \int_0^h \rho (u - \bar{u})^2 dz + \frac{1}{\rho h} \frac{\partial}{\partial y} \int_0^h \rho (u - \bar{u})(u - \bar{v}) dz$$

4.40

The production term P_h is given by

$$P_h = \nu_t \left[2 \left(\frac{\partial \bar{u}}{\partial x} \right)^2 + 2 \left(\frac{\partial \bar{v}}{\partial y} \right) + \left(\frac{\partial \bar{u}}{\partial y} + \frac{\partial \bar{v}}{\partial x} \right)^2 \right]$$

and is due to interaction of turbulent stresses with horizontal mean velocity gradients. The additional production terms, P_{kv} and P_{tv} are due, again, to non-uniformities in the vertical distribution of flow

variables. They are most important at the top and bottom of the layer
and in many circumstances, especially for a dense gas, may be neglected.
The adaptation of the k-ϵ model to depth-averaged flows is certainly
of a rather empirical nature, as evidence by choise of programme constants so far adopted for applications. However, results of applications have been encouraging and stimulate further use and investigation of this approach to turbulence modelling.

5. CONCLUSIONS

A model for the dispersal of a dense gas release has been formulated.
The model uses variable density equations as opposed to contempory
models which involve either Boussinesq or anelastic constant density
equation.   The turbulence modelling used also allows for transport
of turbulent energy throughout the flow and permits large variations
of density due to motion as observed in practice (Puttock et al 1982).
The final model requires solution of nine equations and has eight
disposable model constants.

Such a model has both advantages and disadvantages. The equations are
probably applicable closer to the source than three used previously ,
since this is where large density differences occur. However, after
a significant amount of air entrainment, even for the densest and
coldest gases, the equations are expected to show predictions differing
little from those established using the Boussinisq or anelastic
equations.

A much more sophisticated turbulence model has been created for turbulent transport. This still involves the concept of eddy viscosity
but two transport equations are used to compute this quantity which
is then a function of flow rather than a prescribed quantity as in
existing models. The model is more likely to describe the recirculating
flows occuring near the advancing front of the cloud. It is also much
more appropriate for unsteady flows or flows in which temporal changes
are rapid such as those occuring when a quantity of dense gas is
suddenly released. This model with changing model constants, has predicted transport for a wide range of sheer or recirculating turbulent
flows.

The model does display weaknesses, however, there is significant anisotropy of the turbulent flow. This can occur when there is a preferred direction to the flow, such as when gravity is important or there
is significant streamline curvature. However, no simple model involving eddy viscosity is adequate for flows of this type. Only a higher-
order modelling technique - stress-equation modelling, is applicable.

Such formulations require transport equations for all the turbulence
quantities and a full three-dimensional model would require a solution
of $\sim$ 15 such equations. Thus, computational needs are excessive and
would seem beyond the requirements of a three-dimensional dense gas
model.

A simplified form of stress-equation modelling can, however, be used
to extend a k-ϵ formulation and give effective anisotropic eddy
viscosities. In such a technique, the transport equations for the turbulent fluxes are approximated by algebraic equations which are more
amenable to solution. Thus, the k-ϵ model developed with the aid of
algebraic stress-equation modelling could provide the structure of
anisotropic turbulence.

The k-ϵ formulation does, as has been noted earlier, mean solving two
additional transport equations, which places additional burdens on
computation needs. A further hole deals with possible numerical methods
but it is likely that such considerations will restrict application
of such a model more than turbulence modelling criteria. For this
reason, a simplified model has also been presented which places fewer
demands on numerical techniques. This model is set out in depth-
averaged form and is a quasi-2D formulation. The implementation of
this model would allow for an assessment of the k-ϵ model and density
weighted averaging technique in predicting dense gas releases and ac-
cumulation of experience in using such techniques.

In conclusion, it would seem that a first stage in the development of
a numerical code for dense-gas modelling using k-ϵ turbulence modelling would be a quasi-2D depth-averaged version. This would avoid any
computational difficulties and allow an assessment of the probable
advantages of this formulation over currently-prescribed eddy-
viscosity modelling.

REFERENCES

Chan et.al., Simulation of three-dimensional, time-depen-
    dent, incompressible flows by a finite element method,
    Proc. of the AIAA, 5th Computational Fluid Dynamics
    Conf., Palo Alto, CA, june 1981
Chan et.al., A three-dimensional, conservation equation
    model for simulating LNG vapor dispersion in the at-
    mosphere, Lawrence Livermore National Laboratory Report
    UCID-19210, Sept. 1981
Chatwin, P.C., The influence of basic physical processes
    on the statistical properties of dispersing heavy gas
    clouds, Proc. 7th Biennial Symposium on Turbulence,
    Rolla, MO, 1981
Colenbrander, G.W., A mathematical model for the transient
    behavior of dense vapor clouds, 3rd International Symp.
    on Loss Prevention and Safety Promotion in the Process
    Industries, Basle, Switzerland, Sept. 1980
Cox, R.A. and R.J. Carpenter, Further development of a
    dense vapour cloud dispersion model for hazard analysis
    in S. Hartwig (Ed.) Heavy Gas and Risk Assessment,
    D. Reidel Publishing Co., Dordrecht, Holland, 1980,
    p. 55
Eidsvik, K.J., A model for heavy gas dispersion in the
    atmosphere, Atm.Env., 14 (1980) 769
Ermak, D.L., et. al., ATMAS: A three-dimensional atmos-
    pheric transport model to treat multiple area sources
    Lawrence Livermore Nat. Lab., Livermore, CA, UCRL-
    52603, 1978
Foussat, Andre; Modele de Dispersion Atmospherique Non-
    Isotherme d'un Pulluant Gazeux de Densite Quelconque
    en Presence de Non-Uniformites Orographiques, von
    Karman Institute for Fluid Dynamics, 1981
Fryer, L.S. and Kaiser, G.D.; DENZ - A computer program
    for the calculation of the dispersion of heavy toxic
    or explosive gases in the atmosphere, 1979, SRD R152,
    UKAEA
Havens, J.A., A description and assessment of the SIGMET
    LNG vapor dispersion model, U.S. Coast Guard Report
    CT-M-3-79, 1979
Havens, J.A., A description and computational assessment
    of the SIGMET LNG vapor dispersion model, Journal of
    Hazardous Materials, 6 (1982) 181-195, Elsevier
    Scientific Publishing Company, Amsterdam
Puttock, J.S., Blackmore, D.R., Colenbrander, G.W.; Field
    experiments on dense gas dispersion, J. Haz.Mat. 6
    (1982) 13-41
Zeman, O., The dynamics and modeling of heavier-than-air
    cold gas releases, Lawrence Livermore National Lab.
    Report UCRL-15224, April 1980, Atmos. Environ.16 (4)
    (1982) 741-751.

INVESTIGATION OF ENERGY FLUXES IN HEAVY GAS DISPERSION

G. Schnatz, J. Kirsch, W. Heudorfer
Battelle-Institut e.V., Frankfurt a.M.

1. INTRODUCTION

Since the beginning of the 1970's a multitude of models
has been developed in order to investigate the diffusion
behaviour of heavy gas clouds.
As there was no experimental material to explain the
occuring effects in the initial stages of this research, the
investigations were made with the aid of existing dispersion models from the field of tracer diffusion. These
so-called Gaussian models are conceived for long-term
statistical surveys of air pollution processes. However,
the diffusion of accidental heavy gas spills into the
atmosphere should be treated as a highly unstable process
that cannot easily be classified or parameterized.
The first models (Burgess et al. 1970, 1972, 1974; Drake
et al. 1974; Fay and Lewis 1975) are still more or less
modified and stationary Gaussian-Type models (Table 1).
The use of these models led to erroneous results which
contrasted with actual diffusion behaviour. Subsequently,
the results led to wrong statements about possible consequences (toxic areas, explosion effects). In a comparison
drawn by Havens (1977) Gaussian-type models show results
that differ considerably from each other and from the improved models (Germeles and Drake 1975; Eidsvik 1980;
Fryer and Kaiser 1979; Flothmann and Nikodem 1980; Colenbrander 1980) which also use a high degree of parameteri-
zation of the physical processes involved.
In order to improve the quality of these models, single
effects, such as smooth transition between the phases,
have to be considered to obtain more qualified values of
the employed parameters.
Apart from this development in the field of box-models,
other (numeric) models based on basic principles (SIGMET,
1978; HEGAS, 1979, 1982; and improved versions of the
SIGMET Models: SIGMET N, MARIAH) have been generated. The

S. Hartwig (ed.), Heavy Gas and Risk Assessment - II, 53–65.

Table 1: Types of Models

Early phase of modelling:
— Burgess et al. (1970, 1972, 1974)
— Drake et al. (1974)
— Fay and Lewis (1975)

Advanced Box Models:
— Germeles and Drake (1975)
— Eidsvik (1980)
— Fryer and Kaiser (1979)
— Flothmann and Nikodem (1980)
— Colenbrander (1980)

Numerical Models:
— SIGMET (1978)
— HEGAS (1979)
— improved versions of SIGMET N, MARIAH
— LLL model
— von Karman Institute/Foussat (1981)

essential problem of these models is the adequate treatment
of the vertical turbulent diffusion.
Further explanations concerning this problem will be given
in the report of Hartwig.
In contrast to numerical models just mentioned, the energy
fluxes within the heavy-gas/air system are often not
easily understood in parameterized box-models. This means
that it is not possible to check model quality by taking
an energy balance in the different phases of the diffusion
process.

In order to understand the physical process of heavy gas
diffusion and dispersion, the different reservoirs must
be analysed in their chronological sequence (temporal
development). In the following the energy flux in the
heavy gas/surrounding air system will be discussed.
Desirable and more realistic modifications of the "Gravitational spreading phase" result from the exact consideration of the energy balance of this system in its temporal
development. The changes that influence the results of
the model calculations will be demonstrated with the aid
of examples.

2.   ENERGY BALANCE AND ENERGY FLUXES IN THE INSTATIONARY
     SYSTEM HEAVY GAS FREE ATMOSPHERE

The process of heavy gas spills into the atmosphere, of
spreading and diffusion is related to the occurence of
variable energy fluxes.
In order to make an overview of the possible energy fluxes
easier, it shall be differentiated from the following
cases:
- release of heavy gas, without pressure but ambient
  temperature
- release of heavy gas, under pressure and ambient
  temperature
- release without pressure, refrigerated

The energy fluxes will be considered within a defined,
limited system which consists of a confined part of the
atmospheric boundary-layer and the heavy gas.
The limitation of the atmospheric boundary layer has to
be chosen in such a way that disturbances of the important
parameter which are caused by heavy gas spills are small
outside the system. We assume that the energy fluxes
quoted below over the top of the observed system shall
be constant in time so that the overall energy content of
the system will not change:

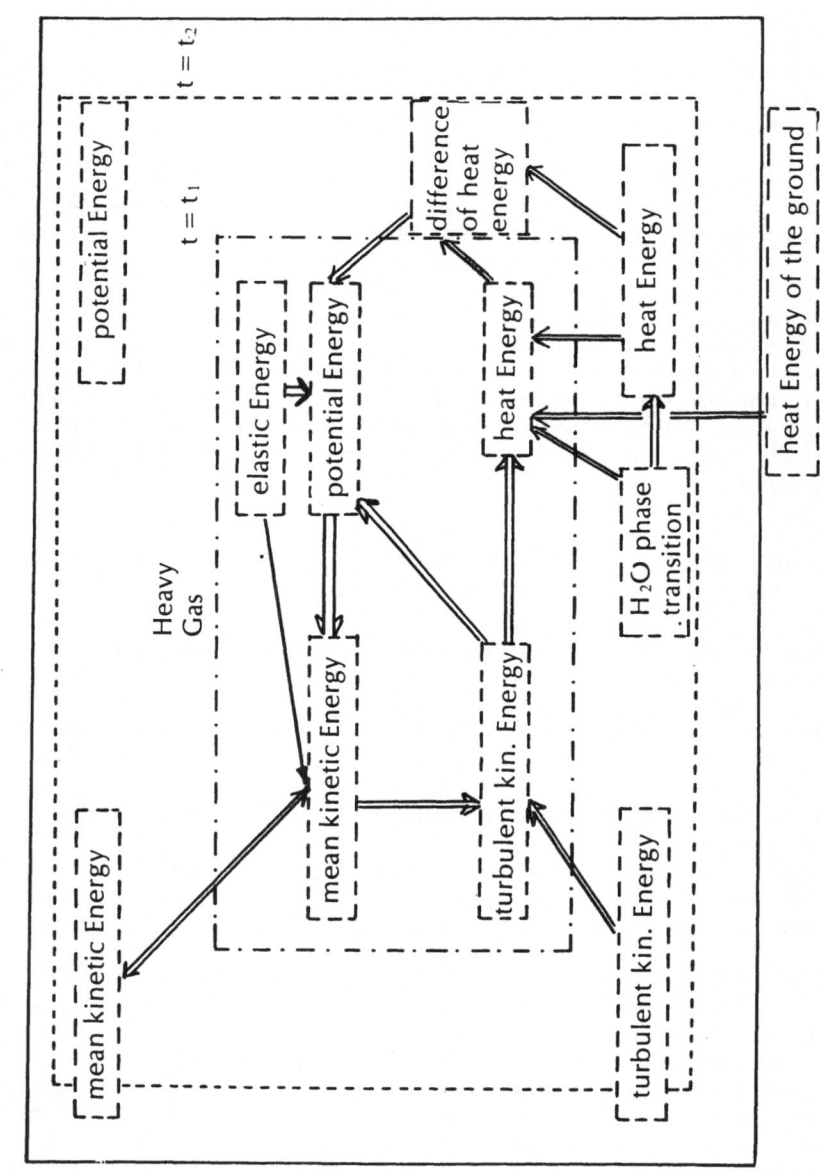

Fig. 1: Energy Diagram; Time > 0

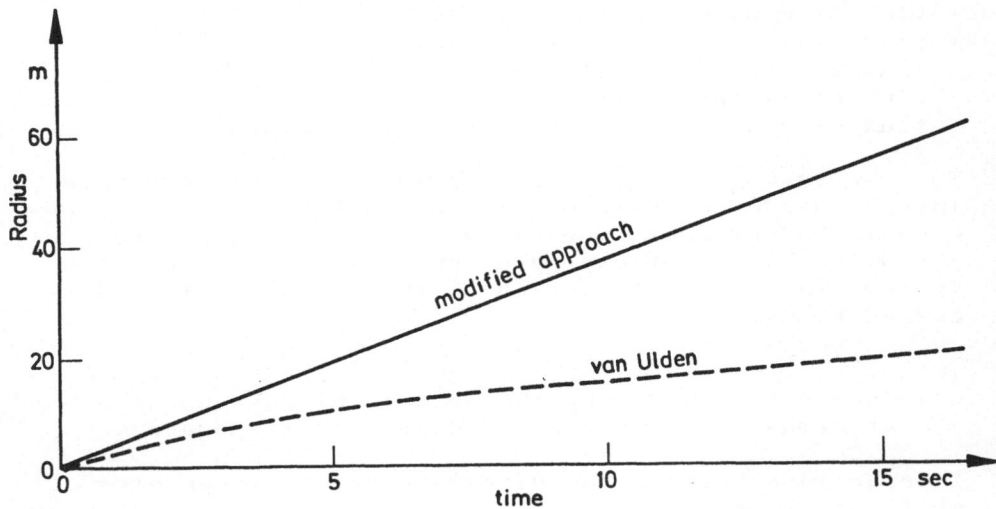

Fig. 2: Comparison of the Radius of the Cloud as a function of time for the modified approach without friction effects and the van Ulden formular

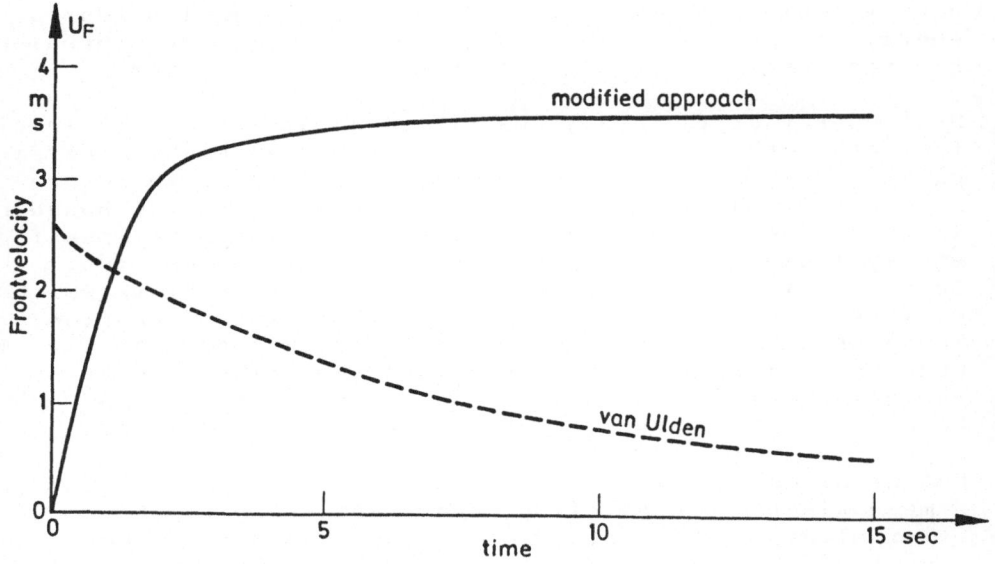

Fig. 3: Comparison of the front velocity of the Cloud as a function of time (models same as Fig. 2)

- flux of medium kinetic energy;
- flux of turbulent kinetic energy;
- **flux of** sensible heat;
- flux of **latent** heat;
- **flux** of **radiation** (long-wave and short-wave).

We also assume that the part of the atmospheric boundary
layer considered is in dynamic equilibrium with the surrounding atmosphere. Theses assumptions suggest that only
a redistribution of energy occures within the observed
system. The energy fluxes which then occur will be discussed below.

2.1 Energy fluxes during the release of substances stored
    at atmospheric pressure and ambient temperature

Interactions between the different energy reservoirs are
still easily comprehensible during the release of a heavy
gas stored at atmospheric pressure and ambient temperature.
The reservoirs involved in the interaction process are
shown in fig. 1. There is a distinction between the reservoirs of the atmospheric boundary layer and the heavy gas
reservoirs.

For t_o=0 the time just when release occurs, the total
energy content of the heavy gas has the form of potential
energy. There are neither mean kinetic energy nor turbulent
kinetic energy available within that closed system.

Heat energy of the heavy gas system is at a constant level:
that means there is no transfer in the system which dissi-
pates energy to heat. As the system is closed, there are
no energy fluxes interacting with the atmospheric boundary
layer around the heavy gas. Within the atmosphere itself
energy fluxes appear between the different reservoirs.
Theses fluxes are not very relevant to the problems which
will be discussed below because of their short duration.
Therefore these interactions have been omitted from fig. 1.
When the heavy gas is released into the atmosphere ($t > t_o$),
various energy fluxes within the heavy gas itself and
across the interface of heavy gas and surrounding air
appear.
Potential energy of the heavy gas is transferred to
- mean kinetic energy of the heavy gas (direct)
- turbulent kinetic energy of the heavy gas (indirect) and
  part of the turbulent kinetic energy is retransferred
  to potential energy, the other part is dissipated
- heat energy of the heavy gas (dissipation) (indirect).

In addition energy cascades from a macroscopic level to a molecular level. The heavy gas gains energy from outside because mean kinetic energy of the atmosphere is transferred to mean kinetic energy of the heavy gas (acceleration of the cloud, drift movement) and mean kinetic energy of the atmosphere is transferred to turbulent kinetic energy of the heavy gas.
While the two systems (heavy gas/boundary layer) can be distinguished easily shortly after release, later the boundary becomes obscured during the mixing of the heavy gas with the atmosphere. This is indicated by the dashed lines in fig. 1.
After a given period of time the different reservoirs namely the atmosphere and the heavy gas can be treated as one reservoir.

2.2 Energy fluxes occuring at the release of heavy gases stored under pressure and at ambient temperature

In contrast to the energy fluxes in section 2.1 an additional energy reservoir is of importance when pressurized heavy gases are released. Pressurized gases contain elastic energy which may be regarded as a special form of potential energy. Part of the elastic energy of the stored gas is transferred to potential energy. The amount depends on the direction of the jet and the discharge velocity. The other part of the elastic energy available is transferred to:
- mean kinetic energy of the heavy gas (direct)
- turbulent kinetic energy of the heavy gas (indirect)
- heat energy of the heavy gas (dissipation) (indirect)

Apart from these energy fluxes withing the heavy gas there are also interactions between heavy gas and atmosphere which are similar to those described in section 2.1.

2.3 Energy fluxes occuring at the release of refrigerated gases at atmospheric pressure

A further complication of energy fluxes and reservoirs is reached when cold gas releases are considered.

When a cold gas is released into the atmosphere the temperature difference between the atmosphere and the heavy gas leads to an additional contribution to the available potential energy which is transferred to
- kinetic energy of the heavy gas (directly)
- turbulent kinetic energy of the heavy gas (indirectly)
- heat energy of the heavy gas (indirectly).

Furthermore, a constant reduction process in the available
potential energy occurs which is caused by the heat fluxes
between atmosphere, ground surface and the heavy gas, i.e.
by heat radiation and heat conduction, respectivly.

3.   Energy balanced heavy gas motion

3.1 Heavy gas motion without friction effects

Energetic aspects of the type just mentioned are often not
strictly applied to the modelling of heavy gas dispersion
as demonstrated below for gravitational spreading.
We describe here the slumping of the heavy gas in some
detail, starting with the simplest case: namely energy
balance of an ideal incompressible fluid (without friction
effects). Furthermore, we assume that no entrainment occurs.
With these constraints, the energy balance can be written
as

$$E_{kin} + E_{pot} = const. = E_{total} \qquad (1)$$

$$E_{pot} = \int dV \; \Delta\rho gz = \Delta\rho g \frac{R^2}{2} H\pi \qquad (2)$$

$$E_{kin} = \int dV \frac{1}{2}\rho u^2 \qquad (3)$$

when t approaches t+dt the following holds

$$\pi R^2 H = \pi (R+dR)^2 (H+dH) \qquad (4)$$

for an arbitrary partial cylinder of radius r $(r < R)$ equation (5) holds:

$$\pi r^2 H = \pi (r+dr)^2 (H+dH) \qquad (5)$$

Elimination of H and dH yields:

$$\frac{R^2}{r^2} = \frac{(R+dR)^2}{(r+dr)^2} \qquad (6)$$

Defining in equ. (6) $\frac{dR}{dt} = u_R$ as frontvelocity and $\frac{dr}{dt} = u_r$ as frontvelocity of a partial cylinder yields

$$u_r = \frac{r}{R} u_R \qquad (7)$$

If R_o, H_o are the radius and height for t=0, than the following holds for time t=0

$$\pi R^2 H = \pi R_0^2 H_0 \qquad (8)$$

Time derivation yields:

$$u_H = -2 R_0^2 H_0 \frac{1}{R^3} u_R \qquad (9)$$

where $u_H = \frac{dH}{dt}$ is the vertical velocity of the top of the cylinder.

Starting from a relation similar to those of equation (4) and (5) it can be shown that the vertical velocity u_h within a cylinder of height h (h<H) decreases linear with decreasing height h.

$$u_h = \frac{h}{H} u_H = -2 \frac{h}{R} u_R \qquad (10)$$

Now we can express the velocity u of equation (3) in terms of the integration variables r and h.

$$u^2 = u_r^2 + u_h^2 = \left(\frac{r^2}{R^2} + 4 \frac{h^2}{R^2} \right) u_R^2 \qquad (11)$$

Substitution of the velocity in equation (3) and integration yields:

$$E_{kin} = \frac{1}{4} \rho \pi R_0^2 H_0 u_R^2 \left(1 + \frac{8}{3} \frac{R_0^4 H_0^2}{R^6} \right) \qquad (12)$$

Now we can derive an equation for the front velocity in terms of the front radius R of the gas cloud. The total energy for the time t=0 is stored in the form of potential energy. From that follows:

$$E_{total} = E_{pot} \, (t=0) = E_{kin} + E_{pot}$$

$$\Delta\rho \, g \, \frac{R_0^2}{2} H_0^2 \pi = \frac{1}{4} \, \rho\pi \, R_0^2 \, H_0 \, u_R^2 \left(1 + \frac{8}{3} \, \frac{R_0^4 H_0^2}{R^6}\right) + \Delta\rho \, g \, \frac{R^2}{2} H^2 \pi \quad (13)$$

by replacing $H = H_0 \dfrac{R_0^2}{R^2}$, the front velocity $u_F = u_R$ can be written

$$u_F = \sqrt{2 \frac{\Delta\rho}{\rho} \, g \, H_0} \, \left(1 - \frac{R_0^2}{R^2}\right)^{1/2} \left(\frac{1}{1 + \frac{8}{3} \, \frac{R_0^4 H_0^2}{R^6}}\right)^{1/2} \quad (14)$$

This equation for the front velocity differs from the equation used by van Ulden and from other models which were derived later.

The equation which is used in the models just mentioned can be written as follows:

$$u_F = \sqrt{c \frac{\Delta\rho}{\rho} g H_0} \, \frac{R_0}{R} \quad (15)$$

The substantial difference between both equations lies in the fact that, when using equation (15), the maximum front velocity appears at the time t=0, whereas in equation (14) the front velocity starts with zero. From an energy point of view this means that van Ulden assumes that the total energy for t=0 appears not only in the form of potential energy, but also in the form of kinetic energy which is even at its maximum for t=0. Both energy reservoirs decrease with increasing time.

Obviously, this contradicts the physical facts. Kinetic energy is zero and increases with increasing time.

If we keep in mind that we assumed no friction, the kinetic energy will approach a constant value. This behaviour is reflected in equation (14).

3.2 Heavy gas motion including friction effects

The modified equation (14) shows a more realistic approach immediately after release of a heavy gas than van Ulden's approach (eq. 15). However the cloud behaviour is not described very well by both equations (14, 15) for large times.

In equation (15) the front velocity decreases with $u_F \sim R^{-1}$.

In equation (14) the front velocity approaches a maximum value for large times. This is caused by the fact that no entrainment and no loss of kinetic energy due to friction is considered.

In this section we include the friction energy in the energy balance.

Because of physical reasoning the differential friction energy dE_R/dt is proportional to the gas density ρ and to the square of the local heavy gas velocity u_r^2:

$$\frac{dE_R}{dt} = \int_0^{2\pi} \int_0^R \frac{1}{2} \alpha \rho u_r^2 \, r \, dr \, d\varphi \tag{16}$$

In eq. (16) α is a proportionality factor which contains the roughness length of the ground surface.

In order to integrate this equation, the time integration is transformed to an integration over the increasing front radius R'.

Using eq. (7) the friction energy E_R is:

$$E_R = \frac{1}{4} \pi \alpha \rho \int_{R_0}^R R'^2 \, u_{R'}^2 \, dR' \tag{17}$$

Adding E_R to equation (1) and differentiation with respect to R leads to a linear differential equation of first order for the square of the front velocity $\ddot{u}_R$:

$$(aR^7 + bR) \frac{du_R^2}{dR} + (\alpha R^9 - 6b) u_R^2 - dR^4 = 0 \tag{18}$$

with

$$a = R_0^2 H_0 \qquad b = \frac{8}{3} R_0^6 H_0^3 \qquad d \approx 4 \frac{\Delta\rho}{\rho} g H_0^2 R_0^4$$

For eq. (18) no analytic solution exists. Nevertheless for
large values of R an   asymptotic expression of eq. (18)
exists

$$u_R(R \gg R_0) \approx \sqrt{2\frac{\Delta\rho}{\rho}gH_0} \left(\frac{2H_0R_0^4}{\alpha R^5}\right)^{1/2} \tag{19}$$

So u_R decreases for large values of R with $u_R \sim R^{-2.5}$ in
contrast to the $u_R \sim R^{-1}$ for the van Ulden model.

4. CONCLUSION

A different approach to gravity spreading is gained from
the sole consideration of energy balance. The modified
approach yields a faster decrease of the front velocity
as a function of the cloud radius, than the van Ulden
model.

To improve the existing box models (which are for certain
areas a very cost-effective tool) energy balance especially
should be checked.
To use the above-mentioned approach the, friction velocity
must be determined by using large-scale experiments.

5. REFERENCES

Burgess, D.S., Murphy, J.N., Zabetakis, M.G. (1970);
        Hazards of LNG Spillage in Marine Transportation,
        U.S.Bureau of Mines, Performed for the U.S. Coast
        Guard (NTIS AD-705078)
Burgess, D.S., Biordi, J., Murphy, J.(1972); Hazards of
        Spillage of LNG into Water, U.S. Bureau of Mines,
        Performed for the U.S. Coast Guard (NTIS AD-754498)
Burgess, D.S. et al. "Volume of flammable mixture resulting
        from the atmospheric dispersion of a leak or spill",
        15. Int.Symp.Combustion, Tokyo, Reprinted by the
        cimbustion institute, Pittsburgh, P. 289 (1974)
Colenbrander, G.W. (Sept. 1980); A Mathematical Model for
        the Transient Bahavior of Dense Vapour Clouds,
        3rd Loss Prevention Symp., Basle, Switzerland
Drake, E.M.; Harris, S.H., Reid, R.C.; "Analysis of vapour
        dispersion experiments", in (American Gas Assicia-
        tion), "LNG Safety Program Interim Report of Phase
        II Work, A.G.A. Project IS-3-1, Battelle Columbus
        Laboratories (1974)
Eidsvik, K.J. (1980); A Model for heavy Gas Dispersion in
        the Atmosphere, Atm. Env. 14, 769

Fay, J.A.; Lewis, D.H.; "The inflammability and dispersion of Hazardous Cargoes by Sea and Inland Waterways, Report CG-D24-76, p. 489 (1975)

Flothmann, D.; Nikodem, H.J. (1980); A Heavy Gas Dispersion Model with Continuous Transition from Gravity Sprea-ding to Tracer Diffusion, in Heavy Gas and Risk Assessment, ed. S. Hartwig, D. Reidel Publishing Co., p. 89

Foussat, A. (Jan. 1981); Modele de Dispersion Atmospherique Non-Isotherme D'un Polluant Gazeux de Densite Quel-conque en Presence de Non-uniformites Orographiques, von Karman-Institute, Technical Note 135

Fryer, L.S., Kaiser, G.D. (1979); DENZ-A Computer Programme for the Calculation of the Dispersion of Heavy Toxic or Explosive Gases in the Atmosphere, SRD R 152, UKAEA

Germeles, A.E.; Drake, E.M.; "Gravity Spreading and Atmo-spheric Dispersion of LNG Vapor Clouds", Proc. 4th Int. Symp. Transport of Hazardous Cargoes by Sea and Inland Waterways, p. 519-539, Report CG-D-24-76 (1975)

Havens, J.A.; "Predictability of LNG Vapor Dispersion from Catastrophic Spills onto Water: An Assessment", U.S. Coast Guard Report CG-M-09-77, April 1977 (NTIS AD/A-040 525)

Havens, J.A. (Febr. 1979); A Description and Assessment of the SIGMET LNG Vapor Dispersion Model, U.S. Coast Guard Report CG-M-3-79

Schnatz, G.; Flothmann, D. (1980); A 'K' Model and its Modification for the Dispersion of Heavy Gases, in Heavy Gas and Risk Assessment, ed. S. Hartwig, D. Reidel Publishing Co., p. 125

Ulden van, A.P. (1974); On the Spreading of a Heavy Gas Released Near the Ground, Proc. 1st. Int. Symposium on Loss Prevention and Safety Promotion in the Process Industries, Delft, Netherlands, pp. 221-226

Ulden van, A.P. (1980); The Unsteady Gravity Spread of a Dense Cloud in a Calm Environment, J. Env. Sci. 20, No. 1. Also presented at NATO/CCMS Conference, Nov. 1979.

ENTRAINMENT MECHANISMS OF AIR IN HEAVY GAS CLOUDS

K. Emblem and T.K. Fanneløp
SINTEF   NTH
7034 - TRONDHEIM-NORWAY

ABSTRACT

A laboratory experiment has been designed to highlight the mechanisms
of turbulent entrainment in a spreading heavy gas cloud due to local
velocity differences. The time dependent volume source is liquid
nitrogen which evaporates through contact with water and the gas
spreads in a two-dimensional channel over an insulating floor. The
most characteristic feature of the cloud is the raised head. This
head may develop into a pronounced front vortex which at times is
observed to detach from the main cloud. The vortex gains intensity
from a continuous feed of heavy material along the cloud bottom to the
front. Detachment occurs when this feed ceases due to friction.
Mixed gas and air appears to be continuously shed from the frontal
vortex. Therefore the main cloud slides forward under a "superlayer"
of negatively buoyant material. The cloud behind the front vortex is
nearly flat and the velocity in the cloud proper is also nearly un-
changed at growing distances aft the gravity current head. These
observations are related to the gradual spill and will probably not
be found after an instantaneous gas release.

1. INTRODUCTION

In making engineering predictions of heavy gas dispersal in the
atmosphere, the greatest uncertainty stems from lack of information
on the turbulent entrainment of air into the cloud. There is dis-
agreement both on the type of entrainment mechanism which is most
important, and on the numerical values of the related entrainment
coefficients. Suggested values differ by an order of magnitude.
It is believed that the entrainment rate is influenced primarily by
the local difference in velocity between the cloud and the air both
at the top interface and at the front, but atmospheric turbulence is
also important. Of secondary importance are, for cold clouds in
particular, convective currents inside the cloud due to bottom surface
heat transfer and water vapor condensation. In a field experiment all
these effects are present simultaneously and it is correspondingly

S. Hartwig (ed.), Heavy Gas and Risk Assessment - II, 67–75.
Copyright © 1983 by Battelle-Institut e.V., Frankfurt am Main, Germany.

difficult to isolate and quantify individual contributions. We have
for this reason attempted a laboratory experiment designed to highlight the mechanism of turbulent entrainment due to local velocity
differences.

2. EXPERIMENTAL LAYOUT

The experiments were conducted in a deep channel, 24 m long, 2.4 m
wide and 1.6 m, with walls of clear plastic draped over a wooden frame
(see Fig. 2.1). Liquid nitrogen could be released at one end in a
controlled and repeatable fashion from a tray spanning the full width.
Along the bottom of the channel there was a water layer, 16 cm deep
mostly covered with material of low heat conductivity (floating
styrofoam plates). Only an area of 1.7 m length adjacent to the tray
had an exposed water surface. Upon release and subsequent contact
between the liquefied gas and the water, vigorous boiling occurred
resulting in a heavy cloud of cold nitrogen gas spreading along the
length of the channel. A typical spill volume was 8 liters of LN_2
and the boiling would last for about 40 seconds.

The purpose of the insulating surface was to minimize heat transfer
to the cloud from the water as it gives rise to thermal convection of
strength sufficient to influence the entrainment. This strong thermal
effect was observed in our earlier experiment (Krogstad and Emblem
1980, SINTEF report No. A80005). One purpose of the present experiment
was to study the effect of the forward (gravity generated) motion on
entrainment, and it was thought important therefore to eliminate the
thermal effects as far as possible.

2.1. Measurements

Velocity measurements: Conventional aerodynamic methods are not
suitable for velocity measurements, and a pulsed hot wire probe was
developed for the present purpose. Only one probe was available and
therefore only a limited number of positions could be conversed. The
measured velocity is the component in the horizontal plane in the
direction along the channel. Temperature measurements: These were
made by use of 25 thermo-couples with rapid response triggered by the
arrival of the cloud front. The thermo-couples were placed at different
heights and at different distances downstream from the point of release
and from the signals we can determine the leading edge velocity and
(with less precision) the height of the dense cloud. The positions
of the thermo-couples were changed during the experiments to get a
more complete set of measurements in certain regions. Concentration
measurements: The gas concentration was measured by a set of four
analyses testing for the paramagnetic properties of the gas. The
sampling probes were placed on a vertical column and the relative
positions of the sampling points could be changed. In the present
experiments this column was positioned in the far half of the channel
relative to the point of release. Visual observations: Timed

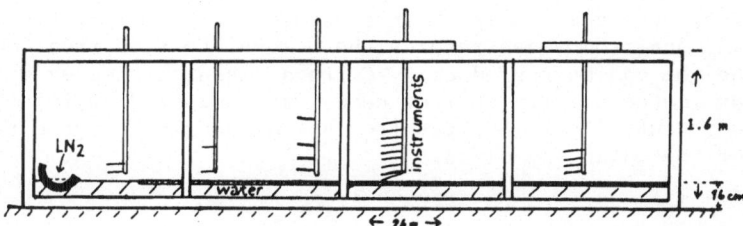

Fig. 2.1 The experimental channel

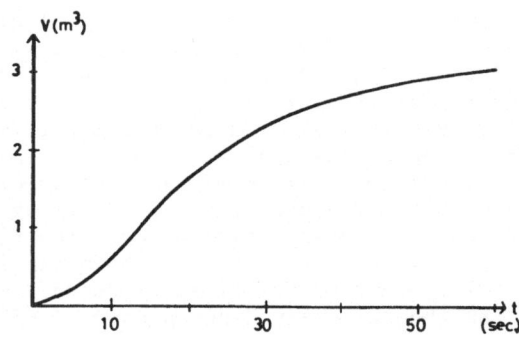

Fig. 3.1 Volume of evaporated gas vs time

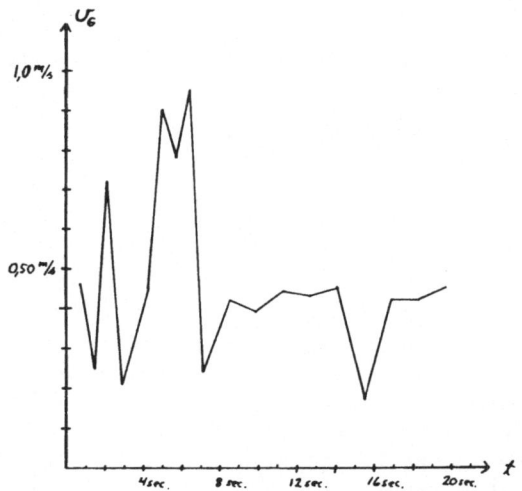

Fig. 3.2 Typical variation of velocity inside the cold
layer as function of time (t = 0 corresponds to
arrival of leading edge at the position of the
velocity probe) Height of probe: 0.03 m,
Position: 19.14 m.

photographs and continuous film coverage are available for a number
of spills and from these records the time-history of front motion and
cloud height can be reproduced. Certain visualization experiments of
the flow in the gravity current head were conducted by injecting
coloured, neutrally buoyant gas from a line source near the floor just
ahead of the advancing cloud.

3. RESULTS AND DISCUSSION

3.1. Source conditions

When the liquid nitrogen is poored onto water a pool of nitrogen will
spread on the water surface. The area of this pool will after a while
reach a maximum. Then the pool area will decay until the evaporation
is completed. Measurements have shown that in the present experiments
the whole evaporation process has occurred through film boiling.
Considerable superheating of the gas has also been measured. A mean
heat flux from the water/ice of 19 kW/m^2 has been calculated from the
measured data of superheating and pool-area. Assuming that the heat
flux is constant through the evaporation process the time-dependent
released volume of gas can be estimated (as shown in Fig. 3.1). A
power curve fit to Fig. 3.1 over the first 24 seconds shows good
correspondence to the theoretically deduced power 1.67 which stems from
the assumption that pool area grows as t to the 2/3 power (Fanneløp and
Waldman, 1972, AIAA J, 10, pp 506-510). The best power fit to the volume
curve is 1,75 up to 17 seconds which decays to 1.67 at 24 seconds.

3.2. Velocity measurements

According to the box model for time-dependent sources the leading edge
position will be given by

$$X_L = (\frac{3}{2+q})^{2/3} K^{2/3} t^{\frac{2+q}{3}}$$

(3.1)

where

$$K = [k \frac{\Delta\rho}{\rho} g V_0]^{\frac{1}{2}}$$

Here $q \approx 1,67$ for $t < 24$ seconds as described in section 3.1. Table
3.1 gives a comparison between the calculated leading edge position
and the positions found in the experiments. A considerable discrepancy
is shown. Two possible reasons for this discrepancy are that viscous
effects are neglected in the box model and that this model assumes a
homogeneous cloud which overestimates the $\Delta\rho \cdot H$ product in the leading
edge region especially in the case of a time-dependent source.
In the same series of releases, the velocity probe for internal
measurements was successively mounted at a height of 0.03 m at distances
7.8 m and 19.14 m. The local horizontal velocity inside the cloud was
found to be highest in the neck region close to the floor and it was

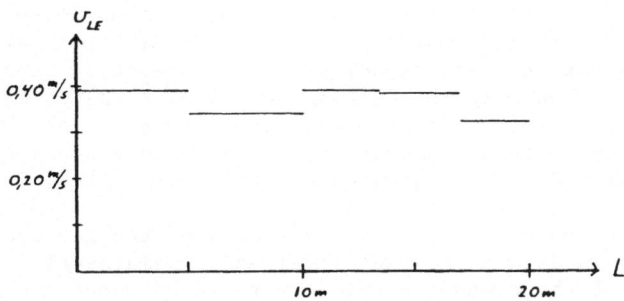

Fig. 3.3 Leading edge velocity averaged over distance between stations at different intervals from the source.

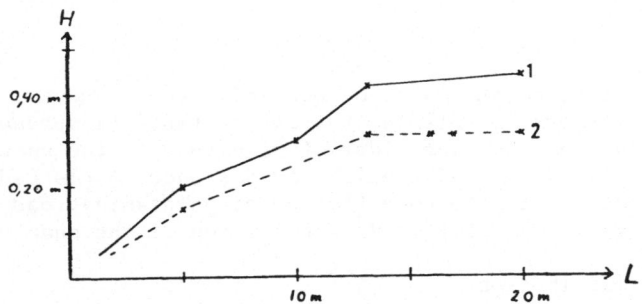

Fig. 3.4 Height of 1) gravity current head, and 2) top of superlayer (in the main cloud) both versus distance from point of release.

Time after release (sec)	14	24
Estimated front position (m)	12	23
Measured front position (m)	5	8

Table 3.1

observed to be more than twice the frontal speed in some cases.
See Figs. 3.2 and 3.3. (Britter & Simpson (1979) (JFM. 94, pp 477-496)
found a 30 % overshoot in their steady state experiment). There is a
corresponding rapid change in the driving potential ($\Delta\rho g$) in the neck
region. This effect is apparent from the temperature traces shown in
Fig. 3.4. The rapid increase in velocity is probably associated with
dynamic effects related to the presence of the front vortex.

Another interesting feature is that the velocity of the gas behind
the neck region would drop to a lower level and remain nearly constant
over a large part of the cloud. A velocity ratio of about 1,0 to 1,5
relative to the leading edge velocity is observed. This can be
explained by the time-dependent source and does not contradict the
general assumption that the gas reaches the maximum velocity near the
leading edge in cases of instantaneous spills.

3.3. Cloud heights

Exact measures of cloud heights are not possible, they depend
also on how the thickness is defined, but the statistical spread in
the data obtained are of the same order of magnitude as the uncertainty
in the measurements (0.06 m). The height as distance is plotted in
Fig. 3.4. It is interesting to note that detailed examinations of
records show zero growth rates towards the far end of the channel.

3.4. Temperature distribution

A selection of temperature profiles inside the cloud at a fixed
position is shown in Fig. 3.5. The profiles across the head show
greater vertical extent of the cold layer than that found in the main
cloud at some distance from the head. Strong entrainment in the vortex
region is evident from the higher temperatures. In the neck region the
warm mixed gas is driven down almost to the floor, but a shallow cold
layer which feeds the vortex along the floor is always present except
after detachment. The main cloud is nearly "tophat" in character as
far as the temperature is concerned, but a thick superlayer provides
a gradual transition to the temperate air layer above.

The mean temperature in the cloud increases almost linearly. This is
illustrated by the temperatures plotted in Fig. 3.6 which are measured
at a fixed height of 0.02 m. The heat flux from the floor is due to
conduction through a thin air layer (< 0,002 m) above the floor. The local
increase rate of the   temperature in the gas cloud due to conduction
is 1.6 K/sec, 0,4 K/sec, 0.2 K/sec and 0,04 K/sec at distances 5 m,
10 m, 13 m and 20 m, respectively. The measured increase in temperature
is more rapid than this should indicate, due to air entrainment. A rough
estimate shows that about 2/3 of the energy required to heat the cloud
to the recorded temperature comes from mixing.

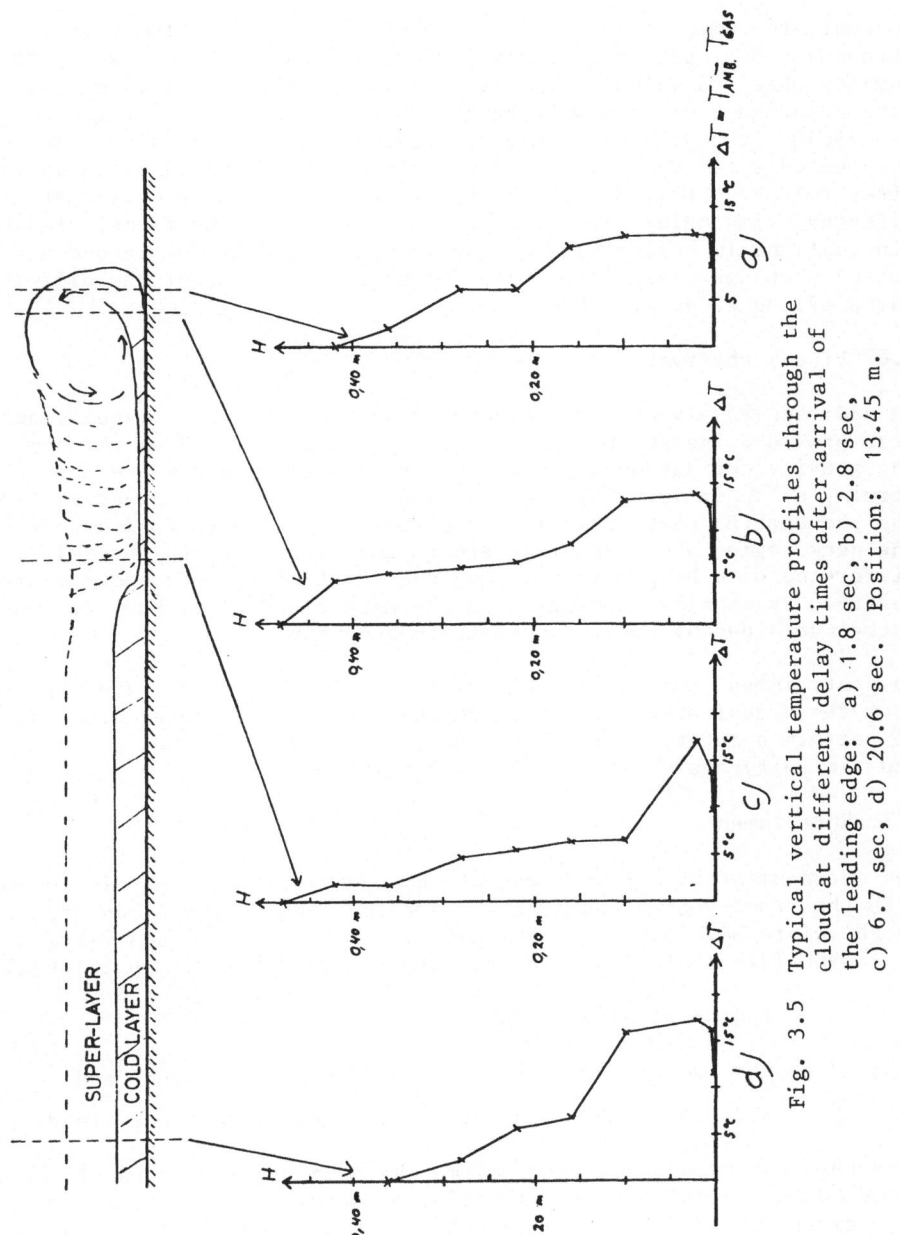

Fig. 3.5 Typical vertical temperature profiles through the cloud at different delay times after arrival of the leading edge: a) 1.8 sec, b) 2.8 sec, c) 6.7 sec, d) 20.6 sec. Position: 13.45 m.

3.5. Concentration measurements

Concentration measurements have been obtained only at two different
streamwise stations in four vertical positions. Fig. 3.7 shows the
maximum observed values for a typical run at the two stations. The
concentration decreases with height, but this drop is more gradual at
x = 19.10 m than closer to the release, i.e. at x = 12.75 m. Not
unexpectedly the dense layer is clearly more pronounced at early times
after release. Fig. 3.8 shows typical concentration profiles at
different time delays from the time of passage of the front. Rapid
fluctuations in concentration cannot be detected by the method used,
but slow changes associated with the characteristics of the different
parts of the cloud are observed.

3.6. Visual observations

In addition to giving cloud thickness information, the visual records
also provided insight into the overall flow patterns. From the top of
the gravity current head pockets of the gas/air mixture are shed,
forming a "super layer" above the cloud proper. It is observed from
the photographs that identifiable pockets are slowly losing height in
the neck region while being stretched and tilted due to the velocity
difference with height in this region. Part of the gas pockets are
observed eventually to merge with the main cloud, i.e. they are
entrained aided by their own negative buoyancy.

The entrainment and flow processes in the frontal region are apparent
from the visual observations where the entrained coloured smoke is seen
to perform a rotary motion from the leading edge, around to the neck
and (in part) again forward to the leading edge.

3.7. Entrainment

The entrainment of air into the cloud is observed to occur in connection
with the front vortex and in the sheer layer above the cloud proper.
We therefore have estimated the entrainment by a model described by
Fay, 1980 (Sec. Int. Symp. on Stratified Flows, Vol. 1, pp 471-494).

$$dV/dt = \pi R^2 U_z + 2 \pi RHU_r$$

adopted to the two dimensional case. Here $U_z = U_{LE} \cdot 10^{-3}$ and
$\dot{U}_R = U_{LE} \cdot 5.3 \cdot 10^{-2}$ and U_{LE} is the front velocity of the cloud.

From the estimated entrained volume the mean concentration of the
cloud can be calculated as a function of leading edge position.
This gives $\bar{C}$ = 83 % for X_L = 13 and $\bar{C}$ = 66 % for X_L = 19. Although
the concentration measurements are scattered and few, this seems to
correspond reasonably well to concentrations recorded in the experiments.

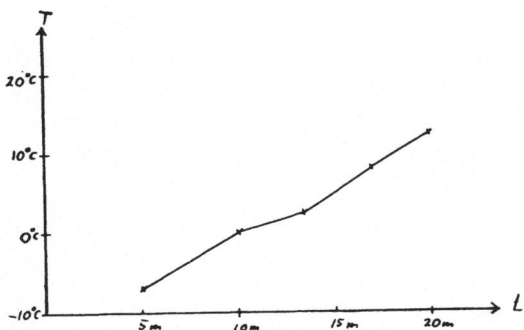

Fig. 3.6 Mean horizontal temperature profile at height 0.02 m
at the time when the leading edge reaches the far
end of the channel.

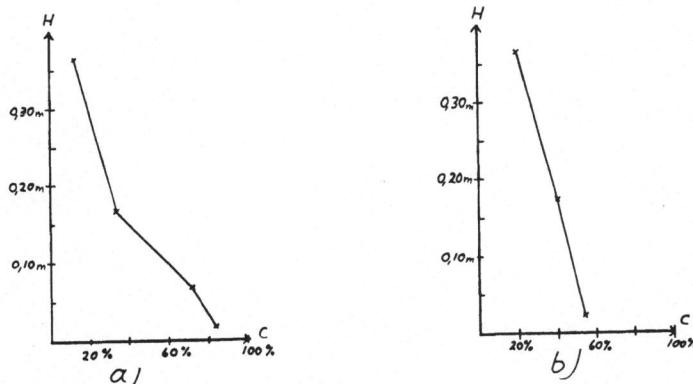

Fig. 3.7 Peak recorded values of concentration at
different heights.
(Data shown averaged over a large number of tests)
Position: a) 12.75 m, b) 19.10 m.

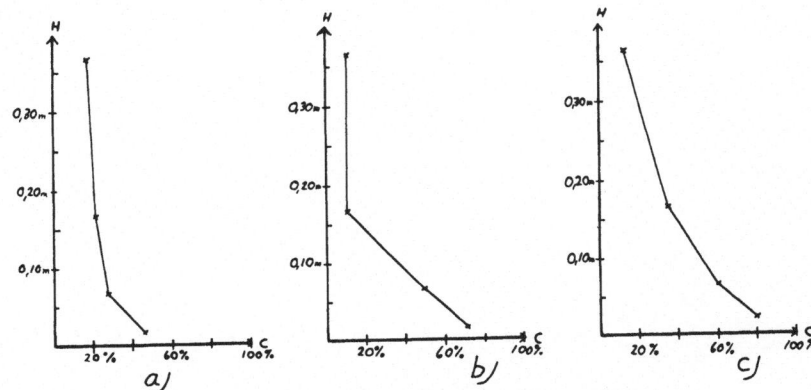

Fig. 3.8 Typical vertical concentration profiles at different
times after arrival of the leading edge: a) 7,0 sec,
b) 13.5 sec, c) 19.0 sec. Position: 12.75 m.

RECENT DEVELOPMENT OF A SIMPLE BOX-TYPE MODEL FOR DENSE VAPOUR CLOUD
DISPERSION.

C.I. Bradley, R.J. Carpenter
British Gas Corporation, Midlands Research Station.

P.J. Waite, C.G. Ramsay, M.A. English
Cremer and Warner

ABSTRACT

Hazard and risk analyses of terminal, storage and processing facilities
storing toxic or flammable liquids and gases often requires the
evaluation of many postulated release scenarios involving the evolution
of dense vapour clouds. If large sophisticated dispersion models are
used then the cost of this evaluation can become prohibitive. However,
simple models, based upon sound physical principles, can be applicable
to the evaluation of the dispersion from many release situations. This
paper describes some recent improvements to a previously published
simple mathematical model (Cox and Carpenter, 1979)[1]. These include
an allowance for vapour cloud momentum and ambient wind profile in
calculating the cloud advection speed. A technique for the application
of the model to transient vapour evolution rates is also described.
The use of the model in hazard and risk analyses is discussed.

1. INTRODUCTION

It is necessary to make careful assessments of the hazards and risks
associated with such activities as the transportation and storage of
large quantities of liquefied gases. In any hazard or risk analysis of
liquefied gas facilities the consequences of accidental releases of
regrigerated and pressurised liquids must be assessed. One of the
possible hazards following such a release is the formation of a dense
vapour cloud due to the vaporisation of the liquid into the atmosphere.
These gases may be either flammable (e.g. LNG and LPG), toxic
(e.g. Cl_2) or both (e.g. acrylonitrile - ACN). An important part of
any risk analysis is the prediction of how far these dense gas clouds
can travel before their concentrations are reduced to a safe level by
mixing with the atmosphere.

Currently, vapour dispersion predictions can be carried out with
essentially two distinct types of mathematical models. The simpler of
these is the 'box' type model which treats the vapour cloud as a single

S. Hartwig (ed.), Heavy Gas and Risk Assessment - II, 77–89.

cell, or box, containing vapour at a uniform concentration. Air is
allowed to mix into the box and dilute the vapour as it is carried off
downwind. The box is allowed to increase in width as it spreads due to
gravity slumping. This type of model can be viewed as a one-cell
Lagrangian model since the box moves with the fluid. The main
alternative to the 'box' model is the large three-dimensional
hydrodynamic model which provides, in principle, a more detailed and
rigorous approach. These models can use a mathematical framework of
the order of 10,000 cells, or boxes, in their computations and the
vapour cloud is assumed to pass through these as it moves downwind.
The larger models are therefore capable of providing a much more
detailed description of the cloud as it mixes with the atmosphere. The
grid in the three-dimensional models is fixed, that is Eulerian, and
the velocity field must be calculated for the whole domain using the
Navier-Stokes equations. The advection and dispersion of the
contaminant gas or vapour is then calculated from the convective-
dispersion equation.

     In a risk analysis, dispersion calculations have to be carried out
for a variety of spill scenarios for a range of wind speeds, wind
directions and atmospheric stability conditions. This implies a large
number of calculations. The three-dimensional codes, although
conceptually superior to the box models, tend to be expensive to run
and therefore their routine use in risk analyses is somewhat
impractical. Three-dimensional codes, however, do in principle offer
the advantage of being able to account for the effects of large objects
(e.g. walls, buildings and storage tanks) on the advection and mixing
of the vapour cloud. Box models can only approximate the presence of
these objects in a global way by adjusting the effective surface
roughness or adding kinetic energy. Consequently where terrain can be
represented as aerodynamically rough but topographically flat, box
models can be used to obtain reasonable estimates of vapour dispersion
distances. On the other hand the use of these models for a site where
large obstructions are present would, in general, result in the
overestimation of vapour dispersion distances. A pragmatic approach to
the hazard assessment of a facility where large obstructions are
present would be to examine the effects of the obstructions with a
three-dimensional model for selected releases but to perform most of
the consequence analyses with the simpler model paying due regard to
these demonstrated effects.

     At the last Battelle Symposium on Heavy Gases a simple method of
the box-type was described (Cox and Carpenter, 1979)[1]. The model was
designed to predict the dispersion of both 'instantaneous' and
'constant continuous' releases of dense cold gas. It allowed for the
following phenomena:-

1. Spreading of the cloud due to gravity

2. Advection of the cloud downwind at a constant speed equal to the
   wind speed at 10m height

3. Air entrainment through the leading edge and top surface of the cloud during the gravity-dominated phase of cloud development

4. Heat transfer from the ground, from the mixing with entrained air, from the mixing with entrained water vapour in the air and from the mixing with entrained liquefied gas droplets

5. Dispersion in a Gaussian manner during the atmospheric turbulence-dominated phase.

It was recognised that there were three major deficiencies in the model as it was then published. First was the assumption that the cloud always travelled at the wind speed at 10m height. Second was the difficulty encountered when applying the idealised release types to real releases since real releases often vary significantly with time. Finally it was found that the model predictions for dispersion of LNG vapour during stable weather conditions were unrealistically sensitive to the assumptions governing the transition from gravity-dominated dispersion to atmospheric turbulence-dominated dispersion. (The dispersion of higher molecular weight gases, such as LPG vapour, is less sensitive to this transition criterion).

In this paper the first two of the above problems are addressed. A momentum equation has been included in the model to calculate the advection of the cloud, and a method is described for application of the model to highly transient gas releases. Finally a description of the use of simple box type models in hazard and risk analyses is given at the end of the paper.

2. CALCULATION OF VAPOUR CLOUD ADVECTION

Hitherto, most box-type models have either assumed that the vapour cloud travels at a constant speed equal to the wind speed at a fixed height (e.g. Cox and Carpenter)[1] or that the cloud travels at the wind speed at some fraction of the cloud height. Clearly both assumptions are, at best, only approximations and in a rigorous treatment of dense vapour cloud advection the conservation of momentum should be taken into account.

For an instantaneous release of a cloud of dense vapour, Figure 1, it is found that the total rate of change of momentum of the cloud is given by:

$$\frac{d}{dt}(m_t V) = \frac{d}{dt}\left[(m_s + m_e + m_v)V\right]$$
$$= D - F + \dot{m}_e U_{av} + \dot{m}_s U_H \qquad (1)$$

which is simply a statement of momentum conservation in the frame of reference moving with the cloud. The first term on the right hand side

a) INSTANTANEOUS RELEASE

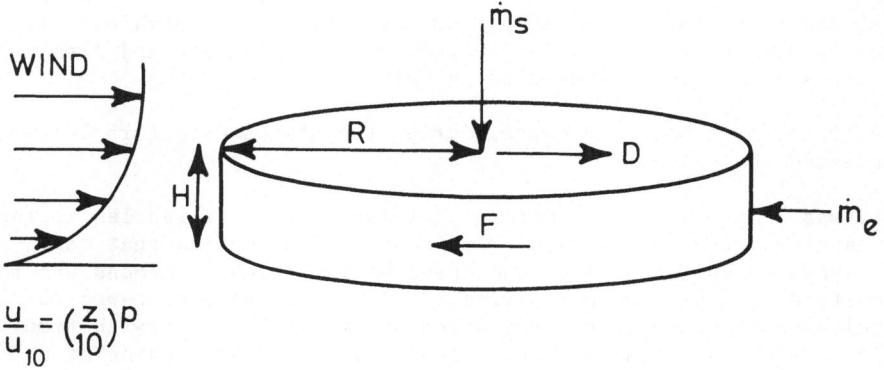

$$\frac{u}{u_{10}} = \left(\frac{z}{10}\right)^p$$

b) CONTINUOUS RELEASE

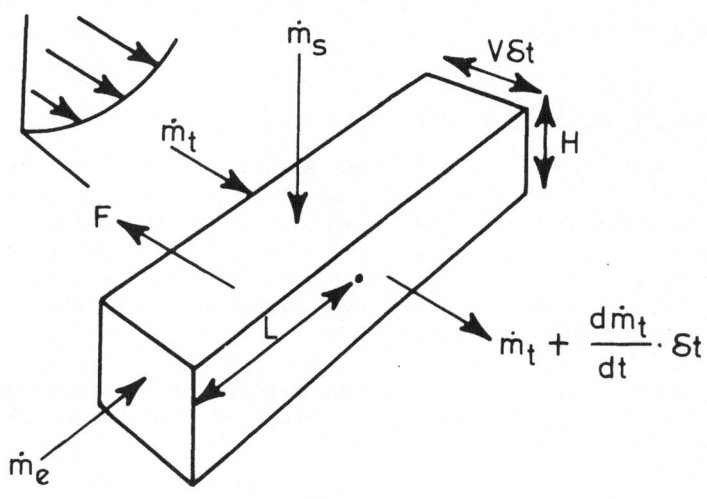

FIG. 1. PLUME / CLOUD GEOMETRY

of equation (1) is the aerodynamic drag experienced by the cloud due to
the air moving over it. (For description of symbols used in the equations,
see below.) The second term is the friction ional drag from the ground
and the third and fourth terms allow for the increase in momentum due
to air entrainment at the edge and top surface of the cloud, respectivily.
The aerodynamic drag, D, is calculated by using a drag coefficient, C,
which is assumed to take the value 1.0 such that

$$D = \int_0^H C R \rho_a (u_{(z)} - V) |u_{(z)} - V| \, dz \qquad (2)$$

The friction force, F, is calculated by analogy with the shear stress
in the atmospheric boundary layer giving

$$F = \tau \pi R^2 = \rho_c \left(\frac{u_*}{u_{10}}\right)^2 V^2 \pi R^2 \qquad (3)$$

Re-arranging equation (1), the acceleration of the cloud is given by

$$\frac{dV}{dt} = \frac{1}{\dot{m}_t} \left[\dot{m}_e (u_{av} - V) + \dot{m}_s (u_H - V) + D - F \right] \qquad (4)$$

The average velocity over the height of the cloud, u_{av} can be defined
in terms of the exponent, p, of the velocity distribution.

$$\frac{u}{u_{10}} = \left(\frac{z}{10}\right)^p \qquad (5)$$

Whence,

$$u_{av} = \frac{u_H}{p+1} \qquad (6)$$

For a continuous release of dense vapour the total rate of change
of momentum in the element shown in Figure 1, which is moving with the
plume, is given by:

$$\frac{d}{dt}(\dot{m}_t \, \delta t \, V) = \frac{d}{dt}\left[(\dot{m}_s + \dot{m}_e + \dot{m}_v) \delta t \, V \right]$$

$$= \frac{d}{dt}(\dot{m}_s \delta t \, u_H) + \frac{d}{dt}(\dot{m}_e \delta t \, u_{av}) - \tau 2 L V \delta t$$

$$- g\left[(\rho_c - \rho_a) \frac{d(LH^2)}{dt} + LH^2 \frac{d\rho_c}{dt} \right] \delta t \qquad (7)$$

The first two terms in equation (7) account for momentum addition due
to air entrainment, and the third term is due to frictional drag. The
final term has no analogue in equation (1) since it allows for the
interaction of one 'slice' with another due to the variation in cloud
dimensions and density as the plume is advected downwind. There is no
aerodynamic drag term in equation (7) since the main effect of this
force for a continuous release is to reverse the upwind flow and give
the plume its initial acceleration.

Re-arranging equation (7), the acceleration of the plume is given by

$$\frac{dV}{dt} = \frac{1}{\dot{m}_t}\left[\frac{d\dot{m}_s}{dt}(U_H - V) + \frac{d\dot{m}_e}{dt}(U_{av} - V) - 2\tau LV - g\left((\rho_c - \rho_a)\frac{d(LH^2)}{dt} + LH^2\frac{d\rho_c}{dt}\right.\right.$$ (8

Equations (4) and (8) provide the information required to calculate the cloud advection speed for instantaneous and continuous releases.

The other equations incorporated into the model have been described fully in a previous publication (Cox and Carpenter, 1979)[1]. Briefly these are as follows:

1) Gravity spreading equations:

$$\left.\begin{array}{l}\dfrac{dL}{dt} \quad \text{(continuous plume)} \\[2mm] \dfrac{dR}{dt} \quad \text{(instantaneous cloud)}\end{array}\right\} = \left(k g H \frac{\Delta\rho}{\rho_a}\right)^{1/2}$$ (9)

2) Mass conservation equations:

$$\frac{dm_t}{dt} = \dot{m}_e + \dot{m}_s = \gamma 2\pi R H \frac{dR}{dt} + \frac{\alpha U_i \pi R^2}{R_i}$$ (10)

for an instantaneous cloud, where $R_i = g\frac{\Delta\rho}{\rho}\frac{L}{U_i^2}$

and $\quad \dfrac{dm_t}{dt} = \gamma 2 H V \dfrac{dL}{dt} + \alpha U_i \dfrac{2LV}{R_i}$ (11)

for a continuous plume.

3) Enthalpy conservation equation:

$$m_{v_0} h_v(T_0) + m_{L_0} h_L(T_0) + m_a h_a(T_a) + m_{wv} h_{wv}(T_a) + q$$
$$= m_v h_v(T) + m_L h_L(T) + m_a h_a(T) + m_{wv} h_{wv}(T) + m_{wL} h_{wL}(T)$$ (12)

This equation allows for the presence of aerosols, both water and liquefied gas, in the cloud.

The mathematical model also allows for heat transfer from the surface of the earth if the cloud is cold and, in addition, includes an equation of state to allow calculation of the cloud density.

The inclusion of the momentum equations in the mathematical model necessitates that the model be re-calibrated against suitable experimental data. This is presently being carried out. It is to be expected that the modified model will yield better predictions of cloud height and the time of passage of the cloud than the earlier model and,

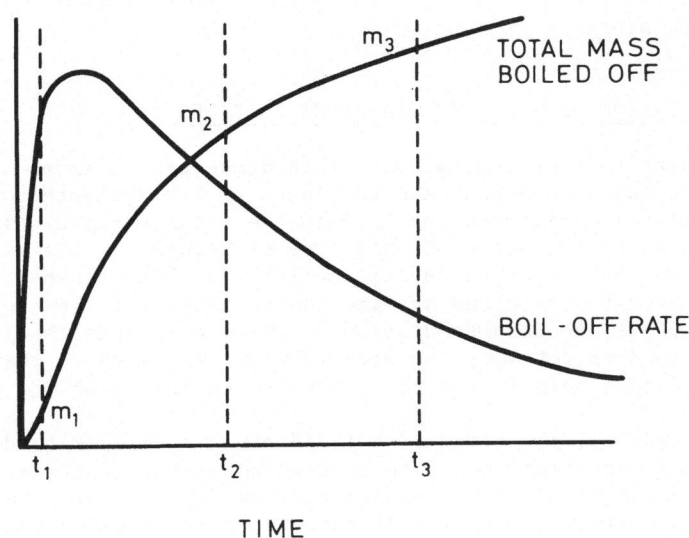

FIG. 2. TYPICAL BOIL-OFF PLOT FOR A SPILL OF CRYOGENIC FLUID INTO A BUND

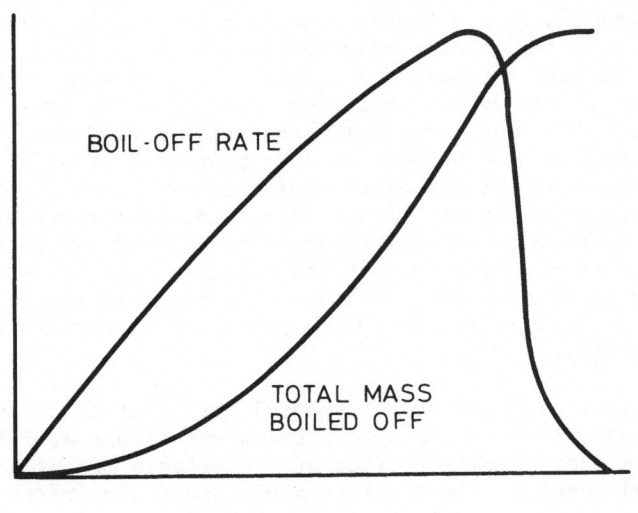

FIG. 3. TYPICAL BOIL-OFF CURVE FOR A SPILL OF CRYOGENIC FLUID ONTO UNBOUNDED WATER

as such, provide a more accurate description of vapour cloud advection
and dispersion.

3. APPLICATION OF MODEL TO TRANSIENT RELEASES

The present mathematical model treats dense gas releases as either
instantaneous or constant and continuous. The instantaneous cloud
description is quite reasonable when the total gas evolution time is
short. The best example of this type of release is the rupture of a
vessel containing pressurised liquefied gas which flashes quickly to
form a large vapour cloud at atmospheric pressure. The constant and
continuous release representation is reasonable when the rate of vapour
evolutiuon does not vary too greatly with time such as that for a small
release from a hole in a storage/transportation vessel.

In many release situations it is often difficult to decide when
the vapour evolution should be treated as instantaneous and when
continuous, and this decision can have an important bearing on the
predicted maximum dispersion distance. It is becoming accepted that
for large, effectively instantaneous, releases of dense gas the maximum
distance to a fixed concentration increases as the wind speed
increases. However continuous releases would not be expected to behave
in this manner, and this has been demonstrated experimentally by
Hall[2].

As indicated above many release situations could produce vapour
evolution rates which are highly transient in nature. The most notable
example would be the release of a reasonably large quantity of
cryogenic liquefied gas into the secondary containment bund surrounding
a storage tank. The vapour evolution rate from this type of release is
neither short-lived nor reasonably constant, as depicted in Figure 2,
and can be seen to rise quickly to a peak value before tailing-off
steadily with time. On the other hand the unrestricted spillage of a
large quantity of cryogenic liquefied gas onto water produces quite
different vapour evolution characteristics as depicted in Figure 3.
For this case it can be seen that the vapour evolution rate
continuously increases with time up to a point when the liquid pool
reaches its maximum size after which point the vapour evolution rate
drops off sharply.

These examples demonstrate the need for a method of approach to
those releases which produce highly transient vapour evolution rates
such that due account can be taken of the effects of spill size, spill
type and wind speed in the prediction of vapour dispersion distance.
Such a method of approach is described below with reference to the
spillage of a large quantity of cryogenic liquefied gas into a bund.
For the spill into a bund, Figure 2, the initial peak in the
vaporisation rate curve is often sufficiently fast to be treated
adequately as an instantaneous release and the latter stages of the
boil-off often varies slowly enough to be treated as a series of

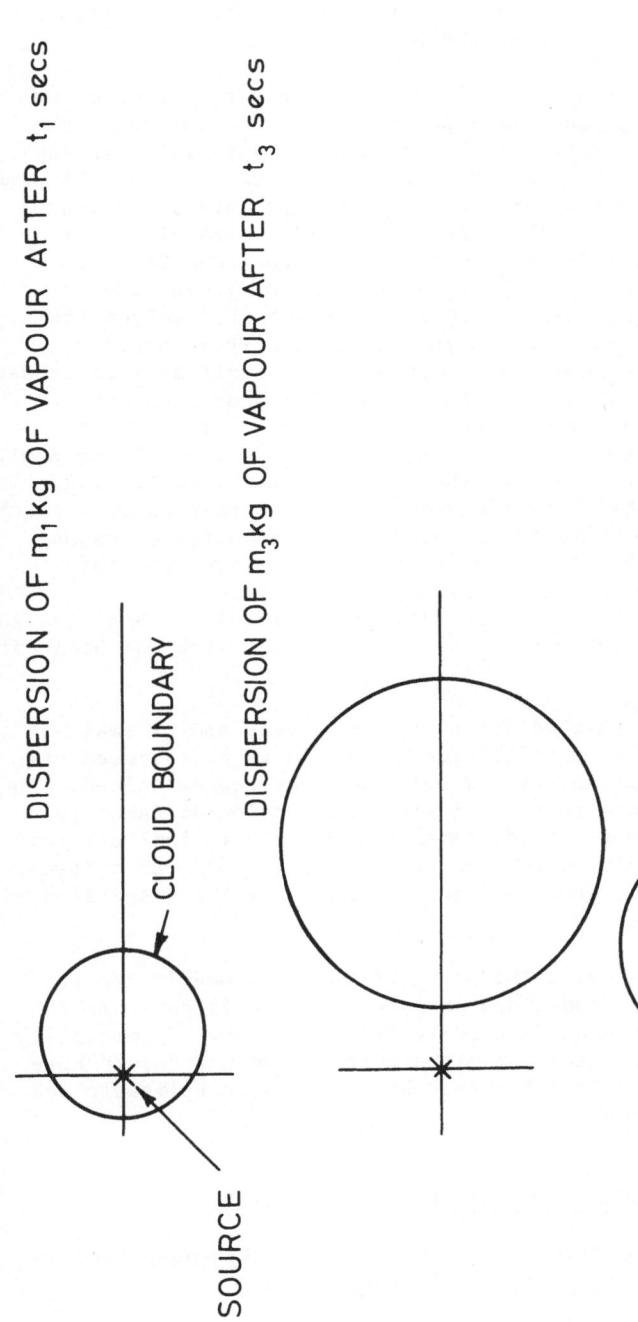

FIG. 4. DISPERSION OF INSTANTANEOUS RELEASE OF VAPOUR

continuous releases. In this case a method of approach is required in order to decide at what point in the vaporisation history one should change from one type of release to the other.

During the initial stages of boil-off the evolving vapour cloud will stay over the liquid pool for a period of time whilst the dense vapour spreads in all directions including upwind (especially at low wind speeds). Eventually the vapour cloud will be advected by the wind off the source whence it will begin to be stretched into a continuous plume. In order to determine the effective initial mass of vapour the instantaneous dispersion model is run to predict the behaviour of different amounts of vapour M_1, M_2, M_3 etc. (Figure 2) over the period of time it takes for each of those amounts to be evolved from the liquid pool (t_1, t_2, t_3, etc). Typical results from these types of calculations are shown in Figure 4. If a small mass is chosen (M_1) then at the time it would have taken for that amount to have boiled-off it is found that the dispersion model predicts that the cloud is still over the source and still being fed by the boiling pool. If a large mass is chosen (M_3) then the cloud has had time to drift off the source and for this case the wind would have started to stretch the cloud into a plume. There is an intermediate quantity of vapour (M_2) that has only just had time (t_2) to clear the source and this is the amount that is assumed to be present in the initial instantaneous release. The continuous dispersion model is then used to deal with the subsequent behaviour starting initially with the boil-off rate at time t_2.

The use of the above method for high wind speeds and/or small releases makes it naturally possible for the cloud to be advected off the source before the peak in the boil-off rate has been attained. In these instances it is found that the maximum dispersion distance is given by the continuous model predictions using the peak boil-off rate. At low wind speeds the initial release is much larger and, as a result the maximum dispersion distance is then controlled by the dispersion of the instantaneous release.

Application of the above method to continuous flammable vapour releases in very low wind conditions can show that it is possible for the initial cloud to have been diluted to below the lower flammability limit before it has drifted away from the source. Such a result here provides an indication that the box-type of model is inapplicable and therefore should not be used.

4. THE ROLE OF SIMPLE MODELS IN HAZARD AND RISK ANALYSIS

The usefulness of a simple dispersion model such as that described in this paper in hazard and risk analyses is three-fold;

(i)   It allows the assessment of a large number of individual failure cases without incurring major costs. This is beneficial in many

applications when it is important that the full range of failure cases
are considered rather than only certain cases modelled with great
accuracy. For example, this can occur in the planning stages of a
proposed new development when the layout and design are being evolved
and the hazard and risk analysis will need to be reworked several times.

(ii) It allows a rapid and economic means of examining the sensitivity
of dispersion distance to important variables such as wind speed,
atmospheric stability, vaporisation rate, relative humidity and surface
roughness.

(iii) It provides dense vapour cloud dispersion information in a
readily usable form insofar as they provide a simple description of the
cloud geometry from which estimates of the important parameters such as
downwind travel distance and crosswind cloud width to specific
concentrations can be obtained.

The results from simple consequence models can be combined and
used in risk contour plotting techniques to summarise the integrated
consequences and probabilities for many hundreds or thousands of
failure cases considered in the analysis of a modern process plant.

Risk contour plotting techniques and their application to the
planning and conceptual design stages of the St. Fergus Coastal Gas
Reception Terminal proposed in 1980 as part of the UK public sector
gas-gathering scheme have been illustrated by Ramsay et al[3]. In the
planning stage of that proposed development at St. Fergus in north-east
Scotland more than 600 postulated failure cases involving the
dispersion of flammable materials were evaluated. In the subsequent
conceptual design stage approximately 3000 postulated failure cases
were considered. The authorities involved in considering this proposed
development utilised the risk contour plots also as a 'safety
criterion' whereby the succession of designs and modifictions were
judged, in part, by their effects on the respective risk contour
plots.

Finally simple models provide an inexpensive means of assessing
the risk from installations where that risk is to be evaluated in terms
of fatalities, either as an average rate of death or by a curve
delineating the frequency of events causing N or more fatalities
against N fatalities. An example of such an application is that
performed by Cremer and Warner for the Rijnmond area[4].

5. CONCLUSIONS

A simple dense vapour cloud dispersion model has been improved so as to
allow better allowance for the effects of vapour cloud momentum and
ambient wind profile on the cloud advection speed. Recalibration of
the modified model is underway and it is to be expected that the new
formulation will yield better predictions of cloud height and the time

of passage of the drifting cloud. A technique for the application of
the model to those practical releases which generate highly transient
vapour evaluation rates has been described so as to provide a
methodology for the prediction of maximum dispersion distance. The
usefulness of simple dispersion models in hazard and risk analyses has
been summarised with reference to their range of applicability and ease
of use.

6. ACKNOWLEDGEMENTS

The authors wish to thank the British Gas Corporation and the partners
of Cremer and Warner for permission to publish this paper. Thanks are
also due to E.B. Winn for his valuable contribution to this work.

7. REFERENCES

1. Cox, R.A. and Carpenter, R.J., 1979. Further development of a dense
   vapour cloud dispersion model for hazard analysis. Battelle
   Institut Symposium, 'Schwere Gase'. Frankfurt.

2. Hall, D.J. Further experiments on a model of an escape of heavy
   gas. Warren Spring Laboratory Report No. LR 312 (AP)

3. Ramsey, G.G., Sylvester-Evans R. and English, M.A. 1982. Siting and
   layout of major hazardous installations. I. Chem. E. Symposium
   Series No. 71.

4. Rijnmond Public Authority, 1982.
   Risk analysis of six potentially hazardous industrial objects in the
   Rijnmond area. A pilot study. Published by D. Reidel.

Notation

Symbol	Meaning
D	Form drag on cloud (N)
F	Friction force on cloud (N)
g	Acceleration due to gravity (m/s^2)
H	Cloud height (m)
h_a	Specific enthalpy of air (J/kg)
h_l	Specific enthalpy of saturated liquid (J/kg)
h_v	Specific enthalpy of saturated vapour (J/kg)
h_{wv}	Specific enthalpy of water vapour (J/kg)
h_{wl}	Specific enthalpy of liquid water (J/kg)
k	Gravity spreading coefficient
L	Cloud semi-width (m)
m_a	Mass of entrained air (kg)
m_e	Mass entrained through edge of cloud (kg)
m_{l_0}	Initial liquid mass (kg)
m_l	Mass of liquid (kg)
m_s	Mass entrained through top surface of cloud (kg)
m_t	Total mass (kg)
m_{v_0}	Initial vapour mass (kg)
m_v	Mass of vapour (kg)
m_{wl}	Mass of liquid water (kg)
m_{wv}	Mass of water vapour (kg)
p	Wind profile exponent
q	Heat flux (J/m^2 sec)
R	Cloud radius (m)
Ri	Richardson number
T	Temperature (K)
t	Time (sec)
U(z)	Wind speed at height z (m) (m/s)
U_{av}	Average wind speed over cloud height (m/s)
U_H	Wind speed at top of cloud height H (m/s)
U_{10}	Wind speed at 10m height (m/s)
U_*	Friction velocity (m/s)
U_1	RMS turbulent velocity in longitudinal direction (m/s)
V	Cloud velocity (m/s)
α	Top surface mixing coefficient
γ	Edge mixing coefficient
$\Delta\rho$	Change of density (kg/m^3)
ρ_a	Density of air (kg/m^3)
ρ_c	Density of the cloud (kg/m^3)
τ	Turbulent shear stress (N/m^2)

N.B. over-dot denotes rate of change

Application of a turbulence flow model to heavy gas dispersion
in complex situations

D.M. Deaves, Atkins R & D, Epsom, Surrey, UK

Introduction

In a recent review of mathematical models for prediction of heavy gas
dispersion in the atmosphere (1), the following categorisation was
suggested for the currently available models:

- box models
- advanced similarity models
- K theory models

The review of box models centred on the different approaches to modelling
entrainment, and the uncertainties involved (2),(3). It was stated that
box models may be unreliable because they depend heavily on entrainment
parameters and other empirical input data. This often means in practice
that the application of such models is limited to situations similar to
those in which the parameters governing entrainment etc. were obtained.

Similarity modelling was discussed with reference to Colenbrander's (4)
model, which is essentially an advanced box model. Real similarity
models obtain analytical solutions to the simplified momentum and continuity
equations (eg. Fannelop (5)), which are derived from a transformation to
similarity variables. Such models show encouraging results compared with
box models, although they suffer from the same inability to deal satisfactorily
with the effects of irregular topography, buildings etc., as illustrated in
Figure 1, and some uncertainties in representing entrainment still exist.

Models in the third category (K theory models) use the gradient transfer
assumption to close the full momentum and continuity equations. In
principle, such models remove the need for most of the empiricism necessary
in box models and are therefore applicable to more complex dispersion
situations in which box and similarity models are inadequate. K theory
models are in fact a subset of a wider category of turbulence modelling
which is discussed below.

Turbulence Modelling for Heavy Gases

When equations of fluctuating motion are averaged with respect to time,
certain unknown second-order correlations appear (eg. $\overline{uw}$), which have to
be modelled in order to close the equations. This closure is known as
Turbulence Modelling, and a good early review of such techniques was
presented by Launder and Spalding (6), and later by Bradshaw et al (7).
For the purposes of turbulence modelling for heavy gases, the following
categorisation is appropriate:

- K theory models
- higher order closures

S. Hartwig (ed.), Heavy Gas and Risk Assessment - II, 91–102.

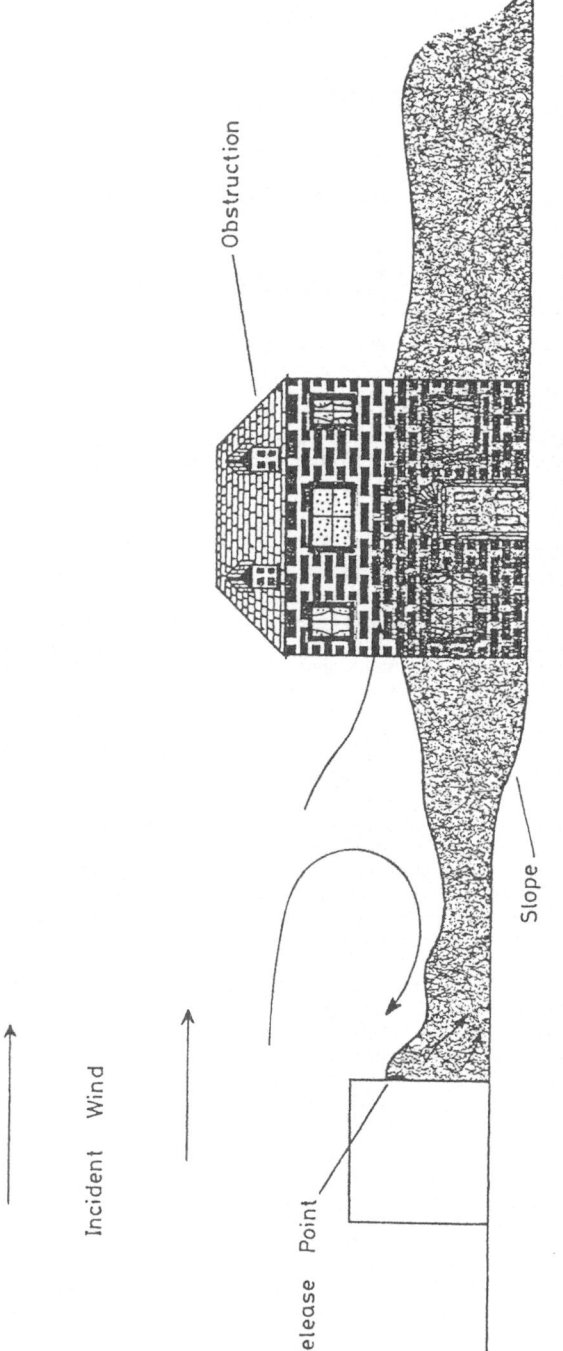

Obstruction

Incident Wind

Release Point

Slope

FIGURE 1   TYPICAL   COMPLEX   DISPERSION   PROBLEM

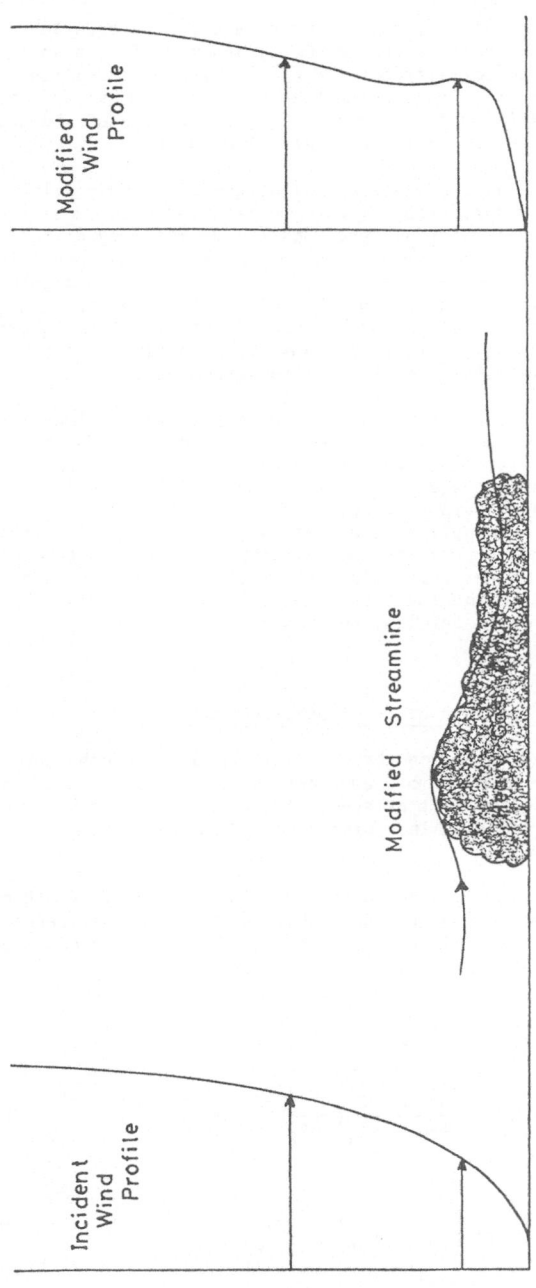

FIGURE 2   MODIFICATION OF WIND PROFILE AND STREAMLINES BY PRESENCE OF

HEAVY GAS

Higher order closures (8) are those in which additional equations are
derived for the Reynolds stresses (eg. $\overline{uw}$), and therefore also for
velocity concentration correlations ($\overline{wc}$ etc.). The new equations
include 3rd order correlations, which then have to be modelled. Such
closure models are still under development, and, because of the additional
equations to be solved, are extremely expensive to run.

The state of the art in turbulence modelling therefore only allows the use
of K theory models for practical dispersion calculations. Most K theory
approaches require the empirical specification of the eddy diffusivities
and eddy viscosities as prescribed functions of space, and possibly of
stability class or Richardson number. This is the approach adopted in
SIGMET (9), which does not allow for the influence of the heavy gas on
vertical buoyancy-driven velocities because of the use of the sigma-
transform in the vertical. Figure 2 shows the effects of a heavy gas cloud
on streamlines and velocity profiles in the atmosphere.

Progress can be made in relating the eddy diffusivities to local effects if
the two-equation (k-ε) model (6) is used. This model, although retaining
eddy diffusivity, allows it to vary in a way which reflects the local
dynamics, since transport equations are also solved for the turbulence
energy, k, and energy dissipation, ε. The constants in the k-ε model are
well documented (8) and apply over a wide range of flows. A slightly
more sophisticated closure can be made by combining with this model an
algebraic prescription for the evaluation of all components of the Reynolds
stress in terms of k, ε and the local mean velocity gradients. This is known
as the algebraic stress model (8), but has not yet been as well proven as
the standard k-ε model.

3.   Development of Equations for a k-ε Dispersion Model

In view of the above remarks on the state of the art in turbulence modelling,
it appears that some form of more advanced turbulence simulation (such as
the k-ε model) offers the best prospects for development of heavy gas
dispersion models. The modelled forms of the equations of motion for such
a scheme are given below.

The computer program HEAVYGAS is currently being developed by Atkins
Research and Development from the CAFE suite (10), which has been used in
a wide range of internal and external turbulent flows. The equations solved are:

$$\frac{\partial \rho}{\partial t} + \frac{\partial}{\partial x_j}\,(\rho u_j) = 0 \qquad\qquad (3.1)$$

$$\frac{\partial(\rho u_i)}{\partial t} + \frac{\partial}{\partial x_j}(\rho u_i u_j) = -\frac{\partial p}{\partial x_i} + \frac{\partial}{\partial x_j}(\mu\frac{\partial u_i}{\partial x_j}) + \Delta\rho g_i \qquad (3.2)$$

$$\frac{\partial(\rho\phi)}{\partial t} + \frac{\partial}{\partial x_j}\,(\rho\,u_j\phi) = \frac{\partial}{\partial x_j}\,(\Gamma_\phi\frac{\partial\phi}{\partial x_j}) + S_\phi \qquad (3.3)$$

Equation (3.3) is the general form of the equation for any scalar ϕ.
The concentration equation takes this form, with $S_\phi = 0$, and Γ_ϕ
the eddy diffusivity; the mass fraction concentration is then used,
and the numerical scheme is found to conserve the overall mass of heavy
gas present. The turbulence energy and dissipation (k and ε) can also
be found as solutions to (3.3), where the term S_ϕ is adjusted to give the
usual form of the k-ε turbulence model equations. Γ_ϕ can then be either
prescribed algebraically (eg. as a function of height for each stability
class), or calculated from values of k and ε, using the equation

$$\Gamma_\phi = c_\phi \; \frac{k^2}{\varepsilon} \; . \tag{3.4}$$

The density, ρ, which is used in the above equations, is the local value,
and is determined from the local heavy gas concentration by

$$\frac{1}{\rho} = \frac{C}{\rho_G} + \frac{1-C}{\rho_A} \tag{3.5}$$

The retention of the local value of ρ throughout the equation of motion
means that the Boussinesq Approximation has not been invoked in its usual
form. However, correlations involving density fluctuations (rather than
concentration fluctuations) have been omitted.

The buoyancy force, $\Delta\rho g_i$, depends on the difference between the local
density, ρ, and a reference density, ρ_o, which is generally taken as the
average along a horizontal grid line. The inclusion of this term allows
the heavy gas to 'drive' local vertical motions and thus to modify the
ambient wind field.

4.   Range of Applicability of the Equations

Equations (3.1) - (3.3) are for the mean quantities ρ, u_i and ϕ.
They have been derived from the fluctuating Navier-Stokes equations,
with the Reynolds stress terms $\rho \overline{u_i' u_j}$ (etc) modelled in the usual (first
order) manner using the eddy viscosity/diffusivity approach. However, the
resultant equations, with second order correlations only, have omitted any
density fluctuations, which, from equation (3.5), will result from
concentration fluctuations.

The additional correlations which appear when the density fluctuations
are considered are of the form $\rho' u_i'$ which could be modelled in the
usual way as $\Gamma_\rho \frac{\partial \rho}{\partial x_i}$. They therefore represent additional diffusion
terms, which are not likely to be significant in highly convective flows,
which are the only ones to have been considered so far using the current
version of the program. It is intended, however, to investigate the effect
of introducing these density fluctuation terms as the next stage in the
program development.

A further restriction to the current version of the model is to isothermal
releases, or at least to releases where temperature effects are not large.
However, an enthalpy equation, of the form (3.3), is available for use
in the program, although evaporation, ground heat transfer and related
thermal effects have not yet been fully investigated.

At present, only a 2D version of the code is in use, with resultant obvious

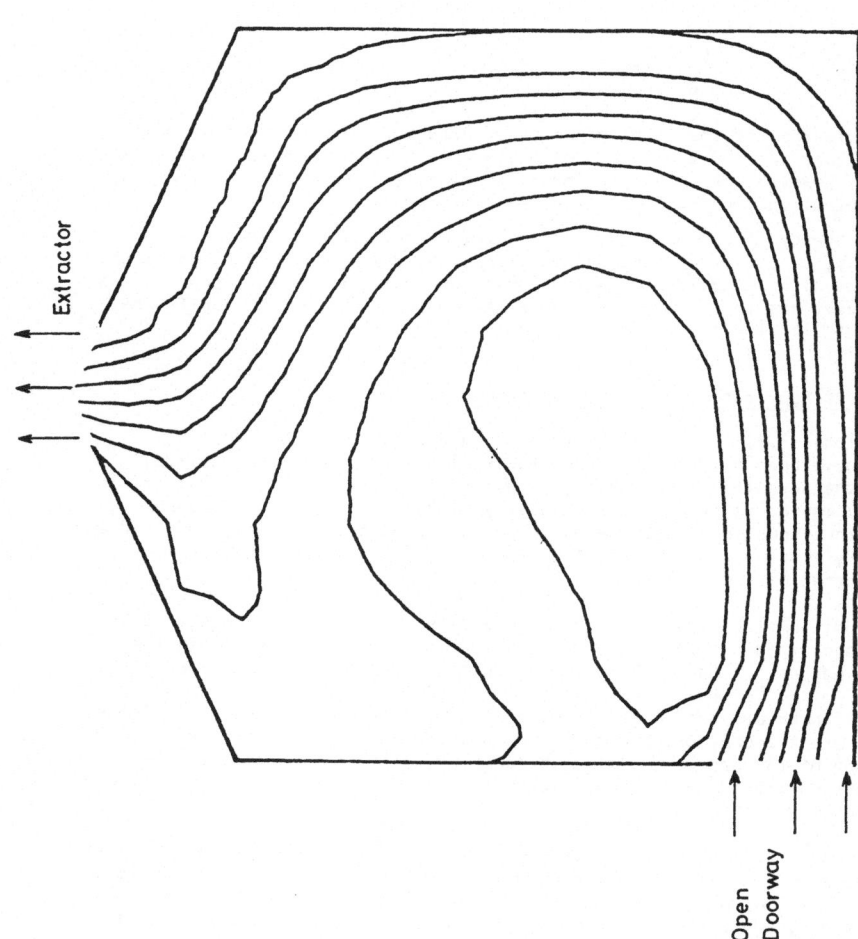

FIGURE 3   STEADY STATE VENTILATING FLOW IN PROCESS BUILDING

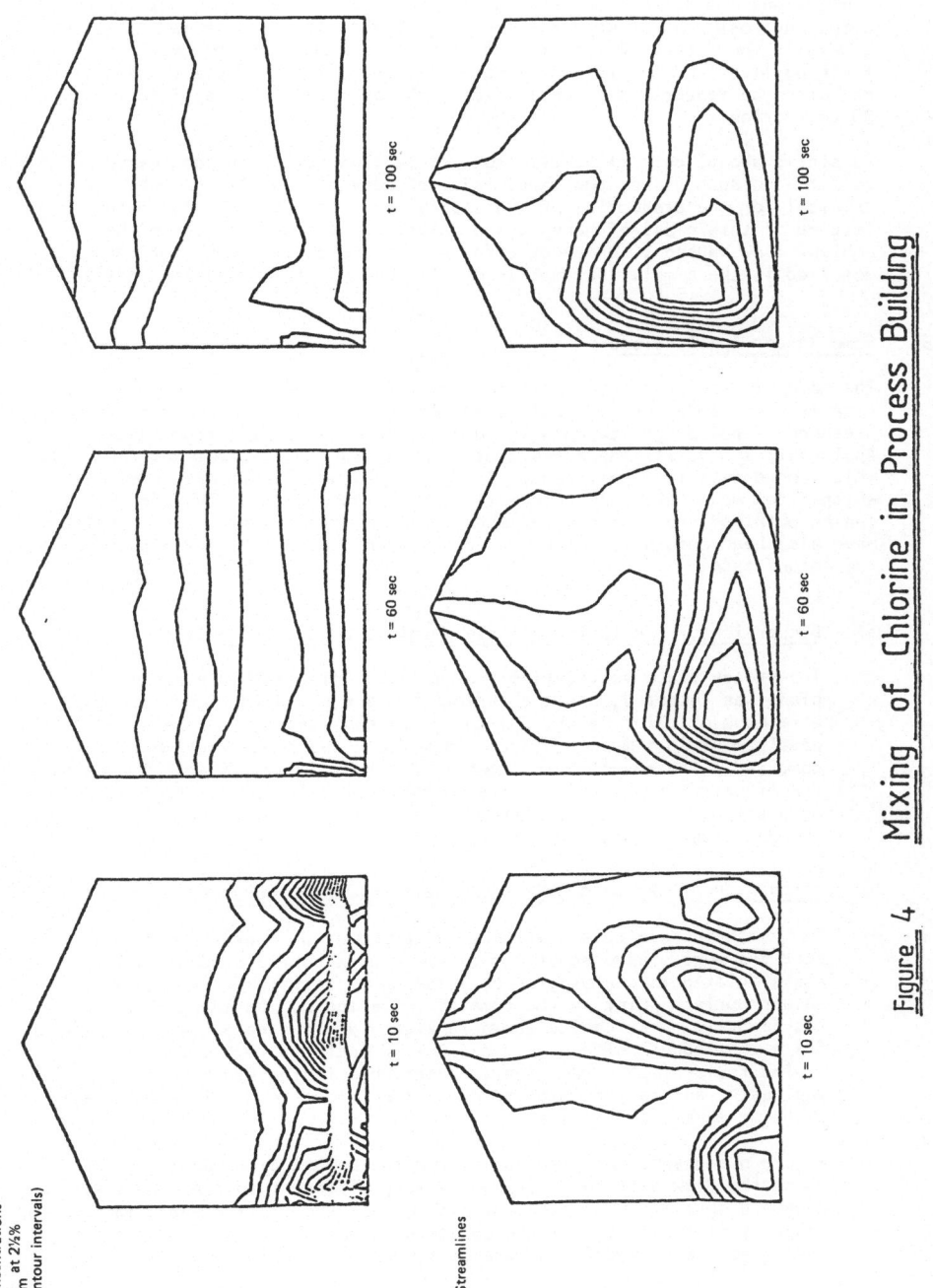

Concentrations
(C_m at 2½%
contour intervals)

t = 100 sec

t = 60 sec

t = 10 sec

Streamlines

t = 100 sec

t = 60 sec

t = 10 sec

Figure 4

Mixing of Chlorine in Process Building

limitations. A 3D version is in preparation and it is hoped that
computing costs can be kept to within reasonable levels. It is
generally found that 2D results are adequate for most of the near
field applications for which the model is appropriate; these results
can often be rendered more realistic by the application of empirical
3D corrections.

As stated above, only flows with some convection have been considered
so far. In such flows, the exact value of eddy diffusivity is not
critical for the production of reasonably accurate results. The main
feature of this class of heavy gas problem is the modification to the
ambient flow caused by buoyancy effects, and these effects cannot be
modelled in the simpler box models or, in general, in similarity models.

5.   Some Recent Applications

The model in its present form is most useful in situations which are
clearly unsuitable for the application of 'box models', eg. in the
presence of buildings, terrain features or other unusual circumstances.
Applications have also been restricted so far to those situations in
which convection is significant. The cases discussed here are those in
which a box model gives no information, but a good qualitative picture
can be obtained from the present model. Approximate quantitative results
have also been obtained, and compared favourably with full scale data for
the second case.

a)   Dispersion of a chlorine release within a ventilated building

The steady state ventilating flow in a building within a process
plant was computed, and the streamlines are shown in Figure 3. Then,
a release of chlorine at time t = 0 was considered, and the
development in time of both the flow field and concentration field
computed. The results are shown in Figure 4, which clearly indicates
the effect of the gas in reversing the recirculation of the flow.
Some estimates may be made from these results of both the time until
any gas escapes, and the rate at which it will escape.

b)   Dispersion of carbon dioxide by a vertical water spray barrier

In the full-scale tests, vertically upward or downward water spray
barriers were placed about 15m downwind of a source of CO_2. This
was simulated in the program by assuming an entrainment velocity at
an appropriate point in the domain, indicated by the large arrows in
the figures. The results shown in Figure 5 are steady state computations
for a continuous release with a downward facing spray. The strong
upwind recirculation was clearly observable in the video records of
the tests, and the downwind concentration profiles are in good agreement
with the experimental data.

Figure 6 shows comparable results for the upward directed spray, in
which the flow pattern is evidently very different to that for the
downward spray. The concentrations reflect these differences and the
tendency of the concentration contours to follow the streamlines
indicates the convective nature of the dispersion in these cases.

Comparison of Figures 5 and 6 indicate the superiority of the upward

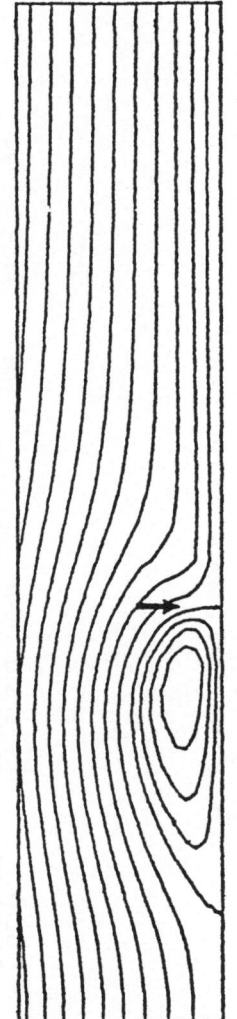

STREAMLINES IN DOMAIN

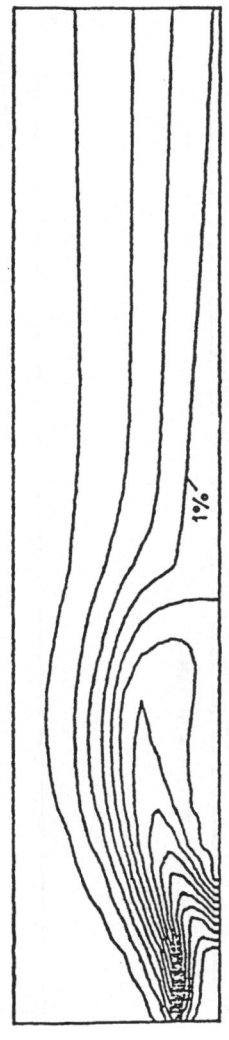

CONCENTRATION CONTOURS IN DOMAIN   (0·25% contours)

FIGURE 5   DISPERSION OF CO$_2$ BY DOWNWARD WATER SPRAY

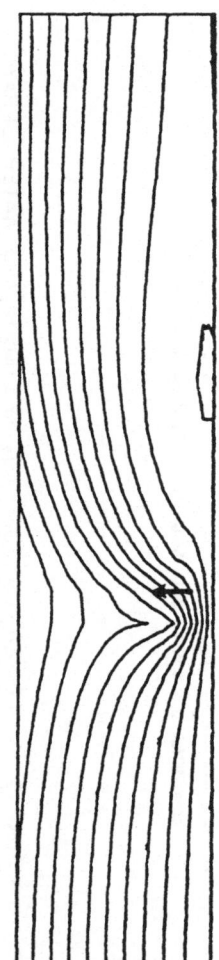

STREAMLINES IN DOMAIN

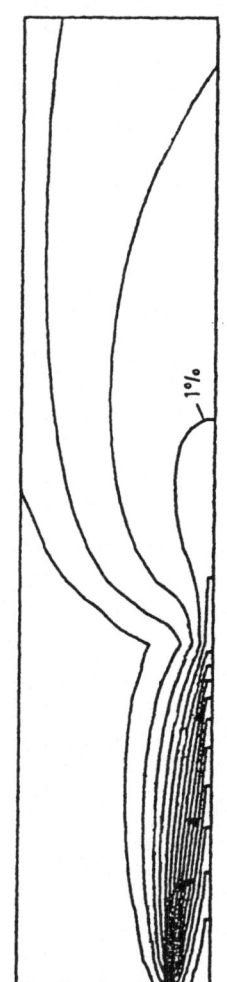

CONCENTRATION CONTOURS IN DOMAIN   (0·25% contours)

FIGURE   6   DISPERSION   OF CO_2 BY   UPWARD   WATER   SPRAY

spray for downwind dispersion, although the enhanced mixing near the source shown in Figure 5 may be preferable in some circumstances. These results clearly demonstrate the usefulness of the modelling approach used in HEAVYGAS to provide results where other simpler models are not strictly applicable.

6. Conclusions

HEAVYGAS, in its present form, is able to deal with 2D heavy gas dispersion problems of the following types.

a) Transient convection and diffusion from a low momentum (ie. non-jet) source, with either an instantaneous or continuous release, (eg. chlorine within a building).

b) Steady state nearfield predictions with low or high momentum sources (eg. spray curtain dispersion).

c) The treatment of more general cases of heavy gas dispersion, which could be described as turbulent buoyant natural convection, can be undertaken. However, such treatment is most satisfactory from the point of view of the stability of the numerical methods employed in the code, when there is a non-zero ambient wind.

The program is at present undergoing further development to widen its range of applicability and improve the modelling of the physics as much as possible. However, it is expected to remain most useful for those situations in which alternative (simpler) methods are invalid.

REFERENCES

1 .   Havens, J.A.   A review of mathematical models for prediction of
      heavy gas atmospheric dispersion. I.Chem.E. Symposium 'The Assessment
      of Major Hazards', Manchester 1982.

2.   Van Ulden, A.P.   On the spreading of a heavy gas released near the ground,
      1st Int. Loss Symposium, The Hague, 1974.

3.   Cox, R.A. & Carpenter R.J.   Further development of a dense vapour cloud
      dispersion model for hazard analysis, in S. Hartwig, Heavy Gas and Risk
      Assessment, D. Reidel, Dordrecht, Holland, 1979.

4.   Colenbrander, G.W.   A mathematical model for the transient behaviour of
      dense vapour clouds. 3rd Int. Symposium on Loss Prevention and Safety
      Promotion in the Process Industries. Basel, 1980.

5.   Fannelop, T.K., Krogstad, P.A. & Jacobsen, O. The dynamics of heavy gas
      clouds. Norwegian Inst. of Tech., Div. of Aero and Gas Dynamics.
      Report No. IFAG B-124, 1980.

6.   Launder, B.E. & Spalding, D.B. Mathematical models of turbulence, Academic
      Press, London. 1972.

7.   Bradshaw, P., Cebeci T & Whitelaw, J.H. Engineering Calculation Methods
      for Turbulent Flow. Academic Press, London, 1981.

8.   Rodi W. Turbulence models and their application in hydraulics State of the
      Art Paper, IAHR, Delft, Holland, 1980.

9.   Havens, J.A. A description and assessment of the SIGMET LNG vapour
      dispersion model. US Coast Guard Rep. CG-M-3-79, 1979.

10.   Moult A & Dean R.B.   CAFE - a computer program to calculate the flow
      environment. CAD80, 4th International Conference on Computers in Design
      Engineering, Brighton, 1980.

DISPERSION OF VERTICAL FREE JETS

H. Giesbrecht, H. Seifert
BASF AG, D-6700 Ludwigshafen, Germany
W. Leuckel
Technical University, D-7500 Karlsruhe, Germany

1. INTRODUCTION

Vent lines of pressure relief systems such as safety valves or rupture discs usually are directed vertically upwards into the atmosphere. The dilution of the frequently
inflammable vent gases is effected primarily by the turbulence of the gas jet itself and, further by atmospheric
turbulence. In case of fluids that are heavier than air,
which is typical for the chemical industry, the exit momentum of the jet is consumed by gravity forces, i.e. the
fluid rises up to a certain height only, called the culmination height, and then sinks down.

It is important do design the outlet in such a manner that
the vent gases are diluted below the limit of inflammability before they reach the ground or some other region
where they can accumulate. In considerations of hazards
from leaks in chemical plants, the break of a connecting
piece on top of a vessel can be regarded under similar
aspects. For the sake of safety, the assumption must be
made that no wind is blowing during venting. In this case
the heavy gas jet behaves like a water fountain, as can be
seen from fig. 1, where a carbon-dioxide jet has been made
visible by adding aerosols and illuminating it from above.

In contrast to the great number of experimental and theoretical investigations about buoyant jets and plumes of
light gases, an essential problem in meteorology,
only few results have been published about jets of heavy
gases. Sato, Osuka and Inoue /1/ measured the culmination
height of cold water jets issuing vertically into heated
water and Turner /2/ performed similar measurements with
saltwater injected upwards into fresh water. Seban, Behnia

103

S. Hartwig (ed.), Heavy Gas and Risk Assessment - II, 103–126.
Copyright © 1983 by Battelle-Institut e.V., Frankfurt am Main, Germany.

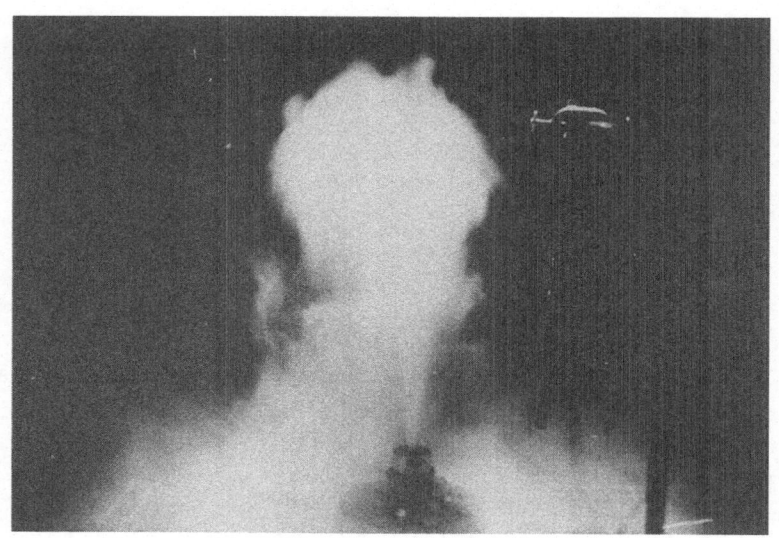

Fig   1: Culminating CO_2-jet
         d_0 = 10 mm, V_0 = 0,85 m^3N/h

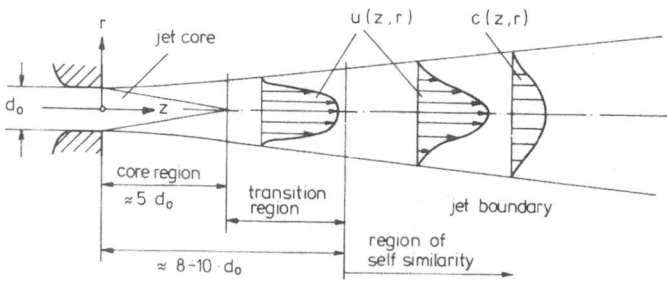

Fig   2: Characteristic regions and profiles of a turbulent
         free jet

and Abreu /3/ investigated the concentration fields in
heated air jets discharged vertically downwards. In this
system they measured the jet penetration length which corresponds to the culmination height of heavy gas jets discharged vertically upwards. They found that the penetration length increases with the square root of the exit
Froude number. Similar Froude number dependency was observed by Hoot and Meroney /4/ and Vignes /5/, who performed
their measurements of culmination height with isothermal
heavy gas jets (Freon 12, CO_2 etc). However, they could
not describe their results exclusively as a function of
the Froude number, but obtaines an additional effect of
the jet density itself.

In 1973 Hess, Leuckel, Stoeckel /6/ used the theoretical
model, presented in chapter 2, to assess the extent of explosive gas/air clouds above pressure relief outlets. In
a subsequent paper, Giesbrecht, Leuckel, Stoeckel /7/ extended the model upon vaporizing fluids. Since the experimental proof of this model was still missing, experiments
with the heavy gases carbon dioxide and Freon 12 were performed. The results of these together with an improved
version of the theoretical model are presented below.

No experimental results about vertical two-phase jets
exist as get. Gärtner, Giesbrecht, Leuckel /8/ presented
some measurements about horizontal nitrogen/air jets; the
essential results are quoted briefly in chapter 5.

2. MODELLING OF VERTICAL TURBULENT JETS WITH BUOYANCY

Gas jets emerging from pressure relief devices are usually in a rather high Reynolds number range ($5 \cdot 10^4$ up to >
10^6), due to exit diameters of at least 20 mm and discharge velocities in the order of 50 to 300 m/s. Mathematical
modelling can be based, therefore, upon the flow and mixing laws of the fully turbulent jet in stagnant surroundings. Due to the fact that the density of the discharged
matter is, in general, different from, in most cases higher
than the ambient air density, buoyancy of the jet in the
atmosphere must be considered, resulting for vertical up-
flowing jets, in increasing or mostly decreasing axial
jet momentum in the downstream direction.

The question arises whether the turbulent entrainment
rate is dependent upon the density difference. Such influence has been found, e.q., by Era and Saima /9/,
Lenze /10/ and Schlanzke /11/ from experiments with gas

jets of very high exit Froude number ($> 10^5$), where buoyancy effects are of minor importance.
There are 3 exit flow parameters of a jet, which are essential for characterising jet propagation and mixing over
large distances:
- exit mass flow rate $\dot{M}_0$
- exit momentum flux I_0
- stagnation density of jet material ρ_0

These parameters apply to incompressible (very low Mach
number) as well as compressible flow. In the case of supercritical relief pressure, momentum flux is to be evaluated as the sum of the exit velocity term, the velocity u_0
being larger than or equal to the critical velocity a^*_0,
and the residual pressure drop term:

$$I_0 = \dot{M}_0 \cdot u_0 + (p_0 - p_{at}) \cdot A_0$$

The residual pressure drop is transformed during a
postnozzle expansion into increased velocity.

Starting from the origin of a fully turbulent jet, one may
distinguish 3 succeeding regions (fig. 2): a "core zone"
where the exit velocity persists on the axis, the turbulent exchange being limited to an annular zone around the
jet core; a "transition zone " where turbulent exchange
has reached the axis; and finally the "region of self-similarity" for radial profiles of local velocity and concentration values. Since the transition zone ends at approximately 8 to 10 exit diameters downstream of the nozzle (with supercritical pressure ratio at about 16 diameters) it seems justified, as an approximation, to treat
the entire jet under the assumption of self-similarity.

With constant density and, hence, in the absence of buoyancy, the mass flow rate of a turbulent jet increases
linearly with axial distance z; this implies a constant
mass entrainment rate per element Δz of axial distance,
which can be expressed by the equation:

(1) $$\frac{d\dot{M}}{dz} = \mu' \cdot \sqrt{\rho_{at} \cdot I}$$

μ' being a dimensionless constant with the value 0.26 from
experiment. In a previous publication /6/ an entrainment
constant μ from a different definition was used, both related by a pure number factor:

$$\mu' = 2 \cdot \sqrt{\frac{\pi}{\ln 2}} \cdot \mu \quad ; \quad \mu \approx 0.06$$

This entrainment law (1), which was previously set up by
Taylor /12/ in a somewhat different form, is also considered valid

for buoyant vertical jets, where $\dot{I}$ stands for the local jet momentum which varies with axial distance z. Hence, the jet mass flow rate M is then no longer a linear function of z.

The z-variation of $\dot{I}$ is described by the momentum balance condition applied to a jet element Δ z, leading to another differential equation:

(2) $$\frac{d\dot{I}}{dz} = -B$$

with the buoyancy integral:

(3) $$B = \int_{r=0}^{\infty} g \cdot [\rho(r,z) - \rho_{at}] \cdot 2\pi r \cdot dr$$

B is positive in the case of jet density higher than atmospheric density, resulting in decreasing jet momentum along vertically rising jets.

Free jet experiments with different densities of the jet and the ambient medium, as performed by several authors /10, 13, 14/ have suggested that the radial profiles of momentum flux density (ρu^2) and of mass flux density of the jet fluid (ρuc) can be approximated by the Gaussian error function:

(4) $$\frac{\rho \cdot u^2}{(\rho \cdot u^2)_{ax}} = \exp\left[-\ln 2 \cdot \left(\frac{r}{r_i}\right)^2\right]$$

(5) $$\frac{\rho \cdot u \cdot c}{(\rho \cdot u \cdot c)_{ax}} = \exp\left[-\ln 2 \cdot \left(\frac{r}{r_{\dot{m}}}\right)^2\right]$$

wherein r_i and r_m stand for the half-value radii of the momentum and mass flux profiles, respectively, and fulfill the proportionality:

$$r_{\dot{m}}/r_i \approx 2/\sqrt{3}$$

It is assumed that these relations hold for buoyant jets as well.

Further calculation is based upon the integral mass and momentum flow rates through any jet cross section, as defined by:

(6) $$\dot{I}(z) = \int_{r=0}^{\infty} (\rho \cdot u^2) \cdot 2\pi r \cdot dr = \frac{\pi}{\ln 2} \cdot \rho_{ax} \cdot u_{ax}^2 \cdot r_i^2$$

(7)
$$\dot{M}_0 = \int_{r=0}^{\infty} (\rho \cdot u \cdot c) \cdot 2\pi r \cdot dr = \frac{4\pi}{3 \cdot \ln 2} \cdot \rho_{ax} \cdot u_{ax} \cdot c_{ax} \cdot r_i^2$$

(8)
$$\dot{M}(z) = \int_{r=0}^{\infty} (\rho \cdot u) \cdot 2\pi r \cdot dr$$

Considering finally the general relationship between local concentration and local density:

(9)
$$\frac{1}{\rho} = \frac{c}{\rho_0} + \frac{1-c}{\rho_{at}}$$

this set of equations in r and z as independent variables can be integrated numerically. A computer program has been developed for this, which also enables the treatment of jets with vaporization or condensation of a liquid/vapor component or the jet fluid.

A closed form solution, which is sufficiently accurate for many application systems, is obtained from the assumption $c \ll 1$, which always is true for the far-downstream jet region. In this case, Gaussian-type radial profiles can be shown to apply for the velocity and concentration themselves:

(4')
$$\frac{u}{u_{ax}} = \exp\left[-\frac{\ln 2}{2} \cdot \left(\frac{r}{r_i}\right)^2 \right]$$

(5')
$$\frac{c}{c_{ax}} = \exp\left[-\ln 2 \cdot \left(\frac{r}{r_c}\right)^2 \right] \qquad \text{with } r_c \approx 2r_i$$

Hence, the integral terms (3) and (8) from above can be solved:

(3')
$$B(z) = g \cdot \frac{\rho_0 - \rho_{at}}{\rho_0} \cdot \frac{4\pi}{\ln 2} \cdot \rho_{at} \cdot c_{ax} \cdot r_i^2$$

(8')
$$\dot{M}(z) = \frac{2\pi}{\ln 2} \cdot \rho_{at} \cdot u_{ax} \cdot r_i^2$$

With $\mathcal{S}_{ax} \approx \mathcal{S}_{at}$, relations (1), (2), (6), (7), (3'), (4') represent 6 equations for 6 functions of z only: u_{ax}, c_{ax}, r_i, $\dot{I}$, $\dot{M}$, B.

Eliminating the axial length z by combining the mass entrainment (1) and the buoyancy balance (2) equation, and substituting B, u_{ax}, c_{ax} and r_i by means of equations (3'), (6), (7), (8') leads to the differential equation:

$$\frac{d\left(\frac{\dot{I}}{\dot{I}_0}\right)}{d\left(\frac{\dot{M}}{\dot{M}_0}\right)} = -\frac{3}{4 \cdot \mu} \cdot \sqrt{\frac{\ln 2}{\pi}} \cdot g \cdot \frac{\rho_0 - \rho_{at}}{\rho_0} \cdot \frac{\dot{M}_0^3}{\rho_{at}^{1/2} \cdot \dot{I}_0^{5/2}} \cdot \frac{\dot{M}}{\dot{M}_0} \cdot \left(\frac{\dot{I}}{\dot{I}_0}\right)^{-3/2}$$

Integrating and considering the boundary conditions for a point source at z = 0 one obtains:

$$(10) \qquad \left(\frac{\dot{I}}{\dot{I}_0}\right)^{5/2} = 1 - \frac{15 \cdot \sqrt{\ln 2}}{32 \cdot \mu} \cdot \frac{1}{Fr_{eq}} \cdot \left(\frac{\dot{M}}{\dot{M}_0}\right)^2$$

Herein Fr_{eq} represents a Froude number, defined from the characteristic exitparameters of an "equivalent jet":

$$u_{eq} = \frac{\dot{I}_0}{\dot{M}_0} \qquad\qquad \text{equivalent jet exit velocity}$$

$$d_{eq} = \sqrt{\frac{4}{\pi} \cdot \frac{\dot{M}_0}{\rho_{at} \cdot u_{eq}}} = \frac{2}{\sqrt{\pi}} \cdot \frac{\dot{M}_0}{\sqrt{\rho_{at} \cdot \dot{I}_0}} \quad \text{equivalent jet exit diameter}$$

$$Fr_{eq} = \frac{\rho_0 \cdot u_{eq}^2}{g \cdot (\rho_0 - \rho_{at}) \cdot d_{eq}} = \frac{\sqrt{\pi}}{2} \cdot \frac{1}{g \cdot \frac{\rho_0 - \rho_{at}}{\rho_0}} \cdot \frac{\rho_{at}^{1/2} \cdot \dot{I}_0^{5/2}}{\dot{M}_0^3}$$

The equivalent jet emanates with the same mass and momentum flow rates as the real jet, but with the ambient density $\mathcal{S}_{at}$ instead of $\mathcal{S}_0$. In case of subcritical exit flow u_{eq} = u_0 and $d_{eq} = \sqrt{\mathcal{S}_0 / \mathcal{S}_{at}} \cdot d_0$.

The culmination condition $\dot{I}$ = 0, introduced into the above relationship $\dot{I} = \dot{I}(\dot{M})$, yields the jet mass flow rate at culmination $\dot{M}_K$:

$$(11) \qquad \frac{\dot{M}_K}{\dot{M}_0} = \sqrt{\frac{32 \cdot \mu}{15 \cdot \sqrt{\ln 2}}} \cdot \left|Fr_{eq}\right|^{1/2} \approx 0.39 \cdot \left|Fr_{eq}\right|^{1/2}$$

In order to determine the height of culmination z_K, the momentum flow rate I(M) from above can be introduced into the mass entrainment equation (1). The resulting integral

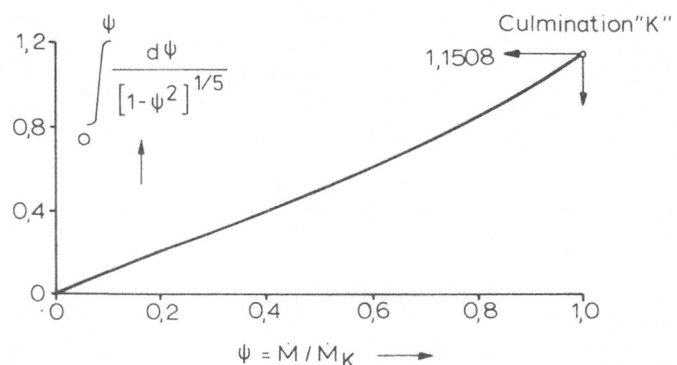

Fig 3: Integral for man/distance relation (equ. 12)

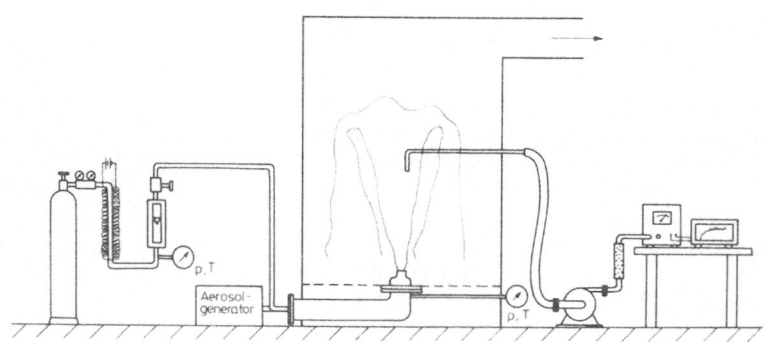

Fig 4: Experimental Set-up

$$(12) \qquad \frac{z}{d_{eq}} = \sqrt{\frac{2 \cdot \sqrt{\ln 2}}{15 \cdot \mu}} \cdot \left| Fr_{eq} \right|^{1/2} \cdot \int\limits_{0}^{\dot{M}/\dot{M}_K} \frac{d\left(\frac{\dot{M}}{\dot{M}_K}\right)}{\left[1 - sign\left(Fr_{eq}\right) \cdot \left(\frac{\dot{M}}{\dot{M}_K}\right)^2 \right]^{1/5}}$$

is a nearly linear function of the mass flow ratio
$\dot{M}(z)/\dot{M}_K$ and can only be solved numerically (fig. 3). The
culmination height z_K is obtained for $\dot{M}/\dot{M}_K = 1$ as the
upper limit of the integral term, leading to:

$$\frac{z_K}{d_{eq}} = 1.56 \left| Fr_{eq} \right|^{1/2} \sim \sqrt{\frac{\rho_0}{\rho_0 - \rho_{at}}} \cdot \frac{u_{eq}}{\sqrt{d_{eq}}}$$

This increase of the culmination height with the square-
root of the Froude number has already been found by several authors experimentally as well as by dimensional analysis /1-5/. Vignes /5/ found some additional dependency
upon the jet density.

Only few results have been published concerning the concentration value in the culmination region which should
vary according to the relation

$$\frac{1}{c_{ax,K}} = \frac{8}{3} \cdot \sqrt{\frac{2\mu}{15\sqrt{\ln 2}}} \cdot \left| Fr_{eq} \right|^{1/2}$$

derived from the equations (7), (8') and (11).

To be complete, it should be mentioned that the theoretical
model is valid for jet fluids lighter than air as well. In
this case, the Froude number is negative and the following
relations can be derived for the far field
($\dot{M}/\dot{M}_K \gg 1$, $\dot{M}_K$ being just a formal quantity without any
physical meaning):

$$\frac{\dot{I}}{\dot{I}_0} = 9 \left(\frac{\mu}{10\sqrt{\ln 2}} \right)^{2/3} \cdot \left| Fr_{eq} \right|^{-2/3} \cdot \left(\frac{z}{d_{eq}} \right)^{4/3}$$

$$\frac{\dot{M}}{\dot{M}_0} = 9 \cdot 2^{5/3} \cdot \left(\frac{\mu}{5\sqrt{\ln 2}} \right)^{4/3} \cdot \left| Fr_{eq} \right|^{-1/3} \cdot \left(\frac{z}{d_{eq}} \right)^{5/3}$$

$$\frac{1}{c_{ax}} = 3 \cdot \left(\frac{4\mu}{5\sqrt{\ln 2}} \right)^{4/3} \cdot \left| Fr_{eq} \right|^{-1/3} \cdot \left(\frac{z}{d_{eq}} \right)^{5/3}$$

$$\frac{u_{eq}}{u_{ax}} = \left(\frac{4\mu}{5\sqrt{\ln 2}} \right)^{2/3} \cdot \left| Fr_{eq} \right|^{1/3} \cdot \left(\frac{z}{d_{eq}} \right)^{1/3}$$

The indicated dependency of the jet properties upon Froude number and height for so-called "vertical buoyant plumes" has been found by numerous authors, e.g., by Chen /15/, Davis et al /16/, Madni and Pletcher /17/. A survey and a comparison of different models are given by Turner /12/ and Seban and Behnia /18/.

3. EXPERIMENTAL ARRANGEMENT AND MEASURING TECHNIQUES

The experimental investigations of vertical heavy gas jets were performed with carbon dioxide (CO_2, M = 44) and Freon 12 (CF_2Cl_2, Difluordichlormethane, M = 121). The test facilities are shown in fig. 4.

The gas was taken from pressure bottles. Since the gas cooled down due to the decreasing vapour pressure in the bottle it was reheated electrically up to the ambient temperature. The mass flow rate of the gas was measured by a floated element flow meter. To protect the jets against perturbations from the surroundings, and for the safety of the personnel, the measurements were performed in a closed chamber with a cross section of 1.5 x 1.5 m^2 and height of 2.5 m.

In order to visualize the flow pattern the gas stream could be tracered with aerosols before entering the testing chamber. At the bottom of the chamber scavenge air could be supplied across the whole cross section in order to carry off the heavy gas through a channel at the top of the chamber after each experiment. The heavy gas was issued from smoothly contoured nozzles with exit diameters of 1.5, 3, 5, 10 and 15 mm. The nozzles were aligned perpendicularly by a bubble level. The flow rate was controlled additionally by measuring static pressure and temperature of the gas entering the nozzle.

The local concentration of the gas in the jet was measured by sampling about 30 l/h through a stainless steel probe with an inner diameter of 3 mm. A traversing mechanism allowing both lateral and axial movement was mounted which provided access to any point in the mixing region. The sampled gas mixture was dried and continuously analysed, in the case of CO_2, by an infrared spectrometer (URAS), and in the case of Freon, by a heat conducting cell (CALDOS). In the near exit region of the nozzle where the gas concentration exceeded 20 Vol.-%, the concentration was determined indirectly by measuring the decrease of

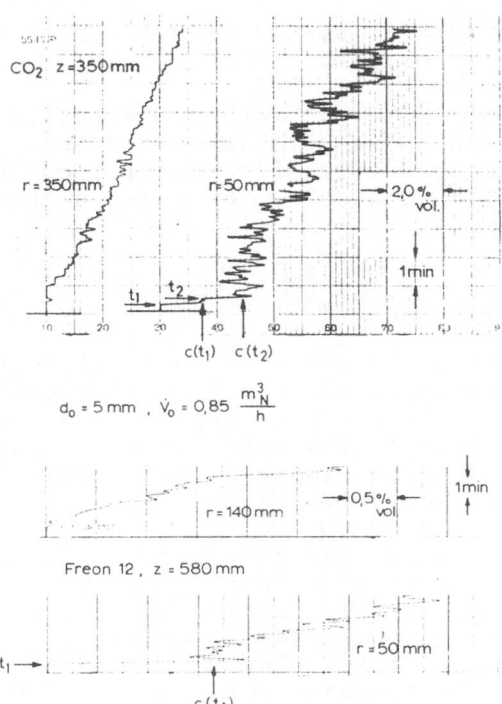

Fig 5: Typical plots of time dependent concentration

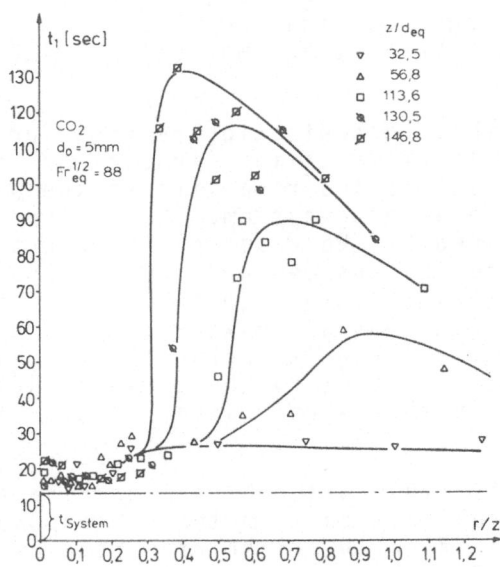

Fig 6: Reduced radial distribution of dead time until first concentration rise

oxygen in the jet with an O_2-analyser based on the paramagnetic property of oxygen.

All measured signals were recorded continuously both on a
curve-drawing recorder and on a tape recorder. The sampling system had a dead time of about 12 sec. In the step
function response the concentration reached the 99 % value
after about 30 sec.

4. EXPERIMENTAL RESULTS

The experimental investigations on vertical jets of carbon, dioxide and Freon have been performed to check the
mathematical model presented in chapter 2. By varying
nozzle diameter and mass flow rate, different values of
Reynolds and Froude number could be adjusted. Values
should be given for the following points:

1. radial distribution of concentration
2. axial decay of center line concentration
3. axial increase of half-value radius
4. concentration of the culmination cloud
5. height of the culmination cloud
6. concentration of the down-flowing gas/air mixture

Since only two sampling systems were available, many experiments had to be repeated. Prior to every test the
chamber was scavenged carefully with air. The intended
mass flow rate of the test gas was set up within less than
two seconds.

Fig. 5 shows typical concentration records for CO_2 and
Freon. After a delay time t_1 the first signal appears.
This time is the sum of the rise time of the jet and the
dead time of the measuring system. The further rise of the
signal is determined by the dispersion process of the
downflowing gas in the chamber and by the response behaviour of the measuring system. In the case of the extremely
heavy gas, Freon, the gas/air mixture of the culmination
region flows down in irregular strands with practically
no dilution nor entrainment by the uprising jet, forming
an increasing but rather stable layer of gas at the bottom
of the chamber.

The mixing of this layer with the surrounding air proceeds
only slowly. As can be seen from the corresponding record,
the concentration after the first rise remains on a short
plateau $c(t_1)$. This value can be assigned to a vertical
jet in unlimited surroundings. Aterwards the concentration
in the jet rises again due to the increasing concentration

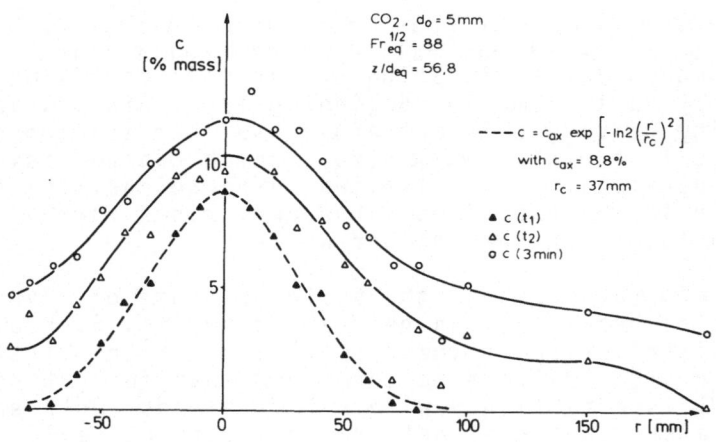

Fig 7: Radial concentration profiles

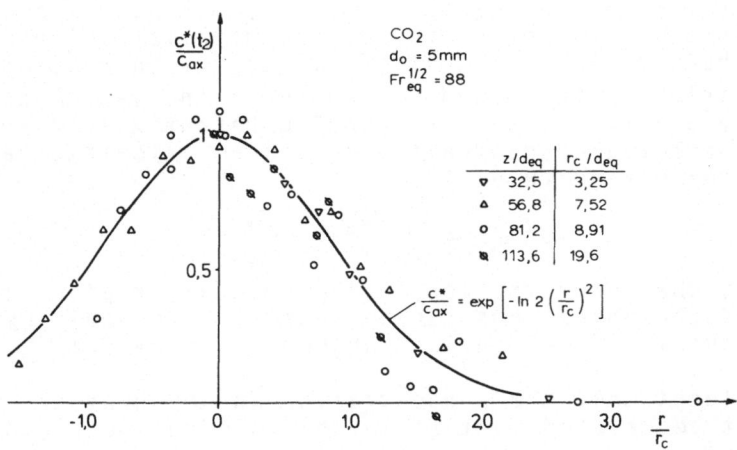

Fig 8: Reduced radial concentration profiles

in the outer region of the test chamber, which can be seen
from the corresponding record for large radius.

In the case of CO_2, the heavy gas jet is dispersed in the
first step by the stream falling after culmination.
As a consequence, the situation of pure air entrainment
exists only up to time t_2. Beginning from this instant,
downfalling heavy gas is entrained, so that the concentration rises to a higher value $c(t_2)$, which can be assigned
to a jet developing in unlimited surroundings. The concentration in the experiment, however, rises further due
to the enclosure of the chamber walls.

Further information about the build-up phase of a vertical jet of heavy gas can be drawn from fig. 6. From the
spatial distribution of the rise time t_1 of the first concentration signal, it can be concluded that the jet region
($r/z < 0.2$) is established within few seconds. Only some
seconds later - but distinct from the central region - the
concentration in the boundary region ($0.2 < r/z < 0.4$) rises
due to gas/air mixture falling from the culmination
region. This gas stream flows along the bottom of the test
chamber and partially up the chamber walls, thus entraining air of the outer region. It is evident that the rise
time increases with rising height.

As can be seen from fig. 7, only the radial concentration
profile at time t_1, i.e. in the period of pure air entrainment, can be approximated by a simple Gaussian
error function, as postulated by the mathematical model.
Later, the profiles are shifted due to the increasing
concentration in the boundary region of the jet. Assuming
that only air from the corresponding height z is entrained, a corrected concentration c^* can be calculated according to the mixing rule

$$c^*(z) = \frac{c(z) - c_{at}(z)}{1 - c_{at}(z)}$$

c_{at} being the concentration in the boundary region. These
corrected concentrations can be approximated again by the
simple Gaussian function, as can be seen from fig. 8

It is evident that the concentration in the boundary region must be related to that in the culmination cloud. As
can be seen from the vertical profiles of CO_2-jets at different Froude numbers, shown in fig. 9, the concentration
c_{at} at corresponding heights are - to a fair approximation - fixed fractions of the corrected centerline concentration c_k^* at the base level of the culmination cloud. As
mentioned earlier, the entrainment of gas/air mixture
flowing down from the culmination region of jets with

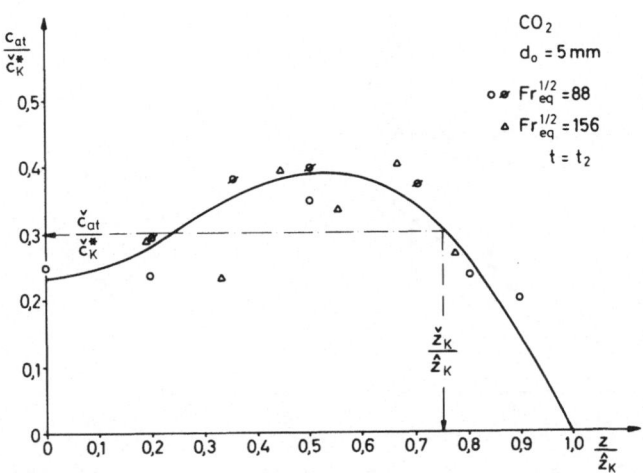

Fig 9: Axial distribution of reduced concentration in the boundary region of the jet

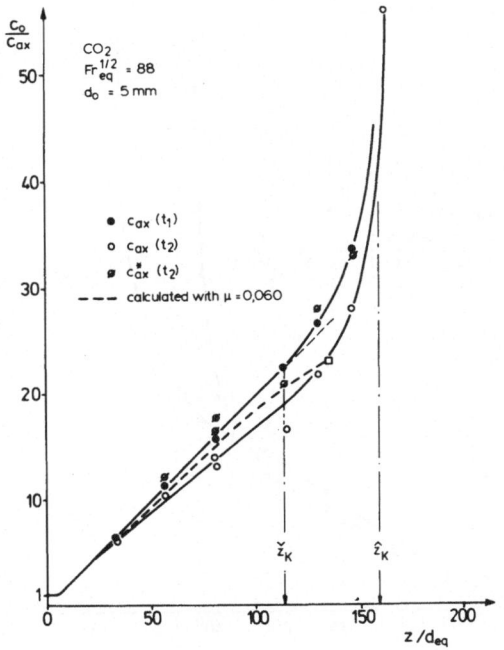

Fig 11: Reciprocal axial concentration as function of reduced height

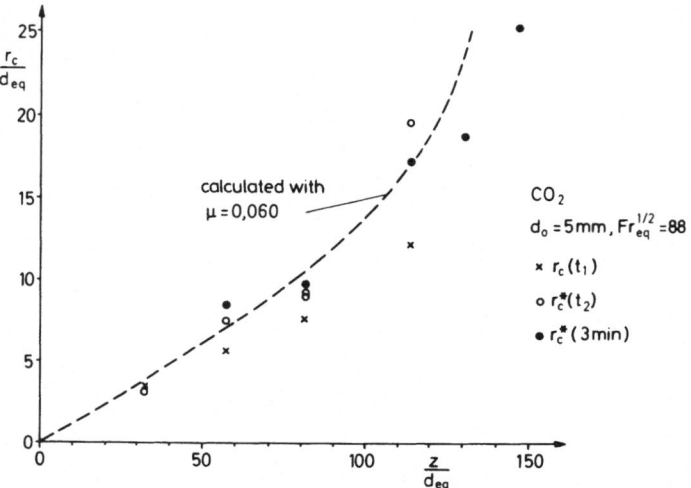

Fig 10: Axial increase of half-value radius of the con-
         centration profile

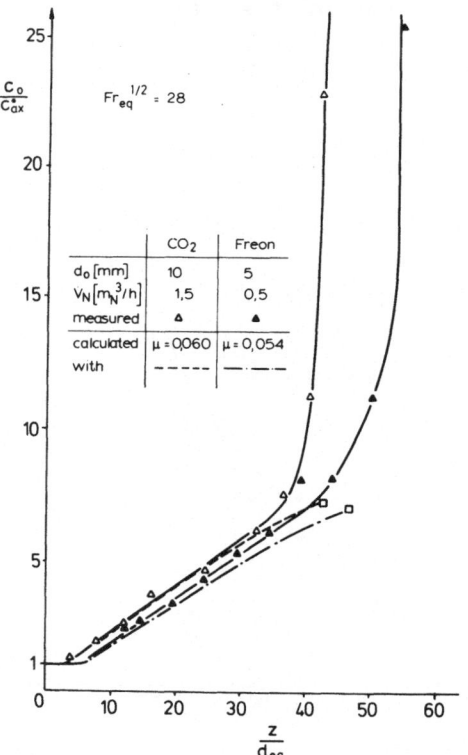

Fig 12: Comparison of reciprocal corrected centerline con-
         centration for different gases

very high density is inferior. As a consequence the concentrations $c(t_1)$, $c(t_2)$ and $c^*(t_2)$ coincide when is involved.

In the mathematical model, presented in chapt. 2, the entrainment of pure air is assumed. A comparison of computed
concentrations is allowed, therefore, only with the measured
values at time t_1 or with the corrected concentrations c^*
at later times. As can be seen from fig. 10, the increase
of the concentration half-value radius r_c with z, computed
with an entrainment coefficient μ = 0.06 is confirmed
fairly well by the experimental results for CO_2-jets. Such
a value for the entrainment coefficient has also been found
by Lenze /10/ for CO_2 jets of similar Reynolds number but
very high Froude number.

The agreement between computed and measured concentrations
on the jet axis is satisfactory only in the jet region
(fig. 11). In contrast to the mathematical model, the realistic culmination cloud has a finite height in which the
concentration diminishes from the concentration at the base level $\check{c}_K$ to zero. The calculated cloud concentration
agrees fairly well with the corrected concentration c^*_{ax}
at the point of deviation from the straight line. The calculated height of the culmintation cloud lies between the
measured heights of base level $\check{z}_K$ and top level $\hat{z}_K$. In the
case of Freon-jets a lower entrainment coefficient of μ =
0.054 has been found more adequate to adjust measured and
calculated culmination heights and concentrations on the
axis (fig. 12). This decrease of entrainment with increasing jet density has been found by Lenze /10/ for jets of
very high Froude number; Schlanzka /11/ even measured the
same entrainment coefficient.

The axial profile of the corrected center line concentration c^*_{ax} discussed for two special Froude numbers is typical for any other Froude number (see fig. 13). By reference of the height z upon the equivalent jet diameter d_{eq}
the reciprocal axial concentrations within the jet region
fall on one straight line for each gas. The point of deviation from that line, i.e. the base level, increases with
growing Froude number.

As predicted by the mathematical model, the culmination
height inreases proportional to $Fr_{eq}^{1/2}$. This is true not
only for the base level $\check{z}_K$ but also for the top level
$\hat{z}_K$ (fig. 14). The latter was measured in different experiments by detecting the height of the first concentration signal from above. The factor of proportionality of
1.9 for $\hat{z}_K$ is the same as that determined by Vignes /5/
from experiments with CO_2-jets.

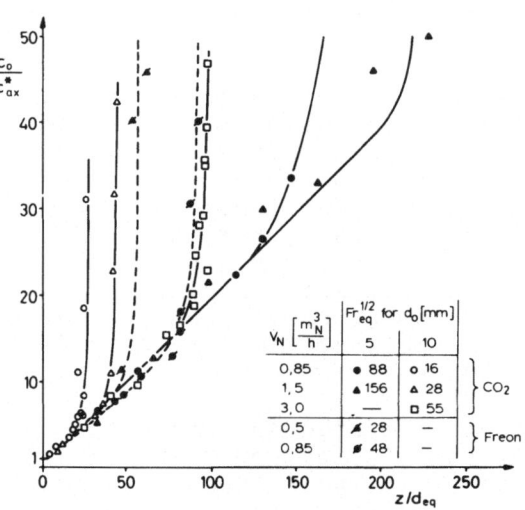

Fig 13: Reciprocal corrected centerline concentration for different gases and Froude numbers

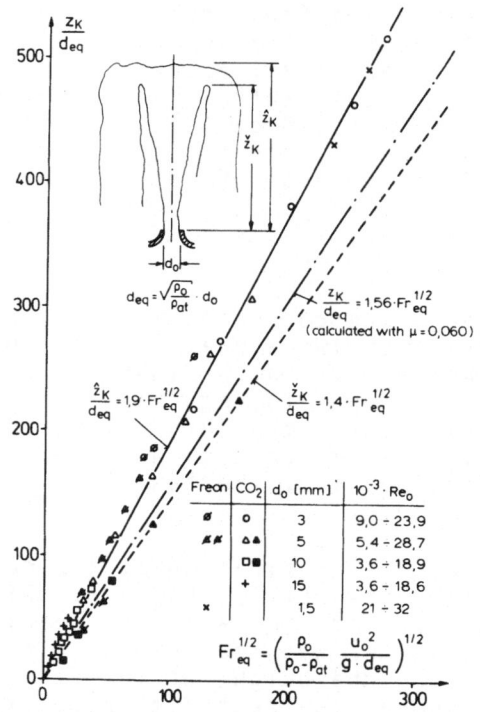

Fig 14: Reduced culmination height as function of equivalent Froude number

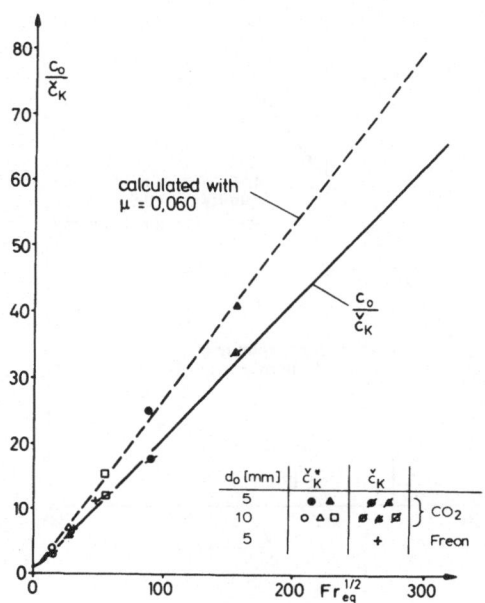

Fig 15: Reciprocal culmination concentration as function of equivalent Froude number

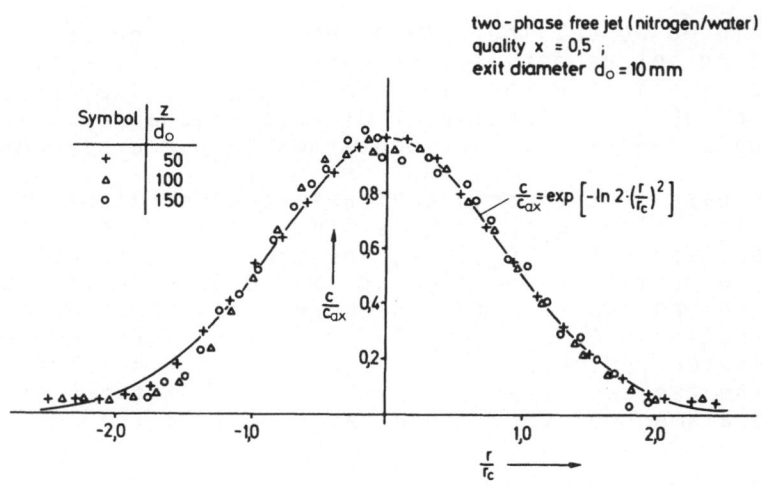

Fig 16: Reduced radial concentration profile for two-phase jets

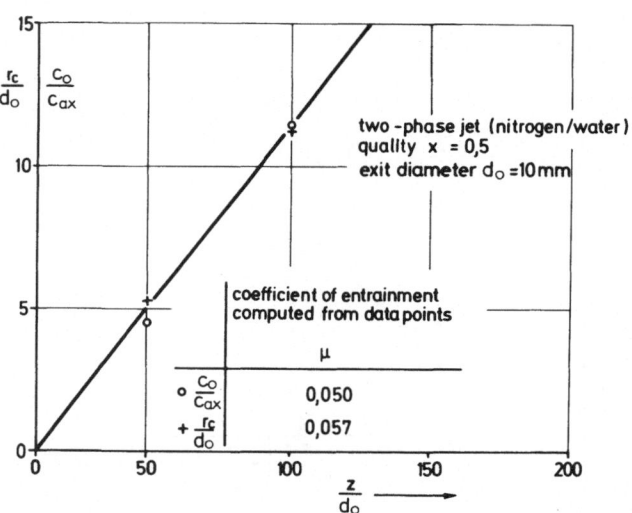

Fig 17: Axial increase of concentration half-value radius
        and reciprocal centerline concentration of two-
        phase jets

As mentioned above, the rate of entrainment decreases with
increasing jet density. Consequently, the culmination height
for Freon-jets is higher than for CO_2-jets. The extreme increase of 50 % for the factor of proportionality as found
by Vignes, however, is not confirmed by our experiments.

The increase of the reciprocal axis concentration in the
culmination cloud with Fr_{eq} is linar as well (fig. 15). As
discussed above for CO_2-jets, the concentration calculated
with an entrainment coefficient $\mu = 0.06$ coincides with
the corrected experimental concentrations $\tilde{c}^*_K$. The actual
concentrations $\tilde{c}_K$ are roughly 30 % higher. For Freon-jets,
the measured concentrations coincide with the corrected
ones. The two experimental points reconfirm the fact that
the entrainment of very dense gases is inferior.

5. EXPERIMENTAL INVESTIGATIONS OF TWO-PHASE JETS

In the liquid is entrained by the gas stream into the vent
line, which is typical, e.g., for the pressure relief of
polymerisation reactors, a two-phase spray jet is ejected

from the outlet. For fluids with high vapour pressure, the
droplets vaporize quickly along with the entrainment of
ambient air, and the jet behaves like a single phase jet.
This has been confirmed by experiments with liquid carbon
dioxide jets.

In the example mentioned above of venting polymerisation
reactors the polymer phase does not vaporize. If blown upwards the liquid phase will reach a lower culmination
height than the gas phase, according to a velocity slip
between both components.

So far, experiments have been performed only with horizontal nitrogen/water jets /8/. As seen from fig. 16 the radial concentration profiles can be described excellently
by a Gaussian-function. Similar to single-phase non-buoant
jets, halb-value radius r_c and reciprocal axis concentration c_{ax} increase linearly with distance from the exit
(fig. 17).

6. SUMMARY

To assess the behaviour of vertical heavy gas jets a theoretical entrainment model has been derived with the following assumptions:

1) Gaussian error functions for the radial distributions of
   the fluid mass and momentum flux density
2) Constant ratio of the half-value radii of the two
   Gaussian functions
3) Mass entrainment proportional to the square-root of the
   local momentum flux

A closed form solution is obtained for the far field
($c_{ax} \ll 1$) yielding the following results:

   Height and reciprocal axis concentration of the culmination region increase proportionally to the square
   root of the Froude number of a socalled "equivalent"
   jet. Such a jet possesses the original mass and momentum flow rate, but leaves the equivalent exit area
   with the ambient density.

The theoretical model has been confirmed by experiments
with CO_2 and Freon jets. In contrast to the model the actual culmination region has an extended height, the theoretical height lying between measured heights of top and
base level. The calculated axis concentration of the culmination region is somewhat higher than the measured base
level concentration, the difference being greater for the

lighter CO_2 gas. In compensation the dilution of the gas/
air mixture flowing down from the culmination region is
higher for CO_2 jets. The result is that for an elevated
source in unlimited surroundings, e.g. an outlet of a
vent line, the concentration of the gas/air mixture flowing down to the ground is, in no case, higher than the calculated culmination concentration. With this in mind, outlets can be designed in such a manner that no inflammable
gas/air mixture can reach the ground.

Acknowledgement
The authors thank Mr. K. Buckel and Mr. M. Vatter for
their assistance in carrying out the experiments.

7. REFERENCES

/1/ Sato, K., H. Osuka, I. Inoue: Maximum penetration
    distance of a vertical buoyant jet. Int. Chem.
    Engng. 21(81)3, 435 - 442

/2/ Turner, J.S.:
    Jets and plumes with negative or reversing buoyancy
    J. Fluid Mech., 26(66), 779 - 792

/3/ Seban, R.A., M. M. Behnia, K.E. Abreu:
    Temperatures in a heated air jet discharged downward
    Int. J. Heat Mass Transfer 21 (78), 1453 - 1458

/4/ Hoot, T.G., N. Meroney:
    The Behavior of Negatively Buoyant Stack Gases
    CEP 73 - 74 RNM - TGH 33, 74 - 210

/5/ Vignes, M.:
    Contribution à l'étude des jets gazeux verticaux dans
    une atmosphère calme. Revue generale thermique,
    83 (68), 1205 - 1217

/6/ Hess, K., W. Leuckel, A. Stoeckel:
    Ausbildung von explosiblen Gaswolken bei Überdachent-
    spannung und Maßnahmen zu deren Vermeidung
    Chem. Ing. Techn. 45 (73)5, 323 - 329

/7/ Giesbrecht, H., W. Leuckel, A. Stoeckel:
    Nachverdampfungs- und Kondensationsvorgänge bei der
    atmosphärischen Strahlausbreitung nach Überdachent-
    spannung. Jahrestagung 1973 des Institutes für Chemie
    der Treib- und Explosivstoffe,
    Pfinztal-Berghausen, 317 - 339

/8/ Gärtner, D., H. Giesbrecht, W. Leuckel:
Effects of two-phase flow upon the emergency relief of
reactor tanks and upon the subsequent atmospheric jet
dispersion, 3rd Int. Symp. on Loss Prevention, Basel
(80), 954 - 967.

/9/ Era, Y., A. Saima:
Turbulent Mixing of Gases with Different Densities
Bull. of the JSME, 20 (77) 139, 63 - 70

/10/ Lenze, B.:
Turbulenter Austausch in Strahlen unterschiedlicher
Dichte
Forsch. Ing.-Wesen 43 (77)3, 75 - 86

/11/ Schlanzke, G.:
Die turbulente Vermischung in Freistrahlen unter Be-
rücksichtigung von Dichtunterschieden, Luft- und
Kältetechnik 2 (75), 71 - 75

/12/ Turner, J.S.:
Buoyancy Effects in Fluids Cambridge University
Press, Cambridge (73)

/13/ Kremer, H.:
Zur Ausbreitung inhomogener turbulenter Freistrahlen
und turbulenter Diffusionsflammen, Diss., Universität
Karlsruhe (1964)

/14/ Günther, R.:
Verbrennung und Feuerungen
Springer-Verlag Berlin-Heidelberg-New York (1974)

/15/ Chen., C.J., C.H. Chen:
On Prediction and Unified Correlation for Decay of
Vertical Buoyant Jets
Transactions of the ASME 101 (79), 532 - 537

/16/ Davis, L.R., M.A. Shirazi, D.L. Slegel:
Measurement of Buoyant Jet Entrainment from Single
and Multiple Sources
Transaction of ASME, 100 (78), 442 - 447

/17/ Madni, I.K. R.H. Pletcher:
Prediction of Turbulent Forced Plumes Issuing Vertically into Stratified or Uniform Ambients
J. of Heat Transfer, Transactions of the ASME,
99(77), 99 - 104

/18/ Seban, R.A., M.M. Behnia
Turbulent buoyant jets in unstratified surroundings,
Int. J. Heat Mass Transfer 19 (76) 1197 - 1204

NOMENCLATURE

A	m^2	area
c	-	mass concentration of jet fluid
d	m	diameter
Fr	-	Froude number
g	m/s^2	gravitational acceleration
$\dot{I}$	N	momentum flux
$\dot{M}$	kg/s	mass flux
p	Pa	pressure
r	m	radial coordinate
t	s	time
u	m/s	axial valocity component
z	m	axial coordinate
ρ	kg/m^3	density
μ	-	entrainment coefficient

Indices

at	in the ambient atmosphere
ax	on the axis
c	concerning concentration
eq	concerning the equivalent jet properties
i	concerning momentum flux
K	in the culmination region
o	concerning the exit conditions
*	concerning state of pure air entrainment
$\widehat{}$	top level of culmination region
$\smile$	base level of culmination region

R&D NEEDS AND OPPORTUNITIES IN LGF SAFETY AND ENVIRONMENTAL CONTROL

J. G. DeSteese
R. E. Rhoads
Battelle, Pacific Northwest Laboratories
Richland, Washington, U.S.A.

ABSTRACT

Mature and safety-conscious industries have been developed to deliver
liquefied gaseous fuels (LGFs) and energy materials to the marketplace.
Accident experience with some of these materials in the past and the
fear of potentially catastrophic releases in the future are the basis
of public concerns about LGF safety and environmental control. Today,
the LGF industries and governments of several countries are conducting
research and development (R&D) programs to address these concerns. A
large part of current R&D is focused on the release phenomenology of
liquefied natural gas (LNG) and liquefied petroleum gas (LPG). Emphasis
is being placed on understanding release phenomena observed in scale
spill experiments and developing models that predict the behavior of
larger accidental releases.

This paper presents a discussion of R&D needs that complement current
phenomenological R&D with the objective of improving LGF release prevention and control (RP&C) techniques. RP&C objectives are considered
to be the primary basis for establishing priorities in LGF safety and
environmental control R&D. The concept of an R&D "relevance matrix"
is proposed as a tool to coordinate and focus R&D activities and is
discussed in relation to improving RP&C strategies for LNG, LPG and
ammonia. A growing body of information on LGF behavior is an asset
that requires management to maximize its usefulness. There are evolving
opportunities for further coordination of R&D efforts between industry
and government both nationally and internationally with the goal of
developing a truly integrated R&D approach, designed to achieve RP&C
objectives.

1. INTRODUCTION

The public, as a whole, derives direct and indirect benefits from the
production and use of liquefied gaseous fuels (LGFs). These benefits
accrue from a wide range of uses including transportation and domestic
fuels, fertilizers, and feedstocks for numerous industrial processes.

127

S. Hartwig (ed.), Heavy Gas and Risk Assessment - II, 127–146.
Copyright © 1983 by Battelle-Institut e.V., Frankfurt am Main, Germany.

The industries that supply these materials have developed sophisticated safety technology and procedures as a result of many years' operational experience and the controlling influence of industry-generated standards and government regulations. The potential hazards associated with accidental LGF releases including cryogenic and chemical burns, asphyxiation, fire, explosions and toxicity have stimulated a high degree of safety consciousness in all aspects of LGF operations. All of these influences have tended to minimize the number of LGF releases and their consequences.

Against this encouraging background, there are a few LGF releases each year that receive wide-spread news media coverage. Accidents involving road or rail tankers containing liquefied petroleum gas (LPG), ammonia and other hazardous liquids are particularly newsworthy because they tend to occur in public places. In general, news media coverage has tended to emphasize casualties, damage and potential hazards of these events out of proportion to the true risks involved. Some special interest groups are also active in making the public aware of LGF release scenarios potentially more destructive than any that have occurred to date. These include, for example, large spills of liquefied natural gas (LNG) that would hypothetically result from the rupture of a large ocean-going tanker or a major storage tank failure in an import terminal or peakshaving plant. Public concern is understandable despite the current expert opinions that such events should have a very low probability of occurrence.

The reaction of industry and government to these concerns has been very responsive. Currently there are several large research and development (R&D) programs and many smaller individual efforts addressing LGF concerns being conducted by industries and government agencies in Europe, Japan, the U.S., and other countries. A substantial body of useful data has already resulted from these efforts and is expected to grow as work continues. However, the emphasis of current work favors LNG with progressively less attention being given to LPG, ammonia and other LGFs. This raises the questions of whether current R&D directions fully reflect the relative risks associated with each LGF and, if not, what other R&D activities are warranted.

The U.S. Department of Energy (DOE) is funding the LGF Safety Studies Project at Battelle's Pacific Northwest Laboratories (BNW) to consider these and other issues related to LGF safety and environmental control.(a) This paper presents R&D recommendations that result from BNW studies on LNG, LPG and ammonia.

(a)This work contributes to the Liquefied Gaseous Fuels Safety and Environmental Control Assessment Program conducted by the Environmental Safety and Engineering Division in the DOE Office of the Assistant Secretary for Environmental Protection, Safety and Emergency Preparedness.

2. GENERAL PERSPECTIVE

There is a continuing need to review the potential benefits of LGF safety and environmental control R&D to maximize the usefulness of current projects and guide the planning of future work. Assessments of this type have been part of the DOE Liquefied Gaseous Fuels Safety Program on LNG, LPG, ammonia and hydrogen.(1) As the number of sources and volume of information increase, the assessment task becomes more difficult.

A major goal of further R&D on LGFs should be to develop improved safety and environmental control practices. R&D planning and priorities should therefore be determined by practical needs and knowledge gaps in release prevention and control (RP&C) techniques. The state-of-the-art in RP&C practice is different for each LGF, and there is a broad range of expert opinion regarding the nature of issues that need to be addressed. For example, one view is that essentially no additional work on release phenomena is necessary. The opposite perspective is represented by those who think that a complete understanding of release phenomena is a prerequisite for planning subsequent R&D projects. Our position is that projects planned on the basis of RP&C priorities are of the most immediate value.

2.1 RP&C Strategy and Priorities

An overall strategy for reducing the frequency and minimizing the consequences of LGF releases includes the following elements:

1. Release Prevention
2. Release Detection
3. Release Control
4. Vapor Control
5. Fire Prevention
6. Fire Detection
7. Fire Control
8. Damage Control

The above order represents the sequence in which each element of the strategy comes into play if the preceding element fails. This order is considered to be the natural priority of RP&C practices and, in turn, the basis for establishing R&D priorities. The Gas Research Institute Program on LNG Safety(2) is an example of R&D based on RP&C objectives and includes projects grouped in release prevention, hazard assessment and hazard control areas.

2.2 The Relevance Matrix as an R&D Planning Tool

R&D planning requires a comprehensive understanding of the extent and usefulness of current knowledge. As more information generated by ongoing efforts adds to the existing data base, it appears both desirable and timely to consider establishing a formal methodology to aid

in defining the need for new projects. As a tool for this purpose,
we suggest the development of a "relevance matrix" which describes
RP&C objectives in terms of net knowledge gaps.

This concept is illustrated schematically in Figure 1. The matrix
identifies the relevance of existing knowledge in various categories
to RP&C problems or objectives with different priorities. For example,
improving the release prevention design of a valve could require, among
many inputs, knowledge of applicable standards and regulations and
human factors that may nullify their effect. Information on vapor generation or flame propagation would probably not be applicable in this
area, whereas it would be essential in the consideration of release control problems. Computerized collection and search techniques would
allow the correlation of information from a large number of sources.
The comparison of existing knowledge with that required for the solution of specific problems permits the identification of the net knowledge gaps associated with particular RP&C objectives. This association would help establish the priority and focus the objectives of
new R&D activities.

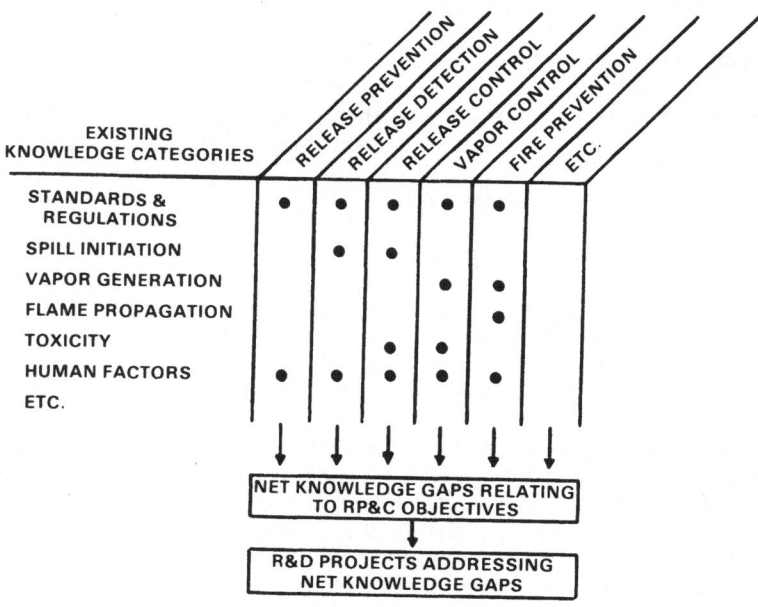

Figure 1. Relevance matrix methodology for R&D planning

The methodology would be amenable to computer accounting for corre-
lating related information on different LGF systems; identifying plans
that have marginal value and providing an up-to-date checklist of
progress toward goals and issues remaining to be addressed. Such a
planning tool could aid the coordination of interagency and international efforts and reduce the risk of conducting duplicate or redun-
dant investigations.

Until recently there has been little need for such a formal methodology
because a reasonable consensus developed on the emphasis of needed
work. Today this area has become controversial. The growing abundance
and detail of available data on LNG and, to a lesser extent, on LPG
offer an initial opportunity to test the value of the relevance matrix
approach. Once developed, it would be applicable to other LGF systems
and could become generally useful to industry as an aid in RP&C planning and decision making.

The balance of this paper contains a discussion of R&D needs and recommendations that relate to current RP&C objectives for LNG, LPG and
ammonia. While the list of topics is not exhaustive, its range illustrates that there is much worthwhile work still to be done and that
important opportunities exist for integrating the R&D initiatives of
the LGF industries with those of other concerned agencies and
institutions.

3. LNG SAFETY AND ENVIRONMENTAL CONTROL

Natural Gas is one of nature's most perfect fuels in terms of energy
density, ease of separation and delivery, and the absence of environmental pollution associated with its use. Natural gas is kept in its
gaseous form whenever possible during distribution and use. However,
the 600-fold reduction in volume resulting from liquefaction is a practical and economical solution for transoceanic shipments and land
transportation to supply satellite storage plants and for storage at
import terminals and peakshaving plants.

In recent years, substantially more attention has been given to LNG
safety hazards than to those of other LGFs. If the characteristics of
various LGFs are compared, it can be argued that LNG is not the most
hazardous. Nevertheless, LNG safety and environmental control is being
advanced by this R&D and a significant portion of the knowledge gained
is applicable to other LGFs, especially to LPG. Some of the more important topics for further consideration are discussed below.

3.1 Risk Assessment

Risk assessment should become an important tool to guide RP&C strategies and facility design, planning and operations. The good safety
record of LNG system operations is such that historical accident data
are not yet statistically significant as bases for meaningful risk
analysis. However, the relative ranking of potential release scenarios

and consequences can be based on existing data using inductive methods
such as failure modes and effects analysis and deductive processes such
as fault tree analysis. These techniques can be used to compare the
order of magnitude of release quantities and frequencies for given
scenarios.(3a, 3b) It is important to periodically update previous
risk analyses as new data become available either as a result of
actual accidents or, preferably, more years of uneventful operation.
Decisions by regulatory bodies should be guided by risk analyses. The
use of formal risk assessment techniques can also be of benefit to the
LGF industries in evaluating current risks, assessing the cost/effec-
tiveness and safety impacts of changes and in the general improvement
of RP&C philosophy and practices.

3.2 Human Factors Effects

The gas industry has developed highly refined systems and procedures to
promote safe and reliable operation of LNG facilities. However, there
is evidence that human factors have contributed to previous accidental
LNG releases. It suggests that additional consideration of human factors could be usefully applied in the design and development of RP&C
systems and practices. Human factors engineering (HFE) is generally
considered to be the domain of military and aerospace systems and,
more recently, nuclear power systems. In general, it appears that HFE
has only rarely been applied, in a systematic manner, to the safety
design of industrial plants such as LNG facilities. Because of this,
HFE enhancement of LNG safety and control systems is a potentially
fruitful R&D area where gains in LNG plant reliability and safety can
be expected for a relatively modest investment. Human factors are
also an issue in the design, fabrication, inspection and maintenance
of facility components and systems. In addition, human factors are
important in release control and emergency response and in injury and
damage control procedures. Human factors are, therefore, a pervasive
consideration over the whole range of RP&C strategies applicable to
LNG systems. As indicated below, similar arguments apply to LPG and
ammonia systems.

3.3 Devices and Systems

There is a continuing need for evolution and development of RP&C
devices that can further reduce the risk of accidental releases.
In the area of release detection, the gas industry would welcome gas
sensors that are more reliable and resistant to poisoning and odorants
applicable at cryogenic temperatures. Emergency shut-off systems
could be improved by designs that reduce the fluid hammer resulting from
their activation. Advanced release and fire control methods are
desirable including the development of new materials and techniques
and improved use of existing approaches such as the use of water
sprays and fire retarding materials. Further work on leak prevention
and arrest technology appears worthwhile.

3.4 R&D on LNG Release Phenomenology

R&D activities to date have produced a vast amount of information on LNG release phenomena. Work has been performed on all aspects of potentially hazardous behavior including spill characteristics, vapor generation and dispersion, and pool and vapor fire phenomena including deflagration and detonations. These phenomena have been evaluated in the laboratory, by simulation in wind tunnels (4a, 4b) and by scale spill tests up to approximately 40 m^3 in size.(5) Much effort has been devoted to the development of computer models which have been generally validated by the scale spill tests. Regulations that govern the safety design and operation of LNG plants are based on the more conservative of these models.

All of this knowledge is directly applicable to LNG release control strategies. There is a need to continue R&D in these areas with the goal of improving abilities to predict the behavior and consequences of larger releases. One approach is to conduct larger scale spill tests. However, the number of tests possible will be limited by both the expense and availability of suitable facilities. As a result, there is considerable value in work to validate scaling parameters with the objective of increasing the confidence that can be placed in laboratory scale experiments and wind tunnel simulations. In the past, spill tests of up to 40 m^3 have been used as exploratory research tools. In the future, exploratory investigations should be conducted at the smallest scale that produces acceptable results. The conduct of larger scale tests should be reserved as a means of confirming the scalability of small scale experimental data. This is an area of opportunity where innovation in dimensional analysis and experimental techniques may be able to resolve current controversies about the scaling relationships that apply to LNG release phenomena.

While the current tendency is to develop more accurate and complex computer programs to describe LNG release phenomena, the gas industry needs uncomplicated but conservative models that can be used quickly to predict the extent of hazards resulting from an accident spill after it has occurred. This information would provide the basis for rapid decisions on appropriate counter measures. There are similar needs for such a tool to guide emergency response procedures for other LGFs and hazardous materials.

A relatively neglected area is the influence of the physical layout of plants, terminals and facilities on the development of spill scenarios. Release characteristics including vapor generation, dispersion and fire behavior may be sensitive to the location of the spill, spill rate and physical obstacles encountered. Potential fire behavior associated with actual plant layouts could correspond to that in semi-confined geometries for which very few experimental results are available. Flame front acceleration caused by obstacles and related effects could be expected to increase explosion hazards and make fire and damage control more difficult. Modeling and scale experiments to evaluate these effects would be an important contribution to improving release control techniques.

Rapid phase transition (RPT) explosions have been observed in LNG spill tests on water.(5) Analysis performed by Lawrence Livermore National Laboratory on data from the China Lake experiments indicates that there may be a critical water temperature range and critical LNG spill rates that control the violence of RPT events. Future research in this area should simulate an LNG spill during loading or off-loading at a marine terminal to determine if an RPT would have the potential of damaging the tanker, loading trestle or other vital equipment in the vicinity.

3.5 New LNG Applications

The final topic in this discussion on LNG is the anticipated growth of natural gas as an automotive fuel. There are already between 15,000 and 25,000 vehicles in the U.S. powered by natural gas. Worldwide estimates are in the range of 300,000 to 400,000 vehicles. The fuel is carried in the vehicle, either as compressed natural gas or LNG. The technology of both fuel containment systems is highly developed and the safety design well engineered. However, as the population of these vehicles grows, additional safety and environmental control concerns may arise. These could involve RP&C considerations during fueling and at the scene of severe accidents. The safety design of LNG tank stations and human factor effects are other possible areas of concern.

4. LPG SAFETY AND ENVIRONMENTAL CONTROL

LPG is used as a petrochemical feedstock and as a fuel. Its use in the home is primarily concentrated in areas where piped natural gas is not available. The LPG industry directly serves about 18 million customers in the U.S. where it is estimated that roughly 60 million people are dependent on LPG for one use or another.

On a statistical basis, the risks associated with LPG production and consumption are less than many voluntary risks taken by the public such as smoking, driving and exposure to medical X-rays. Martinsen and Cavin(6) have estimated that the annual individual risk is about one death per 37 million persons, or 2.7×10^{-8} for LPG land transportation and storage operations in the U.S. The BNW assessment of LPG safety and environmental control developed a total system estimate of annual individual risk in the range 2×10^{-7} to 5×10^{-7}. Risks at the consumer level are the dominant contribution in the BNW assessment. The two estimates are otherwise in reasonable agreement. LPG transportation and storage risks appear to be comparable to the risk

of a person on the ground being killed by an airplane crash. Risks
associated with consumer use of LPG appear similar to the risk of
fatalities resulting from lightning and tornados. Some experts think
that these risks will increase with the expected future increase in
LPG markets and world trade. While existing safety levels appear
acceptable, there are continuing operational and economic incentives
for R&D leading to improvements in LPG safety and environmental
control. The following discussion presents some R&D needs in this area
based on RP&C objectives.

4.1 Accident Data Collection and Analysis

The collection and analysis of accident data are essential first steps
in the development of a sound basis for planning RP&C strategies and
R&D activities, developing codes and standards and making regulatory
decisions. Experience in other areas has demonstrated that accident
records provide a valuable means of identifying hazardous conditions
and accident causes. Unfortunately, there is a general lack of LPG
accident information available as a coherent and consistent body of
data. With some notable exceptions such as LPG transportation, much
of the existing accident information fails to meet many of the basic
criteria of usefulness and is not readily accessible. This lack
precludes the objective assessment of safety-related R&D needs and,
in some cases, even a determination of whether R&D effort is necessary.
The situation arises in the U.S. principally because no single agency
has overall responsibility for LPG safety and environmental control.
There is a need for additional accident data and analysis in each
major area of LPG operations, including production facilities, import/
export terminals, peakshaving plants, all modes of transportation,
consumer storage and appliances, and temporary installations.

4.2 Assessments Supporting Regulatory Decision Making

For an established organization such as the LPG industry, the purpose
of further regulations should be to minimize, at reasonable cost, the
frequency and consequences of accidental LPG releases. As technology
improves and situations develop or change, it appears prudent for
regulatory agencies to review the efficacy of existing regulations before considering additional rules. The following assessments are recommended to provide information that may assist future decisions relating
to the development of both industry standards and new regulations.

Present safety standards and regulations for LNG are far more restric-
tive than those pertaining to LPG. Yet there are indications that
handling LPG involves greater risks. In the U.S. a large amount of
work has been done in recent years establishing the need for and
implementing 49 CFR Part 193 - Liquid Natural Gas Facilities: New
Federal Safety Regulations. No comparable standards have been
prepared for LPG. Much of what has been done in examining LNG
safety is a useful basis for planning improvements in LPG standards
and regulations.

LPG tanks located in large tank farms such as refineries, gas extrac-
tion plants and terminals often follow oil tank safety standards more
closely than LNG tank requirements. These standards do not appear to
adequately account for the hazards of potential LPG releases. LNG
tank design, diking requirements, spacing and isolation requirements,
as well as potential risks of current LPG practice should be reviewed
to determine a basis for upgrading LPG bulk storage practices and
requirements. Similar consideration should be given to storage tanks
at peakshaving plants.

With projected increases in LPG traffic, higher standards may be
required for certain conventional ship systems to minimize the possibility of ship collisions and groundings. Stricter regulations on
navigation equipment, collision avoidance systems, steering apparatus,
and continuity of electric and propulsive power appear to be needed.
Expected increases in the numbers and movements of LPG ships and barges
combined with construction of new terminals and the expansion of existing facilities suggest that greater emphasis needs to be placed
on risk assessment. The majority of risk analyses done so far have
been made in connection with LNG rather than LPG facilities and operations. Many of these studies have been site-specific and have been
quite inconsistent in their methodology. A LPG maritime risk assessment covering a range of site dependent parameters is needed to update
previous efforts and provide a future tool for regulatory decision-
making in this area.

The boiling liquid, expanding vapor explosion (BLEVE) phenomenon
typical of LPG pressure vessels exposed to fire can cause casualties
comparable to those of vapor cloud explosions. There is no information available on how severe this problem is with consumer-type
installations. The frequency of BLEVE occurrence in consumer
storage should be established. Results of such a study should indicate whether changes are appropriate in related codes, regulations
or inspections and enforcement procedures.

Cooperative studies with industry involvement would provide a balanced
perspective on each of these issues. Information that is used subsequently as the basis for establishing standards and regulations should
be guided by formal risk assessment analysis.

4.3 Device and System Improvements

This section presents some of the needs for LPG device and system
improvements that represent current R&D opportunities leading to
improved RP&C practices.

At the present time, structural integrity cannot be determined with
high confidence for either a liquid or gas pipeline. Currently,
there are several methods in use to assess the condition of operating
pipelines that have various limitations or place higher than desirable

stresses on the system. For example, periodic hydrostatic testing applies large pressure cycles that may cause structural flaws to enlarge. It is recommended that a state-of-the-art study be conducted to determine an optimal integrity assessment procedure for LPG pipelines. Appropriate laboratory and field R&D should be conducted to confirm the results of this assessment.

Tank and equipment foundations are not always designed with adequate considerations for soil movement. While tanks and connected components may be built on independent but adequate foundations, soil movements between the foundations may strain the connected piping and devices. Industry experience in various climates and soil conditions and with various industrial piping systems should be examined to determine the extent and nature of movement over time that has occurred in existing refineries, gas plants, and similar facilities. The nature and extent of movement should be examined and techniques for preventing strain in piping and components should be reported. This information would be a basis for recommendations to be incorporated in facility design standards and practices, as required.

Available statistics for LPG tank truck accidents indicate a large incidence of tank leakage after overturns. To reduce the incidence of tank leakage in overturn accidents, R&D efforts should be initiated to improve the design of tank appurtenances. The objectives of this work would be to reduce the leakage potential of fittings, vents and valves and develop a standard overturn qualification test for the whole system. One condition contributing to overturn accidents is sloshing of the LPG in partially filled trailers that interfered with driver control. The severity of this problem should be investigated and anti-sloshing measures developed.

In a highway accident, there is a potential for fire resulting from failure of the truck fuel system even though the LPG tank and trailer are initially undamaged. Fuel system design requirements for LPG trucks should be reviewed with the objective of developing improved systems particularly in more vulnerable areas such as the crossover line.

A number of LPG explosions have been caused by the ignition of LPG leaking from tanks and fuel lines. Studies of safety problems associated with LPG tanks indicate that this is the primary hazard associated with LPG containers used for consumer storage. Currently an odorant is used for leak detection. This, however, is ineffective in many cases. A project should be initiated to develop an economic LPG leak and alarm detector for consumer use. Some types of natural gas alarms are available and should be investigated to determine their adaptability for this purpose.

4.4 R&D on LPG Release Phenomenology

Possibly as much as 80% of our current knowledge of LPG release behavior is based on the extrapolation of LNG information. As mentioned earlier, the continuation of LNG studies can be partially justified by their relevance to other LGF knowledge gaps. In the following discussion of R&D needs in LPG release phenomenology, credit is taken for the related value of the LNG work that is expected to continue. The development and use of the "relevance matrix" tool are also recommended for planning R&D on LPG release behavior.

A model is needed for the gravity- and wind-induced motion of LPG vapor to account for near field flow and dispersion. The model should specifically incorporate, as far as possible, factors relating to the difference in density between the atmosphere and the vapor cloud, and should account for weather factors, topography, obstacles, and vapor generation rates. An improved description of vapor cloud drift and flow could help in defining safe distances between possible LPG accident sites (e.g., transfer and storage sites, railroad grade crossings, etc.) and ignition sources in the neighborhood.

Models used for LPG far field dispersion have been derived from LNG and other gases. Assumptions in some of these models, such as neutral buoyancy, do not apply to LPG. Current studies on LNG should provide far field dispersion models that can be modified for LPG use to give concentrations as accurately as may be required for plant siting and safety considerations.

BLEVEs are perhaps the most serious manifestation of failure in RP&C systems. The rapid vaporization of LPG involves two-phase fluid mechanics and is a very complex phenomenon. In general, work to reduce the occurrence of BLEVEs must first consider changes in tank construction materials, insulation materials and safety valve capacity. These approaches do not require extensive knowledge of LPG behavior. However, critical two-phase flow of a flashing liquid is still poorly understood. The study of two-phase flow in safety valves under BLEVE-initiating conditions could lead to desirable improvements in safety valve design.

Extensive literature already exists on the subject of radiation from LPG pool fires, and fireball formation mechanisms are fairly well understood. While current knowledge appears adequate for establishing isolation distances for structures and materials that may be exposed to LPG fires, existing analytical models do not completely describe the complex phenomena of a free-burning LPG fire. There are additional areas where new knowledge may lead to improvements in fire, injury and damage control strategies. The mechanics of soot production and consumption should be investigated together with the effects of soot on radiation characteristics. Further, pool fire tests on water should be performed for comparison with experiments on land. As a function of time, radiation should be measured in several directions. Other measurements would include fuel consumption rate, vapor and liquid

composition and weather conditions. These and other relevant data
available in the literature should be combined to verify and enhance
existing analytical models and used to model large scale fires. Additional work is needed to determine if available knowledge on radiation
characteristics of test fires is scalable to fires that actually occur.
This effort could be aided by reviewing information on past accidents.
News media films are possible sources of data that have not yet been
fully utilized for this purpose. Data on fireball size, radiation
and duration and other information of the circumstances surrounding
accidents could provide a basis for comparing theory with recorded
characteristics of actual incidents. This comparison could provide
a decision basis for possible experimental verification in latter
phases of this work.

The aerothermochemistry and dynamics of LPG combustion under controlled
conditions are well understood and documented. However, before adequate predictions can be made of flame propagation rates and pressures
that might result from the ignition of LPG vapor clouds, further experimental data are needed under conditions simulating possible or likely
spill scenarios. As with LNG, these experiments should be conducted
in semi-confined geometries to simulate actual spill conditions. Further investigations are required to extend recently performed work on
flame propagation in channels.(7) Flame speed and pressure rise would
be determined as a function of fuel type, fuel/air gradient and external
wind direction and velocity.

It is known that the energy release requirements to initiate detonation
in propane-air mixtures are less than those required to produce detonation in methane-air mixtures. However, there are few specific data on
the energy required for stoichiometric mixtures of propane or butane
in air, and data for oxygen-nitrogen mixtures probably cannot be
reliably extrapolated. Also, no direct data are available on the conditions necessary to detonate LPG-air mixtures exiting from a duct or
in an open system. It is also debatable whether a detonation once
started will be always self-sustaining. Finally, there is considerable
evidence that a disproportionate lowering of the critical limit of a
pure fuel occurs when a small amount of more easily detonated fuel is
added. Experimental approaches are required to investigate these areas
of uncertainty to guide the development of LPG fire and damage control
strategies.

4.5 Human Factors Considerations and Procedure Development

In all aspects of handling LPG as with LNG, ammonia and other hazardous
materials, human factors effects represent a generally underexplored
area in which the application of established HFE principles can further
and profitably minimize risk. Recommendations in this section address
some apparent gaps in RP&C practice relating to human factors effects
in peakshaving operations, transportation and accident response
procedures.

Peakshaving plants may be left unattended during most of the year except during the time LPG is delivered. Many are operated only ten to twenty days per year during extremely cold weather. Deficiencies discovered during plant operations are apt to be tolerated for the short operating period rather than being serviced immediately under possibly severe weather conditions. Service personnel may be required on short notice to leave other work and assist in repairs. This can result in the use of probably less experienced personnel working unsupervised on potentially hazardous assignments. A review of the frequency of these practices may identify the need for improved safety equipment, operating procedures and training of personnel. This review should provide a basis for encouraging the whole industry to adopt state-of-the-art practices.

Errors by tank truck drivers are frequent causes of accidents having the potential of releasing LPG in the public environment. While some educational efforts are underway, these activities do not appear adequate as an organized approach for reaching all persons responsible for safety in LPG truck transportation. Projects should be initiated to develop a broad-based safety education and training and qualification program for LPG truck drivers, maintenance, management and emergency response personnel. Industry initiatives in this area could have positive public relations value.

Spilled LPG is often removed by flushing with water as in the case of gasoline spills. While this method is effective at the scene of the release, there is the risk that LPG may reach ignition sources after flowing through drains and sewers. There is evidence that this has occurred in sewer lines containing vapors from gasoline and LPG spills. The potential hazard of ignition of spilled LPG is a serious problem, particularly since it may expose third parties not involved in or even aware of the spill. Available accident reports should be reviewed to determine the frequency of these ignitions and the time and distances involved in the migration of spilled vapors. Both gasoline and LPG spills should be reviewed to make comparisons of the relative hazards. Based on this comparison, the need for improved practices should be evaluated, including spill containment and treatment methods and emergency response training and procedure development.

5. AMMONIA SAFETY AND ENVIRONMENTAL CONTROL

Ammonia is produced in natural processes and is also widely used in agriculture and industry. Since 1970, ammonia has ranked first in the value of chemicals manufactured in the United States. About 75-80% of the ammonia is used in the production of fertilizer, but there are also about 2500 industrial products derived from it. Ammonia also has a number of energy related uses which may develop in the future.

This common substance in excessive doses is hazardous to all forms of life. It is highly soluble in water and therefore causes skin burns, eye irritations, respiratory discomfort and at high exposure rates, death. Compared to LNG, which under essentially all circumstances is supervised by trained professionals, ammonia is often handled by personnel with limited training in safety problems and limited knowledge of the hazards and behavior of ammonia. There is a higher probability of its accidental release in situations in which members of the public may be affected. It is, therefore, as essential, if not more so, to improve RP&C techniques for ammonia as for other hazardous materials.

The BNW assessment of R&D needs in ammonia safety and environmental control was completed in September 1981.(8) This section provides a summary of recommendations developed in this assessment for improvements in ammonia RP&C techniques.

5.1 Ammonia Accident Data

Currently there is no national system or single responsible agency in the U.S. for gathering and compiling all types of ammonia accident data. With the exception of interstate transportation-related accidents for which, in the U.S., there are federal reporting requirements, much of the data on ammonia releases is collected by various agencies and departments within state and local governments. In many cases, even the identity of ammonia is obscured in a general category such as "alkali" among other material categories used to classify accident data. Without a better accident data base than that currently available, the ability to assess risks associated with distribution and use of ammonia becomes more difficult and impacts the objective planning of further R&D that may lead to improvements in RP&C techniques. A comprehensive accident data collection system should be established at least at the national level, and better still, at a level at which international experience could be assessed.

5.2 Ammonia Release Prevention

Accidental spills occur during production and storage, but most of the accidents involving frequent and significant exposure of individuals to ammonia happen during its transportation and use. Injurious accidents occurring during production are relatively few, and this appears to corroborate that industrial equipment, procedures and training of personnel have been developed on sound safety principles. In the transportation and use of ammonia, accidents are caused predominantly by equipment failure and human errors. It is probable that the estimated large number of releases in agricultural applications that go unreported also involve human factors. Release prevention in ammonia systems and operations relies strongly on individuals observing safety rules and regulations. There appears to be a general need to expand the consideration of human factors effects in training, equipment design and handling, and maintenance procedures that contribute to ammonia release prevention strategies.

It is recommended that additional work be conducted to enhance understanding of failures in ammonia storage and transportation tanks. R&D effort should focus on probable causes such as galvanic corrosion,

stress corrosion cracking and improper maintenance. Results would
provide the basis for mitigation actions and possibly improved standards and regulations.

The effects of galvanically coupled dissimilar metals in both liquid
and vapor ammonia systems should be studied with the goal of developing
improved corrosion inhibitors. Tests should be conducted with and
without welds and stress loads. The stress corrosion cracking (SCC)
resistance of certain commercial grades of steel, in the absence of
corrosion inhibitors, should be considered. The importance of oxygen
in the SCC mechanism should be determined and hydrazine examined as
a material to prevent SCC.

5.3 Ammonia Release Control

Because the toxicity of ammonia is its principal hazard, the ability
to predict vapor cloud generation and dispersion is important in developing and evaluating release control strategies and also emergency
response and evacuation plans. Additional model development and validation is needed to describe the behavior of negatively buoyant ammonia
plumes. Current models can estimate concentrations within a factor of
about 10 whereas with additional effort this might be improved to within a factor of 2 or 3.

Some specific recommendations to be considered include the development
of more sophisticated three-dimensional turbulent mixing models that
provide more accurate dilution profiles for the pressurized release
and carefully controlled experiments to quantify the variables governing the degree of atomization of liquid ammonia during the spill process. Controlled experiments should be performed on wind-induced
evaporation to verify available evaporation models. Spills should be
conducted to determine the partitioning ratios in water and subsequent
cloud characteristics. Model development should include simple and
conservative predictive codes to aid rapid decision-making by those
responsible for emergency response and evacuation actions, and release
control strategies and equipment.

5.4 Ammonia Damage Control

Two major areas need to be addressed in the area of human health effects
from accidental exposure to ammonia. First, more case studies are
needed to further define the long-term health problems associated with
toxic exposure. Second, exposure time/concentration guidelines need
to be developed for use in defining emergency procedures to manage
major spills in population areas.

In the BNW evaluation of ammonia effects on aquatic and terrestrial
organisms, several areas were identified where important information
does not exist. Two important areas are the determination of life
stages of plants and animals that are most sensitive to ammonia, and
the evaluation of toxicity and sublethal effects on these organisms.

Currently, it is not known if spills of ammonia would cause irreversible changes to an ecological community or in what time frame the community could expect to recover from the effects of the spill. The expansion of data bases in these areas would help to define the need for improved ammonia release control and damage control strategies.

Unlike LNG and LPG, there is very little risk of fire or explosion hazards associated with ammonia releases. It is true that ammonia can be burned (its potential as a fuel depends on this property); however, our assessment of the controversial ammonia fire issue is that self-perpetuating ammonia spill fires are practically impossible to initiate. Ammonia flash fires can occur, but they are expected to be short, generally in the order of seconds, duration. Therefore, we offer no general recommendations for fire control, with the exception of one concern as it applies· to emergency response procedures. The hydration of ammonia with water is exothermic. The contact of liquid ammonia with water will increase the evaporation rate as a result, whereas the natural evaporation of ammonia is endothermic, causing the remaining liquid to cool and spontaneously reduce the rate of evaporation further. First contact emergency response personnel, typically a local fire department, use water liberally to wash spilled materials away from the point of leakage. As with LPG, this can be hazardous. Emergency response personnel should attempt to avoid contact between water and liquid ammonia pools with the objective of reducing vapor generation rates.

5.5 New Ammonia Applications

Ammonia has the future potential of being used in several new large-scale applications as a fuel, energy transfer medium or chemical agent. This final section on ammonia discusses safety and environmental problems that might arise in such applications.

If ammonia were used as an automotive fuel, additional RP&C measures would be needed similar to those discussed above for LNG. Ammonia appears to be one of the "cleanest" fuels available and with respect to NO_x, it is ahead of gasoline, methane, propane and methanol. Because the combustion of ammonia does not generate CO_2, it would not contribute to the anticipated danger of CO_2-enrichment in the atmosphere. Overall, from an environmental pollution viewpoint, ammonia would be a highly desirable fuel.

In binary cycle applications, ammonia could be contained under high pressure (up to 2500 psi) as a vapor and as a liquid. As the corrosion behavior of ammonia at high temperatures is not well known, such systems might involve some safety hazards that cannot be estimated very well at the present time. For this application, materials compatibility R&D may be necessary, particularly with respect to stress corrosion cracking. The possibility of hydrogen embrittlement and hydrogen permeation must also be considered as a potential safety problem. As in steam systems, there is the possibility that high pressure ammonia leakage could result from any one of a number of causes. If a major rupture were to occur, the consequences of such an accident could be severe. If the break occurred inside a building, the building would probably be filled with a high concentration of ammonia vapor in a matter of seconds and people in the building might not be able to escape in time. It must be realized, therefore, that the conditions in an ammonia-power-generating facility would basically involve more serious potential hazards than those of steam generating plants and current ammonia storage facilities.

If ammonia were to leak into the main water cycle in a bottoming cycle application its immediate effect would be a change in the water chemistry which in turn could cause long-term corrosion problems. Because the proper treatment of water in steam generator applications has been more art than science, it is hard to predict what the ultimate effect of ammonia contamination would be. It is likely that only practical experience will provide a reliable answer.

Safety requirements and precautions for ammonia containment in ocean thermal-gradient power plants appear to be about the same as those of current ammonia manufacturing, fertilizing, and refrigeration systems. Problems of inadvertent leakage can occur as a result of metal corrosion but do not seem to present a serious safety hazard because of the limited inventory.

Ammonia has been used for refrigeration purposes for such a long time that there has been a substantial amount of experience in managing accidents in ammonia refrigeration plants. As these accidents were reported in the press, it was usually implied that ammonia was the source of the explosion. However, what actually appears to have caused these explosions is the "foul" gas, that collects in certain parts of the system. The "foul" gas originates from a mixing and reaction of ammonia with lubricating oil and contains some hydrogen and hydrocarbons. Present knowledge indicates that because ammonia is such a powerful solvent, any contact or mixing with certain organic materials may result in the development of these highly combustible gases. Therefore, any contact between ammonia and these materials should be avoided. It is recommended that a list be compiled of organic substances which should not be used in contact with ammonia.

The use of ammonia for removing NO_x or SO_2 from flue gases does not seem to involve any basically different or new technological processes which potentially could result in environmental or safety problems. However, the use of ammonia in various methods of coal preparation is still under active development. No standard method has yet evolved, and a hazard assessment at this time is premature, Potentially, however, there exists the possibility that relatively large amounts of liquid ammonia could evaporate and escape into the atmosphere if no adequate precautions were taken. It is recommended, therefore, that developments in this field be monitored in the future so that potential safety and environmental problems can be addressed in a timely manner. The same arguments are applicable to the potential future use of ammonia in water-splitting schemes.

6. CONCLUDING REMARKS

The purpose of this paper has been to discuss R&D topics that can
contribute to improvements in RP&C techniques for LNG, LPG, ammonia
and other LGFs with similar hazard potentials. There are some R&D
areas that appear as common needs for this class of materials including
the improvement of accident data bases, human factors engineering and
hazard prediction tools. Risk assessments are needed in a number of
areas to provide a rational basis for decision making in R&D planning
and the development of standards and regulations. The elements of
an overall RP&C strategy are considered to establish a basic priority
for additional R&D activities. An abundance of information on LGF
release phenomenology has accumulated, and it is now appropriate to
plan R&D with more focused objectives. The concept of the "relevance
matrix" R&D planning tool is suggested as a means of accomplishing this
goal.

As this paper indicates, many of these R&D needs present new opportunities for innovative analyses and techniques. In addition, they should
be the stimulus for increased cooperation and coordination between
industry and government programs. Finally, they show the value of a
collective approach for investigating LGF safety and environmental
control concerns rather than addressing each LGF material as a separate
entity.

7. BIBLIOGRAPHY

1. Liquefied Gaseous Fuels Safety and Environmental Control Assess-
    ment Program: A Status Report, DOE/EV-0036. Prepared by Pacific
    Northwest Laboratory for the U.S. Department of Energy, May 1979.

2. 1982-1986 Five-Year Research and Development Plan, Gas Research
    Institute, April 1981, Chicago, Illinois.

3a. Pelto, P. J., E. G. Baker and R. J. Hall, Assessment of LNG Release
    Prevention and Control. Proceedings of the U.S. Department of
    Energy Environmental Control Symposium, DOE/EV-0C46, September 1979.

3b. Baker, E. G. and P. J. Pelto, "Analysis of LNG Import Terminal
    Release Prevention Systems," PNL-SA-9251. Paper presented at the
    Cryogenic Engineering Conference, August 10-14, 1981, San Diego,
    California.

4a. Liquefied Natural Gas Wind Tunnel Simulation and Instrumentation
    Assessments, SAN/W1364-01. Prepared by R&D Associates and Colorado
    State University for the U.S. Department of Energy, April 1978.

4b. Neff, D. E. and R. F. Meroney, The Behavior of LNG Vapor Clouds
    Final Report, GRI 80/0094. Prepared by Colorado State University
    for the Gas Research Institute, July 1981.

5. Liquefied Gaseous Fuels Safety and Environmental Control Assess-
    ment Program: Third Status Report, PNL-4172. Prepared by Pacific
    Northwest Laboratory for the U.S. Department of Energy, March 1982.
    Reports A, B, C, and D contributed by Lawrence Livermore National
    Laboratory.

6. Martinsen, W. E. and W. D. Cavin, LPG Land Transportation and
    Storage Safety, DOE/EV/06020-T5. Prepared by the Applied Tech-
    nology Corporation for the U.S. Department of Energy, December 1981.

7. Liquefied Gaseous Fuels Safety and Environmental Control Assess-
    ment Program: Third Status Report, PNL-4172. Prepared by Pacific
    Northwest Laboratory for the U.S. Department of Energy, March 1982.
    Report H contributed by P. A. Urtiew, Lawrence Livermore National
    Laboratory.

8. Brenchley, D. L., et al., Assessment of Research and Development
    (R&D) Needs in Ammonia Safety and Environmental Control, PNL-4006.
    Prepared by Pacific Northwest Laboratory for the U.S. Department
    of Energy, September 1981.

MAPLIN SANDS EXPERIMENTS 1980: DISPERSION RESULTS FROM CONTINUOUS RELEASES OF REFRIGERATED LIQUID PROPANE

by J.S. Puttock, G.W. Colenbrander* and D.R. Blackmore
Shell Research Ltd., Thornton Research Centre, P.O. Box 1,
Chester CH1 3SH
* Shell Research BV, Koninklijke/Shell-Laboratorium,
  Postbus 3003, 1003 AA Amsterdam, Netherlands

SUMMARY

In 1980, a series of spills of up to 20 m^3 of LNG and refrigerated liquid propane onto the sea was performed at Maplin Sands in the South of England. Both instantaneous and continuous releases of liquid were made, with data collected from an extensive array of instruments up to 650 m downwind. This paper describes the results of eleven continuous propane spills at rates from 2.0 to 5.6 m^3/min, and in wind speeds from 2 to 8 m/s. A comparison is made of the measured results with those predicted by the dense gas dispersion model HEGADAS. In general, the agreement is good.

INTRODUCTION

A series of experimental releases of liquefied gases onto the sea was performed by Shell Research Ltd. in 1980. A total of 34 experiments included spills of liquefied natural gas and refrigerated liquid propane in quantities up to about twenty cubic metres. A description of the overall programme is given by Puttock et al. (1982). Some of the spills were ignited in order to study the combustion of gas clouds; other spills were purely dispersion experiments. Dispersion results from the eleven continuous propane spills are described here.

The site chosen for the releases was on Maplin Sands, an area of tidal sands with a typical slope of 1/1000 on the north side of the Thames estuary in England. Experiments were conducted during offshore winds for reasons of safety. The point of release was 350 m offshore and, when possible, spills were performed at high tide. So that spills could also be performed at low tide, a 300 m diameter dike was constructed to retain the sea water around the spill point. There was a maximum 0.75 m change in level at the offshore edge of the dike. Behind the 5 m high sea wall was flat farmland.

147

S. Hartwig (ed.), Heavy Gas and Risk Assessment - II, 147–161.

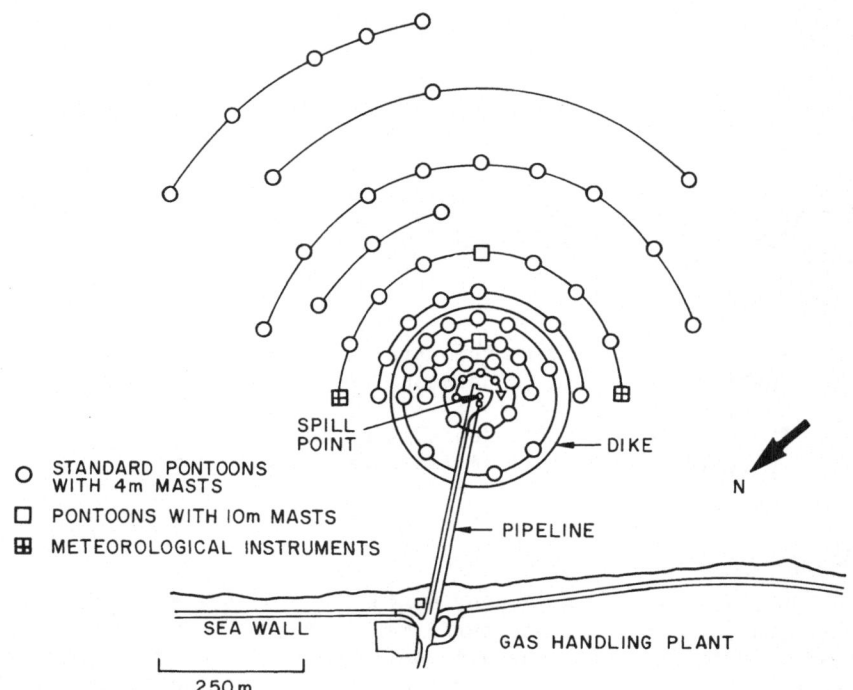

Figure 1. The layout of the instrument pontoons as used for most of
           the continuous propane spills

       In most of the continuous propane experiments the end of the spill
pipe was flared out in a vertical-axis cone with a horizontal plate
below it at the water surface. The liquid emerged from the slot
between the cone and the plate with negligible vertical momentum. In
two cases the liquid was delivered from the open end of a vertical pipe
terminating just above the water surface; and on two occasions the open
pipe-end was below the surface.

       Instruments were deployed on 71 floating pontoons most of which
had 4 m masts. The pattern of deployment for the continuous propane
spills differed from that used earlier (Puttock et al., 1982); more
pontoons were concentrated in the near-field, particularly in the
direction covered by the prevailing westerly wind (Figure 1).

       There were about 360 instruments in the array. Mounted on a
standard pontoon, in addition to combustion instruments, were three gas
sensors (at 0.5 to 0.9 m, 1.4 m and 2.4 m above the sea surface) and
one fast-response thermocouple close to the lowest gas sensor. The gas
sensor is a device based on measurement of the heat loss from a
filament under free convection. Two special pontoons each had ten-
metre masts, six gas sensors, one thermocouple and two three-axis sonic

anemometers. Two further sonic anemometers were deployed elsewhere.

Another two special pontoons were devoted to meteorological measurements. These provided vertical profiles of temperature and wind speed up to ten metres, together with measurements of wind direction, relative humidity, insolation, water temperature, and wave height.

The signals from all these instruments were sampled ten times per second (30 Hz for the thermocouples), digitised on the pontoons and relayed by cable via multiplexers to the computers onshore.

The spills were photographed from three locations, two land-based towers providing orthogonal views, and a helicopter overhead. At each location still photographs and video recordings were taken.

METHOD OF ANALYSIS

A problem is posed in these continuous spills by the meandering of the plume. The array of measuring stations (Figure 1) was set out as described to cover any offshore wind direction. Restriction to a smaller range of wind directions would have further reduced the number of spills which could be performed in the time available. The steady-state plumes in all but low winds were so narrow that they could pass completely between sensor stations occasionally, although rarely for a whole experiment because of variations in wind direction. Thus any sensor might see the plume only intermittently and there were times when the plume did not pass over any stations at a given distance.

An example of this is shown in Figure 2, a gas concentration measurement from a spill (46) in a 8.1 m/s wind speed; there were wide fluctuations as the plume intermittently reached the sensor. This can be compared with a signal obtained during spill 54 in a 3.8 m/s wind (Figure 3) where the plume was wide and steady and the sensor remained near the plume centre.

If model predictions are to be compared with the data, it must be decided exactly what gas concentrations the models are intended to predict. The long-term average concentration at a fixed location downwind of the source is reduced by meandering of the plume. However, meandering of the plume does not change the average maximum concentration inside the plume.

A model dealing with flammable gas dispersion should therefore not predict long-term average concentrations at fixed locations. On the contrary, the value of particular interest in flammable gas plumes is the maximum concentration in the plume at a given distance from the source, whatever its angular position may be. This concentration can conveniently be referred to as the "centreline concentration", though it does not necessarily occur at the centre of the meandering plume. If model predictions are in terms of mean concentrations, then the

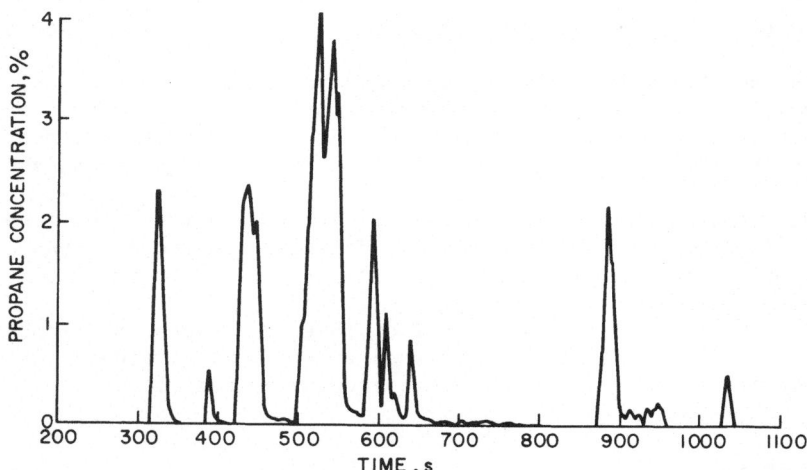

Figure 2.   A gas sensor signal from spill 46, at height 0.9 m and 88 m
            from the source. In these figures, time is measured
            relative to the arbitrary start of data collection for the
            spill; gas concentrations are smoothed using a 3s moving
            average

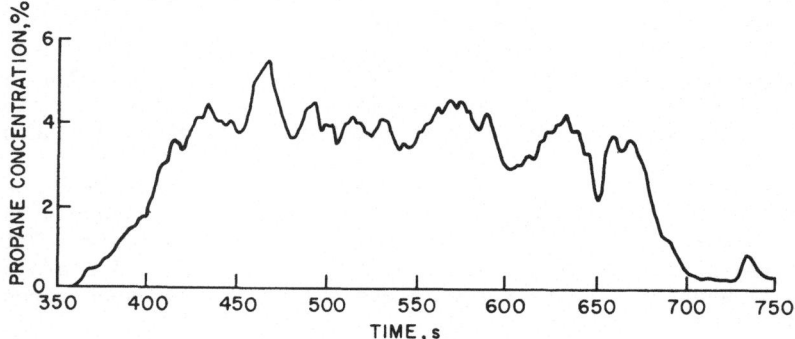

Figure 3.   A gas sensor signal from spill 54, at height 0.6 m and 181 m
            from the source

predicted plume-centre concentration should be compared with the mean
of this "centreline concentration", measured in the experimental plume.
The long-term average concentration at any fixed point may be much
lower, but that fact has little significance for the assessment of
flammable gas hazards. The crosswind variation in concentration
predicted by the model is then assumed to be taken relative to a
moveable centreline.

   To measure the relevant mean concentration it is necessary to
obtain measurements of the "centreline concentration" at each instant,
and subsequently to average the values. This is possible only if there

were several sensor stations across the plume at each distance,
requiring perhaps an order of magnitude increase in the number of
sensor stations. Averaging the intermittent signal at any sensor does
not give the required results. It is more useful to take the maximum
observed concentration, since the "centreline" concentration would be
observed as the "centre" of the plume meandered over a sensor. The
sensor stations were close enough that the plume centreline rarely
failed to pass over one for at least part of the spill time. In fact,
there were usually sensors in the plume for a large part of the spill
time. Then the maximum signal obtained can be assumed to be above the
mean of the "centreline concentration" as defined.

Model predictions

To describe atmospheric dispersion of heavy gas, emitted by area
sources at ground level, the HEGADAS model is being developed by Shell
Research (Colenbrander, 1980).

Steady-state and time-dependent versions of this model exist.
Both versions contain a simple gravitational spreading law and a
Richardson number expression, which quantifies the effect of density
gradients on the vertical dispersion. The description of heavy-gas
dispersion is consistent with the formulation of passive dispersion in
a neutral, stable or unstable atmosphere.

The steady-state version of HEGADAS has been made to simulate the
propane trials described in this paper. The measured air temperature
profiles show a nearly neutral atmospheric stability for all trials.
One would expect, therefore, very similar values of the ratio of the
friction velocity u_* and the 10 m wind speed U_{10}, for all trials. From
the sonic anemometer signals we calculated u^* for each trial (by
definition $\sqrt{\overline{u'w'}}$, where u' and w' are the fluctuating wind speed
components in horizontal and vertical direction, respectively). The
values of u_*/U_{10} we found varied widely between the trials, presumably
owing to the shortness of the averaging times used (usually less than
10 min). For all propane trials we found a mean value of
u_*/U_{10} = 0.034, which is consistent with the estimated surface
roughness at Maplin Sands. This is used as input value for all HEGADAS
simulations; the atmospheric stability was taken to be neutral for all
runs.

A more detailed analysis of the weather data, currently being
undertaken, will result in more accurate estimates of u_*/U_{10} and
atmospheric stability.

TWO EXAMPLES

Two detailed examples of the results obtained will be given: one
spill in a fairly high wind and one in a fairly low wind.

During spill 46, the mean wind speed was 8.1 m/s at an angle of 25° to the shore. The liquid release rate was 2.8 m³/min and the spill lasted for about eight minutes. Further details are included in Table 1. At such a high windspeed this spill was only marginally affected by the density of the gas. Figure 4 is a photograph taken when the plume was fully developed. Very little early lateral spreading of the plume beyond the width of the liquid pool could be seen. An adiabatic mixing calculation shows that, starting at a temperature of -42°C, propane plumes remain visible until diluted to

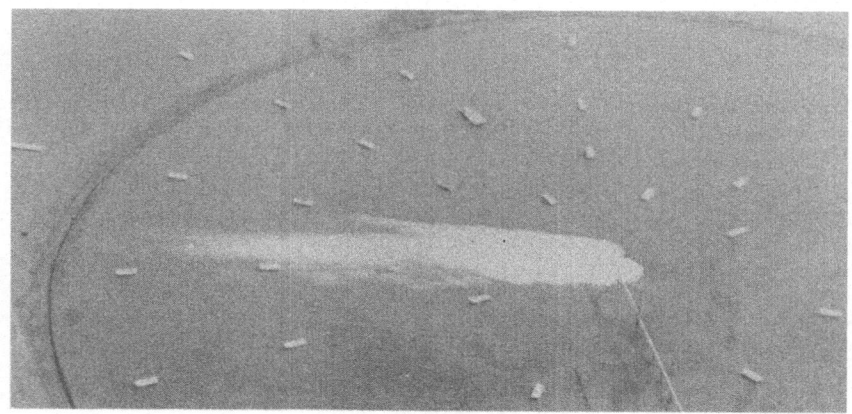

Figure 4.   The fully developed plume of spill 46

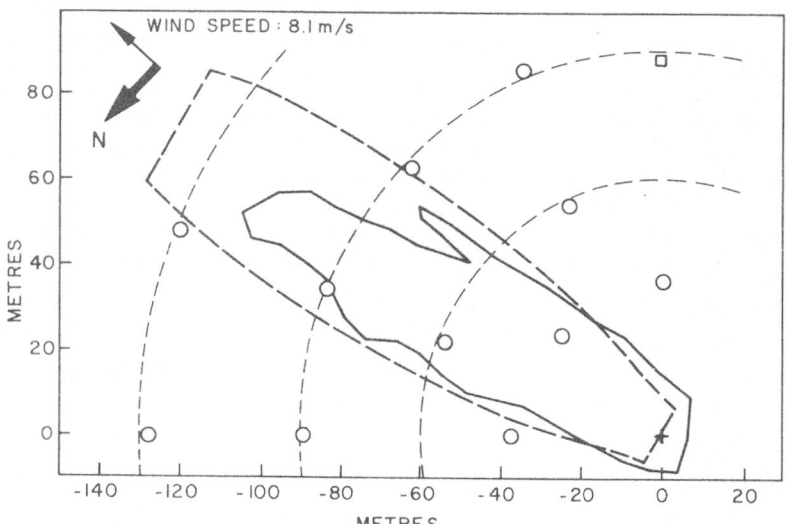

Figure 5.   A plan view of the spill 46 plume, derived from Fig. 4, and the limits of the visible plume predicted by HEGADAS

about 5% concentration for an ambient relative humdity of 70%. Here,
the plume was strongly visible in its centre to about 130 m, with
occasional wisps visible to 200 m.

    The cloud outlines from overhead still photographs such as
Figure 4 have been input to the computer using an image analyser.
After perspective correction, the visible outline can be plotted and
compared with the HEGADAS model prediction (Figure 5).

    Time series of gas concentration at three levels at each sensor
station can be plotted as in Figure 6. The signals have been smoothed
using a three second moving average to eliminate any high frequency
noise present. The gas sensor itself has an exponential time constant
of about 1.1 s (3 s to 90% response).

    The maximum concentration observed at each sensor can be obtained
and then plotted as a function of distance from the source. The model
predictions of mean concentration are shown in Figure 7 for
comparison.

    The gas temperature measurements are also of interest since they
can indicate the extent to which the gas plume has been warmed in its
travel over the sea surface. This heat transfer is not believed to
affect the dispersion of propane significantly but must be modelled
correctly for accurate prediction of LNG dispersion. It is convenient
to plot temperature drop from ambient against gas concentration and
compare this with the relation obtained in the model assuming no heat
transfer to the gas from the surface. The results in Figure 8 are
typical and show that the observed temperatures are higher than the

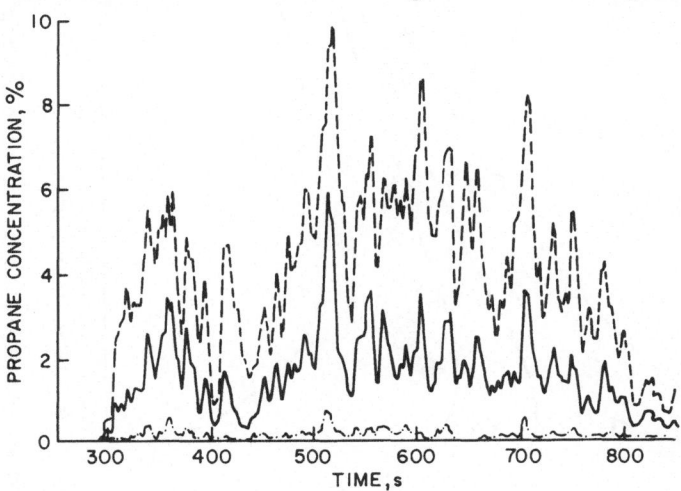

Figure 6. Gas concentrations measured in spill 46 at 34 m from the
           source, -46° from the array axis, at heights ——— 0.9 m,
           ———1.3 m and—·—·—· 2.2 mm

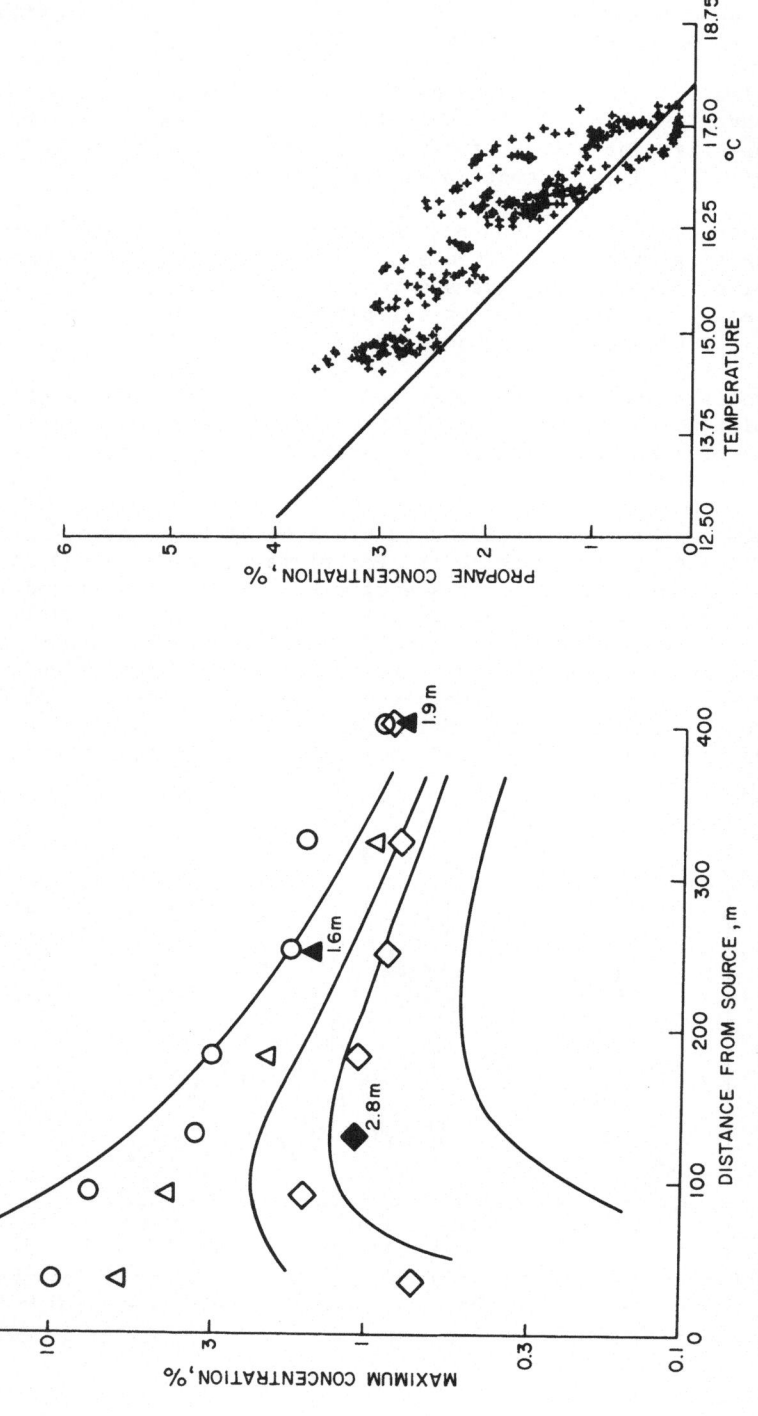

Figure 8. Concentration against temperature at 129 m from the source in spill 46. The line shows the relation given by an adiabatic mixing calculation

Figure 7. Maximum concentrations (smoothed using 3 s moving average) measured at each radius for spill 46, at heights O 0.9–1.0 m, △ 1.3–1.4 m, ◇ 2.2–2.4 m. The curves are HEGADAS predictions of mean concentrations at 0, 0.9, 1.3 and 2.3 m

adiabatic mixing calculation indicates. This is consistent with the
observation (Figure 5) that the visible plume is shorter than predicted
by the model, which does not include heat transfer from the sea
surface.

     Spill 54 was performed in a low wind (3.8 m/s). This produced a
much wider plume than spill 46 (Figures 9 and 10), clearly affected by
gravity spreading in its early stages. The plume was also much lower -
a 1 to 1.5 m high visible plume between 50 m and 100 m from the source,

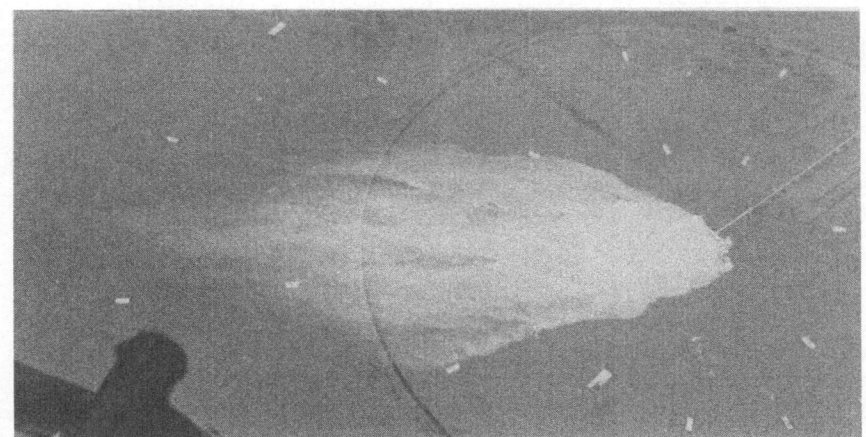

Figure 9.   The fully-developed plume of spill 54

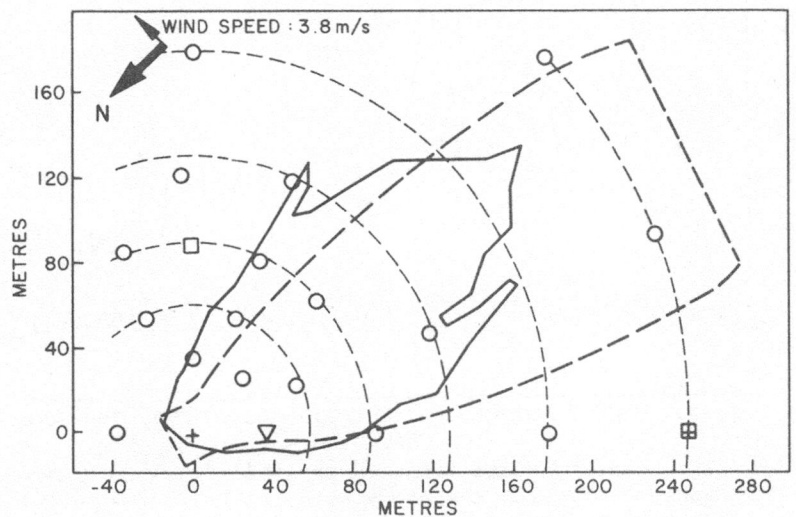

Figure 10.   A plan view of the spill 54 plume, derived from Fig. 9, and
             the limits of the visible plume predicted by HEGADAS

with very little increase in height thereafter. Fortunately, on many
of the pontoons the lowest gas sensor had been moved down from 0.9 m to
about 0.6 m above the surface which enabled useful measurements to be
made although the bulk of the gas may have passed beneath the sensors
close to the source.

The width of the plume ensured that changes in wind direction did
not create significant lateral movement of the plume, and the signals
at most sensors were remarkably steady (Figures 3,11). In this case it
is possible to plot both the maximum concentrations (Figure 12) and the
mean concentrations (Figure 13) as a function of distance.

RESULTS

Details of all eleven continuous propane spills are given in
Table 1. The period of time for which the flow was controlled at the
stated rate is shown, and the wind speed quoted is the mean over this
period. In order to compare the strength of density effects in the
spills, an initial Richardson number is calculated, as defined by
Puttock et al. (1982).

Three of the plumes were ignited, but prior to ignition these
experiments were identical to the dispersion experiments and can
provide useful data. However, on one of these occasions (spill 51) the
plume centre remained steadily between two lines of sensors during the
period of controlled release and so the plume centre was not measured.
Also, in spill 55 the wind direction was outside the range for which

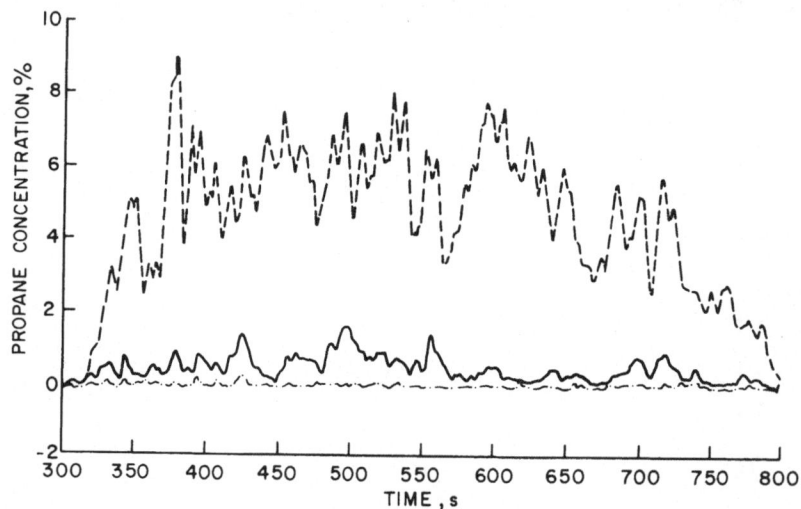

Figure 11. Gas concentrations measured in spill 54 at 88 m from the
source, 45° from the array axis, at heights ————0.6 m,
—————1.3 m and—·——·—2.3 m. The maximum concentration at
88 m, height 0.6 m, plotted in Figure 2, was obtained at
23°

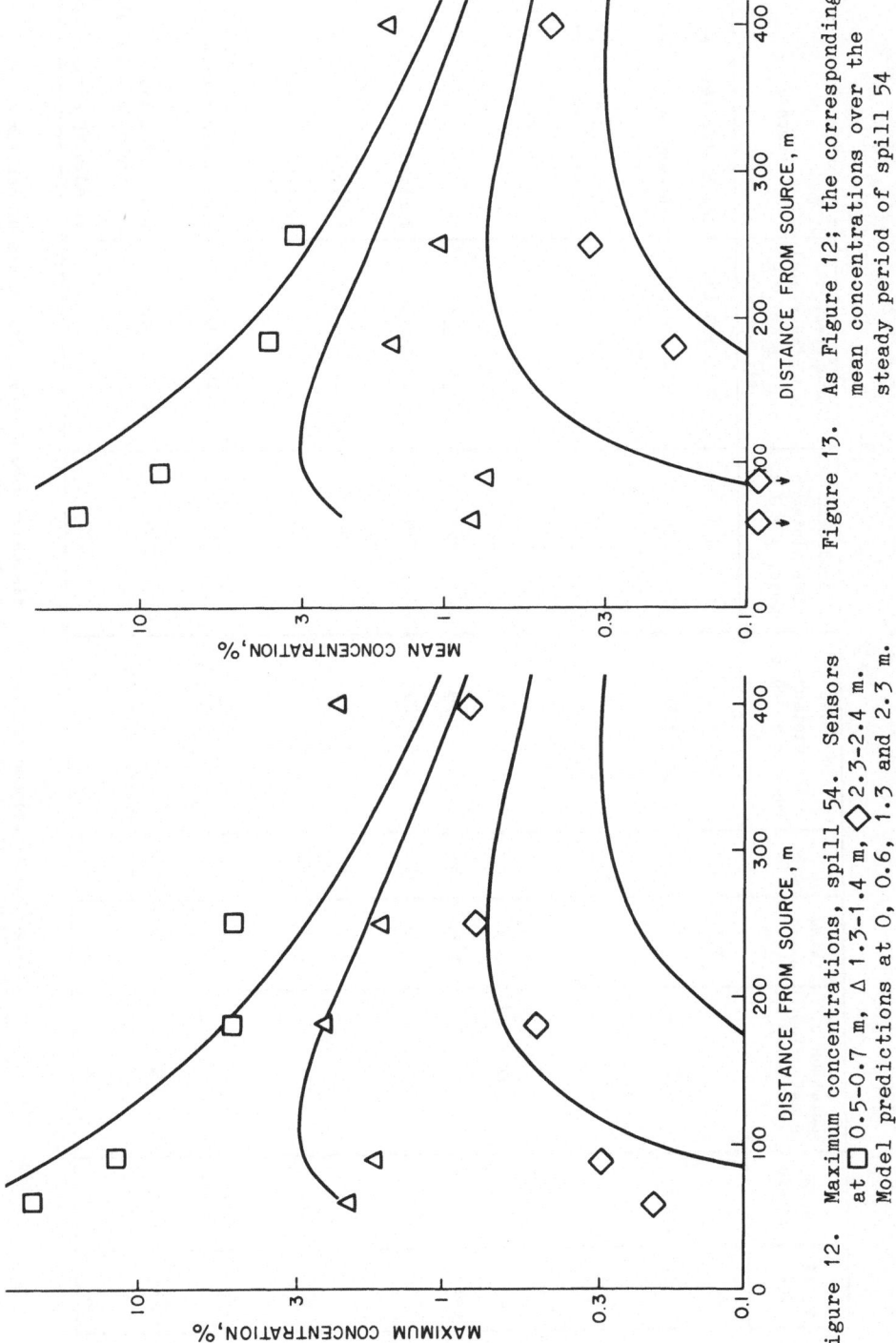

Figure 13. As Figure 12; the corresponding mean concentrations over the steady period of spill 54

Figure 12. Maximum concentrations, spill 54. Sensors at ☐ 0.5-0.7 m, △ 1.3-1.4 m, ◇ 2.3-2.4 m. Model predictions at 0, 0.6, 1.3 and 2.3 m.

Spill no.	Rate, m³/min	Duration of steady flow, s	Wind speed, m/s	Ri_0	Observed LFL distance peak, z = 0.9 m, m	Predicted LFL distance mean, z = 0 m, m	Concentration ratio (far field) peak, z = 0.9 m / model, z = 0 m	Comments
42	2.5	180	3.7	100	220 ± 35	315	0.50 ± 0.10	Underwater release
43	2.3	330	5.5	28	215 ± 20	245	0.79 ± 0.10	Pipe-end just above surface
45	4.6	330	~2	800	-	675	-	Pipe-end just above surface. Wind very unsteady and non-uniform
46	2.8	360	8.1	10	245 ± 35	225	1.18 ± 0.23	
47	3.9	210	5.6	35	340 ± 80	320	1.09 ± 0.27	
49	2.0	90	6.2	18	285 ± 25	210	1.7 ± 0.3	Ignited
50	4.3		7.9	13	210 $^{+50}_{-25}$	280	0.61 $^{+0.17}_{-0.09}$	Ignited
51	5.6	140	6.9	22	-	335	-	Ignited. Plume centre missed sensors when steady
52	5.3		7.9	14	200 ± 30 (mean 130 ± 30)	315	0.45 ± 0.08	Underwater release
54	2.3	180	3.8	90	450 ± 70 (mean 400 ± 100)	295	2.0 ± 0.6	(Data from lowest sensors at z ≈ 0.6 m)
55	5.2	150	5.5	40	-	372	-	At edge of array

Table 1   Details of the continuous propane spills. LFL distances were obtained from plots of maximum concentration against distance; in two cases long-term mean concentrations could also be used

the sensor array was designed and few measurements were obtained. For
two spills (42 and 52) the pipe end was underwater; these will be
considered separately. During spill 45, the wind speed was very low
(~2 m/s) but also unsteady and non-uniform. Consequently, quantitative
comparison with model predictions is not possible for this spill; it
provides a useful qualitative picture of the behaviour of a release in
a very low wind.

For each spill, results were plotted as in Figure 7. A smooth
curve, estimated by eye, was drawn through the data points for the
lowest sensor level. Taking into account the scatter of the data, two
more curves were drawn to estimate upper and lower extremes of
concentration. From the intersection of the first curve with the 2.1%
concentration level an observed distance to the lower flammable limit
(LFL) was obtained. The other two curves provided an estimate of the
uncertainty in this value. A better estimate of the confidence
interval for this parameter may result from an analysis of the
experimental variability of the sensor calibration, which is currently
in progress.

The LFL distance so obtained is derived from the maximum
concentrations observed at approximately 0.9 m (0.6 m for spill 54)
above the surface. This has to be compared with a value derived from
model predictions of the mean centre line concentration at ground
level, as listed in Table 1 (see note 1). In order to estimate the
relation between the experimental peak concentrations and the mean
centre line concentrations at the same level, it is useful to examine
the data from two spills where the plume was wide and the wind steady.
In these two cases (54 and 52) sensors stayed in the same part of the
plume for the whole spill; meandering was not an important effect. The
ratio of peak to mean concentration measured at sensors at the bottom
and middle levels, and not close to the source, is around 1.4. We
shall assume a value of 1.4 for the ratio of peak observed to mean
"centreline" concentration in all the spills.

Some information on the vertical variation of concentration can be
obtained by looking at the ratios between the measurements at 0.9 m and
1.3-1.4 m. If attention is confined to the region near LFL, the
scatter in the ratio of these two concentrations is great, reflecting
both the randomness of the dispersion and the uncertainty of individual
sensor readings at one or two percent concentration. However, it is
more useful to average all the available such ratios for sensors at
250 m or further from the source in all the moderate or high wind
spills which give good data (i.e. 43, 46, 47, 49, 50). The average
obtained in this way is:

(concentration at 0.9 m) = 1.11 × (concentration at 1.3 to 1.4 m)

Linear extrapolation gives a surface concentration higher than at 0.9 m
by a factor of 1.2. In fact, linear extrapolation may be a conservative
assumption since the vertical gradient of concentration must tend to

zero at the surface.

Combining these two estimates leads to the conclusion that for the moderate and high wind spills (i.e. excluding spill 54), the ratio of peak concentration at 0.9 m to mean surface concentration would be very roughly 1.4/1.2 ≃ 1.2 in the far field.

With this in mind, we can compare the measured peak concentrations with the surface mean concentrations predicted by the model. The results are listed in Table 1. A group of five spills have already been mentioned; these were at moderate or high wind speeds, without an underwater source, where there were good observations of the plume centre: spills 43, 46, 47, 49 and 50. For these spills the scatter of the concentration data is 30-40% either side of the expected value, which is reasonable for trials in the atmosphere; some of these variations may yet be explained by a more detailed examination of the meteorological data, particularly measurements of u_*. Spill 54, however, shows a' larger discrepancy. It would have been possible to reach a stronger conclusion about the significance of this if there had been another experiment under similar conditions.

The experimental data on the vertical profile of concentration show a shallower gradient than predicted by the model. The photographic data generally show that the plume remained visible for a greater distance at its centre than at its edges, in contrast to the flat end predicted for the visible plume; this suggests a faster lateral decrease in concentration away from the centreline than predicted. The underwater releases, however, did not give this result.

Underwater releases

The two underwater releases (spills 42 and 52) provide interesting results. It is clear from Table 1 that the gas was diluted faster than in equivalent above-surface spills. These spills also were different in appearance. The plume from spill 42 in a low wind (3.7 m/s) was strongly bifurcated. We suggest this bifurcation is due to the gravity-driven 'slumping' of the gas. The liquid was released below the surface of the water in a downward jet. Most of it probably evaporated below the surface and the result was a stream of gas reaching the surface with considerable vertical momentum and jetting up into the air. So, paradoxically, the gas started to slump from a greater height than in the other spills. It then dropped down, spreading in a cross-wind direction as it moved down wind in the plume. At the edges of the plume were strong gravity-current heads of the type described by Simpson and Britter (1980). It is believed that in the case of spill 42 the slumping was strong enough for much of the mass to be concentrated in the two lateral heads, as was found for the circular head in some instantaneous releases at Porton (Picknett, 1981). In the higher wind speed (7.9 m/s) of spill 52 the effect was not strong enough to produce a visible bifurcation of the plume.

CONCLUSIONS

1. For the well-defined spills with good data, the maximum distance to
   which flammable gas was found to travel was, with one exception,
   between 0.75 and 1.35 times the mean LFL distance predicted by
   HEGADAS (concentrations within ±50% of the expected value). The
   exception was a spill where the discrepancy was about twice as
   great, although with more uncertainty.

2. For two spills in steady conditions (one with underwater release),
   peak/average concentration ratios for the lowest sensors, near the
   LFL region, were found to be around 1.4.

3. Underwater release of liquid propane resulted in a strong jet of
   gas up into the air whose subsequent slumping produced a bifurcated
   plume (in low wind) and increased dilution.

REFERENCES

Colenbrander, G.W.: 1980, A mathematical model for the transient
behaviour of dense vapour clouds, 3rd Intl. Symp. loss prevention and
safety promotion in the process industries, Basle, September 1980.

Picknett, R.G.: 1981, Dispersion of dense gas puffs released in the
atmosphere at ground level, Atmospheric Environment, 15, 509-525.

Puttock, J.S., Blackmore, D.R., and Colenbrander, G.W.: 1982, Field
experiments on dense gas dispersion, J. Hazardous Materials, 6,
13-41; also in Dense Gas Dispersion, ed. R.E. Britter and
R.F. Griffiths, Elsevier, Amsterdam.

Simpson, J.E. and Britter, R.E.: 1980, A laboratory model of an
atmospheric mesofront, Quart. J. R. Met. Soc., 106, 485-500.

NOTE

1. The more obvious approach should be to compare data obtained at
   0.9 m with model predictions at 0.9 m. However, because the
   vertical profile adopted in the model is somewhat arbitrary on the
   part of the plume where gravity effects play a dominant role, the
   experimentally observed vertical profile has been introduced to
   translate the 0.9 m observations to ground level. In this way the
   most important feature of the model is tested, namely its ability
   to predict the maximum concentration at a given distance, which is
   at ground level.

RESULTS OF 40-m^3 LNG SPILLS ONTO WATER

Donald L. Ermak, Henry C. Goldwire, William J. Hogan,
Ronald P. Koopman, and Thomas G. McRae
Lawrence Livermore National Laboratory

ABSTRACT

Lawrence Livermore National Laboratory (LLNL) is conducting safety
research under the sponsorship of the U.S. Department of Energy (DOE)
to determine the possible consequences of liquefied natural gas (LNG)
spills. The LLNL program includes both the collection of data from
various size experiments and development of an ensemble of computer
models to make predictions for conditions under which tests cannot be
performed. In spills of 40 cubic metres (m^3) of LNG onto water done
at the Naval Weapons Center (NWC), China Lake, California in 1980 and
1981, data was collected on gas cloud dispersion and combustion and
rapid phase transition (RPT) explosions. Analysis of the data from
these tests, including comparisons between the predictions of various
models and the data, are presented. The results suggest that large-
scale spills may be more hazardous than would have been predicted
based on earlier small-scale tests. Work performed under the auspices
of the U.S. Department of Energy by the Lawrence Livermore National
Laboratory under contract number W-7405-ENG-48.

INTRODUCTION

With the increase in volume of liquefied gaseous fuels (LGFs) being
transported and stored throughout the world, there comes a growing
possibility of very large spills of flammable materials. The LGF
Spill Effects Program was established at LLNL by the DOE to develop
and validate tools that can be used to predict the effects of a large
LGF spill--including pool spread, vaporization, dispersion, ignition,
combustion, explosion, and damage effects. Because of the complex
interdependence of these phenomena, the development and verification
of predictive tools requires an integrated analytical and experimental
approach which includes an extensive set of small-, medium-, and large-
scale spills of the fuels themselves.

During the summers of 1980 and 1981, the DOE sponsored two series
of field-scale, LNG spill experiments. These tests involved spills
ranging in size from 25 to 40 m^3, at spill rates from 10 to 20 m^3/min.

163

S. Hartwig (ed.), Heavy Gas and Risk Assessment - II, 163–179.
Copyright © 1983 by Battelle-Institut e.V., Frankfurt am Main, Germany.

The first series, called the Burro series, was designed to study the atmospheric dispersion of the resultant LNG vapor cloud, and the second, called the Coyote series, was initially designed to study the combustion of the vapor cloud. During the Burro series of LNG spill experiments in 1980, a number of RPT explosions occurred unexpectedly. The experimental plan for the Coyote series was subsequently modified to include a subseries of experiments to study the RPT phenomena.

This paper presents the major findings from the Burro series of experiments (for more detail see Koopman et al.,1981, and Ermak et al., 1981), and a preliminary analysis of the RPT (also see McRae, 1982) and vapor burn experiments conducted in the Coyote series. The tests were conducted at an existing facility located at the NWC, and were a joint effort by LLNL and NWC.

DESCRIPTION OF EXPERIMENTS

A schematic of the $40\text{-}m^3$ spill facility at NWC is given in Figure 1. In concept, its operation is quite simple. The LNG is forced out of the storage tank by high pressure nitrogen and flows down approximately 100 m of 25-cm-diameter pipe to the spill pond. There it is directed downward onto a spill plate just below the water surface. In the RPT tests the spill plate was adjustable from about 40 cm below the water level to slightly above it. The pond is about 1 m deep at the spill point and 50 m in diameter.

In the Burro series of experiments, we measured the dispersion of the LNG vapor cloud in a variety of atmospheric conditions. Eight LNG spills were done in winds ranging from 1.8 to 9.1 m/s and under a variety of stability conditions. The LNG was spilled at rates from 10 to 20 m^3/min and spill volume ranged from 25 to 40 m^3. Beyond the spill pond, the terrain rose to a height of about 7 m above the water level at a distance of about 80 m downwind and remained relatively level thereafter.

A large array of gas-sensing and wind-measuring instruments was deployed upwind and downwind of the spill point (see Figure 2). One subarray consisted of twenty cup and vane anemometers, each at a height of 2 m above the ground, to map the wind field. Another subarray of 25 stations tracked the cloud using gas concentration sensors at 1, 3, and 8 m. A variety of gas sensor types were used--some measured methane and higher hydrocarbons separately, others simply sensed total hydrocarbons. The turbulence subarray of six stations used propeller bivane anemometers at 1.36, 3, and 8 m and fast gas sensors at 1, 3, and 8 m to measure gas concentration and turbulence effects. Humidity sensors, thermocouples, and heat-flux sensors also collected data at various locations within the array and cameras and IR imagers made remote measurements. All of the stations were battery-powered and micro-processor-controlled, with some onboard memory for collecting data. They communicated with the data-recording trailer by radio telemetry, turning on instruments on command and sending back data when polled.

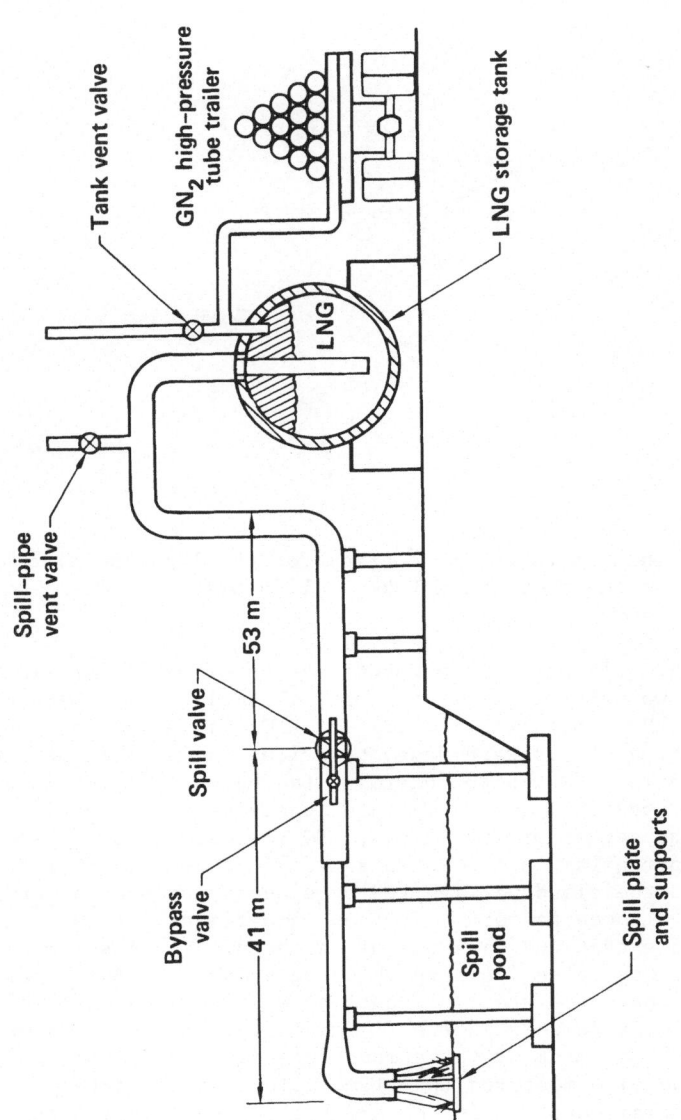

Figure 1. The 40-m^3 LNG spill facility at the NWC, China Lake, California.

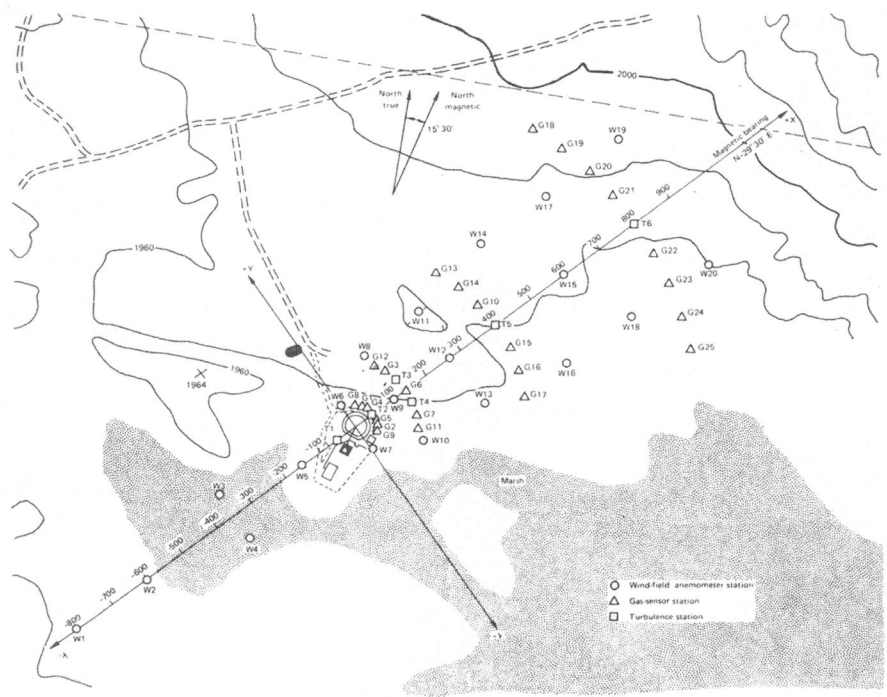

Figure 2.   The instrumentation array for the 1980 LNG dispersion tests
at the NWC, China Lake, California.

As previously noted, the Coyote series of spill experiments consisted of two subseries. In the one, 13 small spills were conducted
to examine RPT explosions. Spills of LNG, liquefied methane, and
liquid nitrogen (LN_2) were done at various spill rates and into
water of various depths to determine the boundary conditions under
which RPTs occur and what influences the size of the explosions. In
the other subseries of five tests, LNG was spilled and allowed to partially disperse (as in the Burro series). However, after the spilling
was stopped the cloud was ignited by either flares or a jet ignition
source and allowed to burn. In these experiments our objective was to
measure the extent and violence of the burn. The instrument stations
used in the Coyote series were the same as for the Burro series with
some additions. The station locations were changed to allow a higher
density of measurements in the flammable region. Besides measuring
the size and movement of the combustible cloud up to the point of
ignition, we also measured the flame velocity, the extent of the fire,
and its thermal output. In the RPT tests, additional instruments
close to the spill point measured overpressure in the air and underwater as well as the composition, velocity, and temperature of the LNG
exiting the spill pipe.

VAPOR DISPERSION

   Dispersion analysis conducted prior to the experiments indicated
that 40-m³ spills at rates on the order of 20-40 m³/min were at
the threshold for observing spill-dominated dispersion effects. This
prediction was essentially confirmed by our experimental series of
vapor dispersion tests. Turbulent processes in the lower atmospheric
boundary layer dominated the transport and dispersion of gas in all of
the experiments except one, Burro 8, in which the wind speed was 1.8
m/s and the atmosphere slightly stable. In the other experiments, the
average wind speed was greater than 4 m/s and the atmosphere was less
stable.

   The shape and dynamics of the Burro 8 vapor cloud was significantly
different than in the other tests conducted under higher wind speeds.
The cloud was much wider and lower in height than from any of the other
experiments and developed a bifurcated structure as shown in Figure 3.
The cloud extended about 40 m upwind of the spill point as well as
beyond both sides of the instrumentation array. In addition, the LNG
vapor cloud remained over the source region for nearly two minutes
after the spill had been terminated. This is in contrast to the other
experiments at higher wind speeds where the vapor cloud was rapidly
dispersed downwind within 10-20 seconds after spill termination.

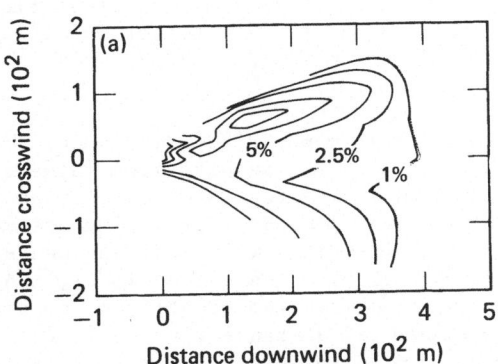

Figure 3.   Burro 8 horizontal LNG vapor concentration contours at 1-m
            elevation and at 200 s, showing the bifurcated structure of
            the vapor cloud. Contours are 1, 2.5, 5, 10, 15, and 25
            percent by volume.

   The low wind speed in the Burro 8 test permitted the gravity flow
of the cold dense LNG vapor cloud to dominate the effects of the surrounding atmosphere. Within the cloud, the mean flow was dominated by
the gravity spreading process with the ambient wind field being essentially displaced above the LNG vapor cloud. The wind speed within the
cloud, shown in Figure 4, dropped essentially to zero. Turbulence

within the cloud was controlled by the mean gravity spread flow and
ground heating. This behavior, observed only under low wind speed
conditions, is likely to occur in larger spills under a variety of
conditions. Additional field-scale gravity spread experiments are
needed in order to evaluate the scaling of these processes.

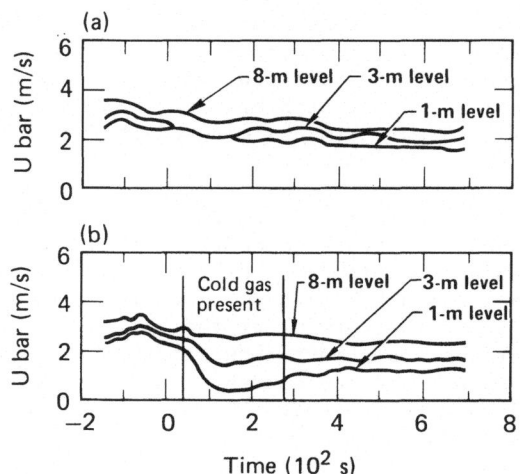

Figure 4. Mean wind speed during Burro 8, (a) at station T-1 upwind
of the spill pond and (b) at station T-2 downwind of the
spill pond and within the gas cloud.

The predictions from three vapor dispersion models for cold dense
gas releases were compared with the results from these tests. The
models vary considerably in the degree to which they approximate important physical phenomena and include restricting assumptions. The
simplest model is the Germeles-Drake (GD) modified Gaussian plume
model (Germeles and Drake, 1975). The second model, SLAB, is a time-
dependent, layer-averaged conservation equation model with one independent spatial variable, downwind distance (Zeman, 1982). The final
model, FEM3, is a fully three-dimensional, time-dependent conservation
equation model (Chan et al., 1981).

The range of applicability of any particular model generally
depends on the degree to which important physical phenomena are
approximated. Of the three models used in this study, the FEM3 model
is the least limited by various approximations and restricting
assumptions, and it did indeed provide the best overall descriptions
of LNG vapor cloud dispersion as observed in the Burro experiments.
Predictions of the vapor concentration distribution in time and space
over the range from 5% to 15% (the LNG flammability limits) were
generally quite good. Estimates of the maximum distance to the lower
flammability limit (LFL) were also quite good or at least conservative
(overestimated X_{LFL}).

A major accomplishment of the FEM3 model was the prediction of the
bifurcated structure of the very wide vapor cloud in Burro 8 (see
Figure 5). Neither of the other two models could have reproduced this
behavior since the shape of the crosswind concentration distribution
was prescribed in both these models and not subject to change by the
conditions in the cloud. Additional model simulations would provide
more insight into the atmospheric and spill conditions under which
this phenomenon would be expected to occur. For more quantitative
agreement with experiment, improvements are needed in the turbulence
submodel, the approximations for the velocity profile near the ground,
and the treatment of terrain.

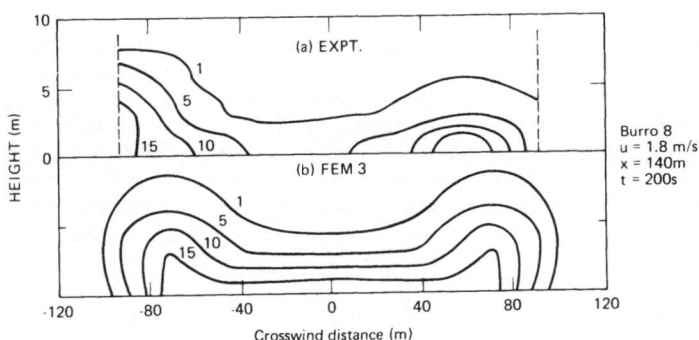

Figure 5. Burro 8 crosswind contour plots of cloud concentration 140
           m downwind at t = 200 s. The dashed lines in plot (a)
           indicate the outer extent of the instrumentation array.
           The contour lines designate the 1, 5, 10, and 15 percent
           levels and the vertical-to-horizontal distance scale is
           1-to-6.

From a safety point of view, it is important to be able to predict
how long the vapor cloud will remain hazardous. In this regard, both
the FEM3 and SLAB models accurately predicted the lingering of the
vapor cloud over the source region after the spill was terminated in
the low wind speed, Burro 8 test (see Table 1). Their ability to make
this prediction was due to the inclusion within these models of the
fundamental principle of momentum conservation in the wind direction.
The GD model, and most other simple models, do not include these
momentum effects, but simply assume that the vapor cloud travels
downwind with the ambient wind speed. As shown by the Burro 8 test,
this assumption can lead to a significant underestimation of the time
for cloud dispersal.

Experiment	Duration of 5% or greater concentration at x = 140 m.
Lobe 1	245 s
Lobe 2	195 s
FEM3	190 s
SLAB	210 s
GD	106 s

TABLE 1.   CLOUD LINGERING AFTER THE SPILL ENDED IN BURRO 8.

The SLAB model provided a fairly good description of the observed concentration distribution, especially cloud width. However, SLAB assumes the vertical concentration distribution is nearly uniform so that the vertical concentration gradient ($\partial c/\partial z$) is essentially zero from the ground up through most of the cloud and then very steep at the top of the cloud. This was generally not the case in these experiments, especially in the high wind speed tests, where the vertical concentration gradient was found to be more gradual throughout the cloud.

The GD model estimates of the vapor cloud concentration distribution were significantly poorer than those of the other two models. The predicted cloud was roughly twice as high and twice as narrow as in the experiments. The main reason for this is that the GD model includes gravity effects only in the initial calculation of the vapor source height and width, but does not include gravity effects on the subsequent downwind dispersion of the vapor cloud. This discrepancy in cloud shape between the GD model results and the experiments illustrates the importance of gravity spread on the crosswind concentration distribution, even when the wind speed is high and turbulence within the cloud is dominated by the ambient atmospheric conditions.

The FEM3 and SLAB predictions of the maximum LFL distance were generally within the experimental uncertainty of + 50 m as shown in Table 2. In the model simulations, the maximum distance to the LFL is quite sensitive to the turbulence level. Typically, a change in the turbulent diffusivity by a factor of two resulted in a similar change in the LFL distance. None of these models considered the effect of ground heating on the vertical turbulence rate. However, theory predicts that the ground heat flux measured in the China Lake experiments is sufficient to have a significant effect on the turbulence level within the cloud. Additional experiments of this size, with both cold and ambient temperature dense gases, are needed to evaluate the importance of ground heating on cloud turbulence level.

	SPILL RATE (m^3/min)	WIND SPEED (m/s)	RICHARDSON NUMBER	LFL(m)			
				Expt.	GD	SLAB	FEM3
Burro 3	12	5.4	−0.22	255	126	215	190
Burro 7	14	8.4	−0.018	200	150	264	210
Burro 9	18	5.7	−0.014	325	235	315	330
Burro 8	16	1.8	+0.12	420	661	418	630

TABLE 2.   MAXIMUM DOWNWIND EXTENT OF THE LFL (IN METRES).

RAPID PHASE TRANSITION EXPLOSIONS

   Two of the eight Burro series spills produced RPT explosions, while six of the 18 Coyote spills produced RPTs. Of these, two distinct types of RPTs occurred—early RPTs and delayed RPTs. The early type of RPTs began immediately with the spill, and in some cases continued for the duration of the spill. These explosions were generally located near the spill point and appeared to be primarily underwater. Most of the RPTs that occurred at China Lake were of this type. The delayed type of RPTs occurred at the end of the spills, and were generally located away from the spill point out on the LNG pool surface. Delayed RPTs occurred on three tests.

   The largest RPTs recorded at China Lake occurred during the Burro 9 spill. A list of the largest of these explosions is given in Table 3. The TNT equivalents shown in Table 3 only correspond to the explosive energy transmitted to the atmosphere. The underwater component of the explosive energy could be significantly larger. As can be seen, the RPTs began immediately and persisted for over a minute. Only those RPTs above the dashed line occurred early enough that they could have been affected by the cooldown-enriched LNG; all below the dashed line occurred while LNG of storage-tank composition was spilling out of the pipe. At some time within the first 30 seconds of the spill, the spill plate was torn loose. The Burro 9 spill was unique in that it was the only spill conducted without a spill plate, at the maximum spill rate, and with the pond temperature greater that 17°C. As will be discussed shortly, these three parameters (spill plate depth, spill rate, and pond temperature) are important ingredients for the occurrence of RPTs during large-scale spills.

   The Coyote RPT tests typically consisted of three short spills at low, medium, and high spill rates. In theory, all the other parameters (LNG composition, spill plate depth, water temperature, etc.) would not change during the three consecutive spills. Spill tests with liquid methane and liquid nitrogen were performed to further examine the effect of composition on the occurrence of RPTs.

Time[a] (s)	Side-on Pressure[b] (psi)	TNT Equivalent[c] (g)
6.5	0.12	65
7.1	0.15	115
9.2	0.27	530
21.4	0.57	3400
35.1	0.72	6300
43.2	0.10	41
46.0	0.12	65
54.1	0.12	65
54.9	0.13	80
66.9	0.19	215
72.7	0.12	65

[a] t = 0 is start of spill-valve opening.
[b] Measured at distance of 30 m.
[c] Equivalent free-air point-source explosion of TNT.

TABLE 3. OCCURRENCE TIMES AND MAGNITUDES OF MAJOR BURRO 9 RPT EXPLOSIONS.

There appears to be some correlation between the occurrence of RPTs and the water temperature. With one exception, all of the RPTs occurred for water temperatures greater than 17° C, suggesting that the colder water inhibited RPTs. The depth of water above the spill plate also appears to affect the occurrence of RPTs. In the last spill, conducted without a spill plate, RPTs did occur, even with a pond temperature of 11.6° C. Thus it appears that the depth of penetration into the water is an important factor affecting RPT occurrence.

One of the main parameters which has been associated with the occurrence of RPTs is the composition of the LNG being spilled. In the early 1970s, Enger and Hartman (1972) of the Shell Pipeline Corp. conducted one of the most thorough and extensive studies of RPTs from spilled LNG to date. Even though the amount of LNG spilled was small (~ 0.1 m^3), the number of spills was large, lending considerable credibility to the accuracy of the results. A summary of these results for the methane, ethane, and propane mixtures is shown in Figure 6. Enger and Hartman concluded that only LNG with compositions within the envelope shown in the figure, or LNG allowed to "age" into this envelope, could produce RPT explosions. The large-scale China Lake RPT results for the early RPTs, also shown in Figure 6, are clearly well out of the Enger-Hartman envelope. Furthermore, there is a lack of consistency for the large-scale spills, i.e., LNG of similar com-

position may or may not produce an RPT. Apparently there are other
mechanisms, besides the prespill composition, which dominate the occurrence of RPTs for these large spills. The asterisk in Figure 6 is an
estimate of the maximum amount of cooldown enrichment for the most
enriched LNG of the Coyote series. This calculation shows that even
the first cubic metre of LNG during that spill was still well out of
the Enger-Hartman envelope.

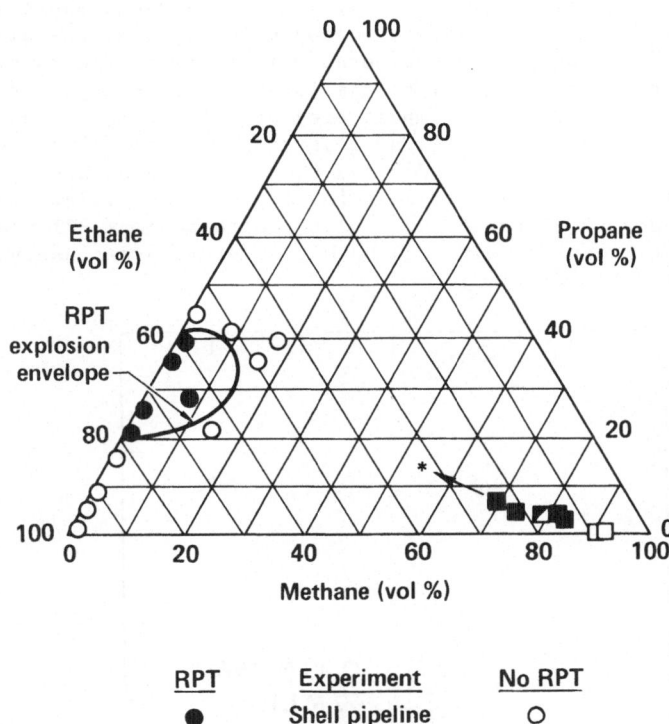

Figure 6. Large-scale RPTs and the Enger-Hartman RPT explosion
          envelope.

This does not mean that enrichment has nothing to do with RPT
explosions in spilled LNG. Certainly the delayed type of RPT must be
a result of enrichment or "aging" on the pond. It is even quite possible that some form of local or transient enrichment may play a part
in the early type of RPTs. Nonetheless, for large-scale spills the
results indicate that prespill composition is not a good indication of
the likelihood of an RPT.

A plot of RPT explosive yield versus LNG impact pressure is shown
in Figure 7. Also included in the figure are the results of the MIT

laboratory-scale impact test results (Jazayeri, 1977). The MIT tests
involved pure liquid methane, ethane, propane, and nitrogen, as well
as some mixtures of these cryogens. The experimental procedure was
such that the water was actually impacted onto the cryogens. The data
from these small tests indicate a gradual increase of explosive
strength with increasing impact pressure. The collapse, due to impact
pressure, of the film-boiling vapor barrier was proposed as the mechanism responsible for RPTs in the case of cryogens which would not
normally explode during a simple spill. As can be seen, there is no
real trend of RPT yield with impact pressure for the large-scale
spills. There is, however, a considerable increase in the magnitude
of the RPTs going from the small to the large spills--approximately
five orders of magnitude at an impact pressure of 0.8 atm. The
observation of a fairly large RPT at less than 0.1 atm and no RPT at
an impact pressure of 0.7 atm indicates that impact pressure may not
be the mechanism responsible for RPTs in large-scale spills. In fact,
spill rate appears to correlate much better with maximum RPT yield
than does impact pressure, although both parameters are closely related
in this type of spill facility.

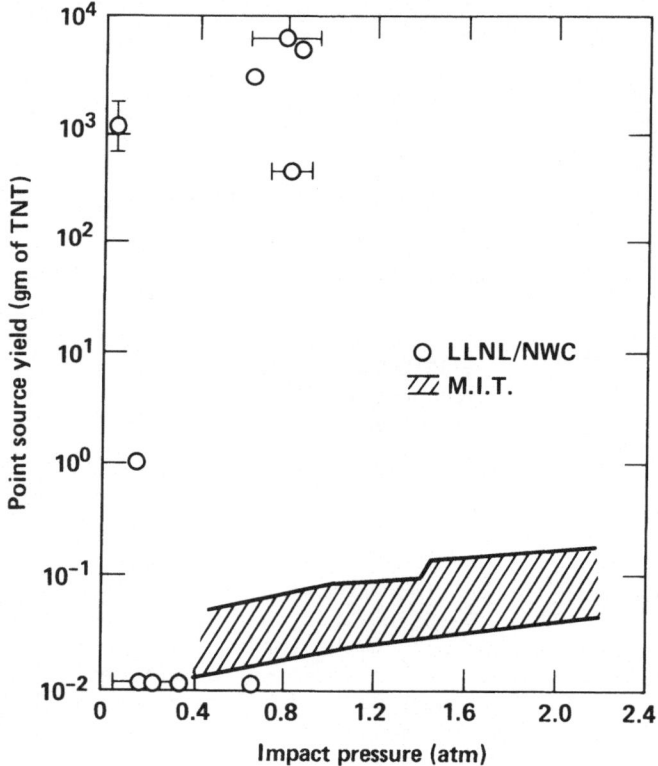

Figure 7.   Comparison of LLNL/NWC and MIT impact pressure data with
             RPT explosive yield.

A plot of RPT yield versus LNG spill rate is shown in Figure 8. Of
all the parameters correlated with RPT yield, the spill rate produced
the best correlation. As can be seen, there is an apparent threshold
or abrupt increase in the RPT explosive yield at about 15 m³/min.
Unfortunately, this rapid increase of RPT yield just begins to occur
at the spill rate limit of the China Lake facility. Consequently, the
scaling of the RPT yield to higher spill rates is uncertain at this
time and needs to be investigated further.

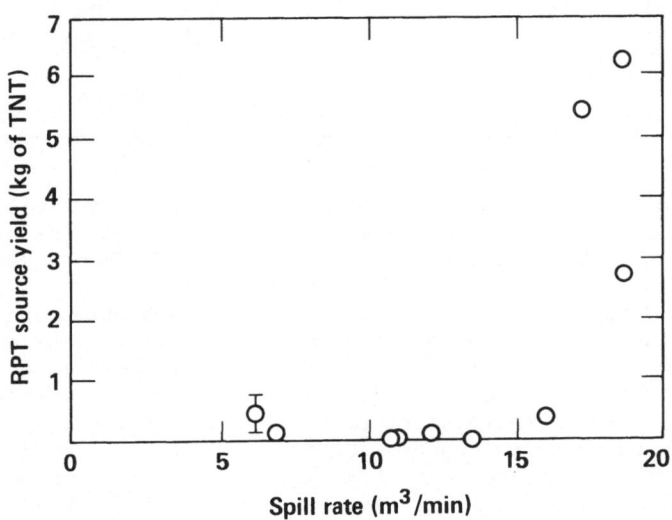

Figure 8. The effect of spill rate on RPT yield for large-scale
           spills.

VAPOR CLOUD COMBUSTION

In the field-scale vapor burn tests, we are attempting to discover
what factors affect flame velocity, to correlate the cloud combustion
region with the gas concentration contours prior to ignition, and to
measure the thermal energy. The analysis of data from these tests is
still in the early stages. However, our field data has shown no
significant large-scale flame acceleration during any of the unconfined
vapor burn tests at NWC. Large-scale tests to investigate the effects
of turbulence or partial confinement have not yet been performed.

The flame speed in the 40-m³ vapor burns is one of the main
variables of concern. Table 4 shows the boundary conditions of each
of the vapor burn tests. Times given are all measured from the
beginning of the spill. Four techniques were used to measure flame
speed: 1) high speed optical cameras, 2) side-on infrared imagery, 3)
overhead infrared imagery, and 4) in-situ flame sensors. All data has
not been analyzed as yet and therefore the results are preliminary.

Coyote Test Number	Date	Material Spilled (% CH₄)	Spill Date (m³/min)	Spill Duration (min)	Wind Speed (m/s)	Ignition Source	Ignition Time (min.)
2	8/20/81	70	16	0.5	5.9	Flare	1.2
3	9/02/81	79	14	1.1	5.8	Flare	1.7
5	10/07/81	75	17	1.6	10.0	Flare	2.0
6	10/27/81	82	17	1.4	4.8	Flare	1.9
7	11/12/81	99.5	14	1.9	6.0	Jet	2.4

TABLE 4.   COYOTE VAPOR BURN TEST SUMMARY

In the Coyote 3, 5, 6, and 7 experiments good flame velocity measurements in the downwind direction were obtained from the side-on infrared cameras. The wind speeds in these four experiments were 5.8, 10.0, 4.6, and 6.0 m/s. The average downwind flame velocities were 12.6, 16.4, 11.9, and 18.9 m/s respectively. If we subtract the wind velocity from these numbers we obtain flame velocities relative to the air of 6.8, 6.4, 7.3, and 12.9 m/s. That is, the velocities were all around 7 m/s, except for the last experiment, when it was almost twice that. It must be noted that the field of view of the infrared imager differed from experiment to experiment, so we should not yet draw general conclusions. In the only flame pressure measurement we performed, the results from Coyote 6 showed pressures of about one millibar. The side-on infrared camera also shows us that the flame height in these tests ranged between 55 and 65 m.

The overhead IR imagery on Coyote 6 and Coyote 7 allows us to follow the flame front in both the upwind and the downwind directions. We see the data plotted in Figure 9. Velocities appear to be very high near the ignition source (whether it is a jet or a flare) and to fall off rapidly as the flame moves away from the source in either direction. Relative to the wind, the flame velocity seems to be symmetric about the source.

It is too early in the analysis of our experiments to draw general conclusions. However, the flame heights and velocities we observed were much larger than those seen in the Maplin Sands tests of Shell. What caused these differences will be the subject of further investigation.

SUMMARY AND CONCLUSION

The 40-m³ experiments produced results that can have a significant impact on assessing the hazards resulting from a large spill of a liquefied gaseous fuel. Our postulate that large clouds might behave

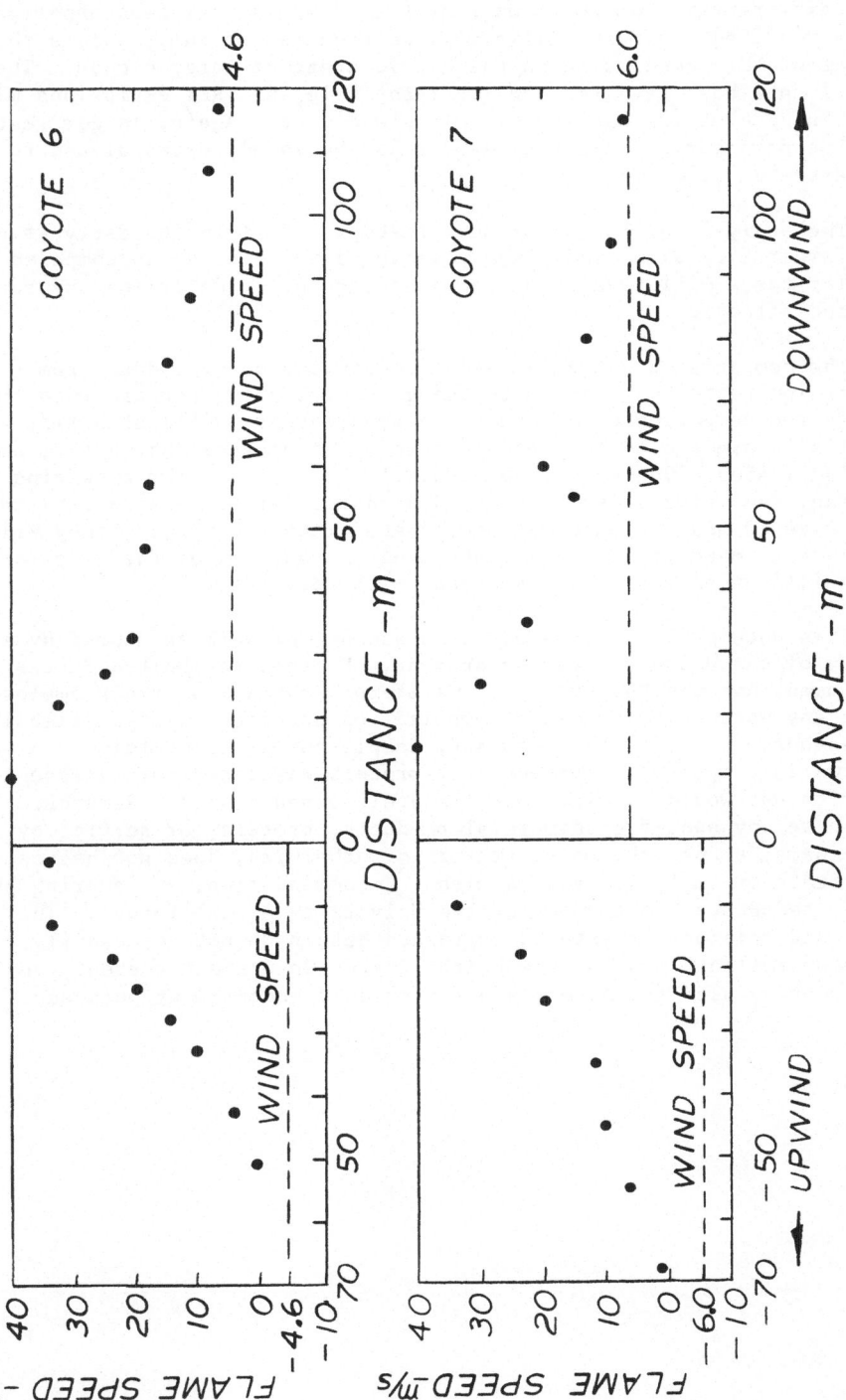

Figure 9. The Coyote 6 and 7 flame velocities as a function of time in
both the upwind and downwind directions.

very differently from small ones as they disperse has been experimentally verified. Larger spills will be required to fully assess the
impact of this conclusion on the hazard areas of large clouds. The
experiments have revealed the important role that RPT explosions might
have in determining the outcome of an accident. Again, larger tests
will be necessary before the results can be safely extrapolated to
large-scale accidents.

The analysis of the vapor burn tests is still in its early stages.
The larger fire sizes and flame velocities we observed as compared to
smaller tests will have to be examined further to determine any scale
related effects.

The comparisons between model calculations and the data from
dispersion tests has been encouraging. Introducing terrain into the
models and comparing with data taken from larger spills at a flat
spill site are the next necessary steps. It appears that data from
spills of about 100 - 200 m^3 of LNG will be critical for answering
the many remaining questions on LNG hazards. After these questions
are answered and the answers incorporated into the models, they will
be ready to make predictions concerning the outcome of the very large
LGF spills which could be associated with accidents.

This document was prepared as an account of work sponsored by an
agency of the United States Government. Neither the United States
Government nor the University of California nor any of their employees,
makes any warranty, express or implied, or assumes any legal liability
or responsibility for the accuracy, completeness, or usefulness of any
information, apparatus, product, or process disclosed, or represents
that its use would not infringe privately owned rights. Reference
herein to any specific commercial products, process, or service by
trade name, trademark, manufacturer, or otherwise, does not necessarily
constitute or imply its endorsemment, recommendation, or favoring by
the United States Government or the University of California. The
views and opinions of authors expressed herein do not necessarily
state or reflect those of the United States Government thereof, and
shall not be used for advertising or product endorsement purposes.

REFERENCES

Chan, S.T., Gresho, P.M., and Ermak, D.L.: 1981, A Three-Dimensional, Conservation Equation Model for Simulating LNG Vapor Dispersion in the Atmosphere. Lawrence Livermore National Laboratory, Livermore, California, UCID-19210.

Enger, T., and Hartman, D.E.: 1972, LNG Spillage on Water. II. Final Report on Rapid Phase Transformations. Shell Pipeline Corp., Research and Development Laboratory, Houston, Texas, Technical Progress Report No. 1-72.

Ermak, D.L., Chan, S.T., Morgan, D.L., and Morris, L.K.: 1981, A Comparison of Dense Gas Dispersion Model Simulations with Burro Series LNG Spill Tests Results. Lawrence Livermore National Laboratory, Livermore, California, UCRL-86713.

Germeles, A.E. and Drake, E.M.: 1975, Gravity Spreading and Atmospheric Dispersion of LNG Vapor Clouds. Proceedings of the 4th International Symposium on Transport of Hazardous Cargoes by Sea and inland Waterways, Jacksonville, Florida.

Jazayeri, B.: 1975, Impact Cryogenic Vapor Explosions. M.S. Thesis. Massachusetts Institute of Technology, Cambridge, Massachusetts.

Koopman, R.P., Cederwall, R.T., Ermak, D.L., Goldwire, H.C., McClure, J.W., McRae, T.G., Morgan, D.L., Rodean, H.C., and Shinn, J.H.: 1981, Description and Analysis of Burro Series 40-m^3 LNG Spill Experiments. Lawrence Livermore National Laboratory, Livermore, California, UCRL-53186.

McRae, T.G.: 1982, Preliminary Analysis of RPT Explosions Observed in the LLNL/NWC LNG Spill Tests, Lawrence Livermore National Laboratory, Livermore, CA, UCRL-87564.

Zeman, O.: 1982, The Dynamics and Modeling of Heavier-Than-Air Cold Gas Releases, Atmospheric Environment 16, p. 741.

FURTHER ANALYSIS OF CATASTROPHIC LNG SPILL VAPOR DISPERSION

J. A. Havens
T. O. Spicer
Chemical Engineering Department
University of Arkansas
Fayetteville AR 72701

This work was supported by U.S. Coast Guard Contract DTCG23-80-C-20029
with the University of Arkansas. The opinions or assertions contained
herein are the private ones of the writers and are not to be construed
as official or reflecting the views of the Commandant or the Coast
Guard at large.

INTRODUCTION

As part of our work under a contract with the U.S. Coast Guard to
develop mathematical models for heavy gas dispersion for incorporation
into the U.S. National Emergency Response System (1), we are conducting
an assessment of currently available heavy gas dispersion models. This
model assessment program involves three phases.

1. Comparison of model predictions with presently available
   accident and field test data, with data to become
   available from large scale test programs such as the
   DOE Burro tests (2) and the SHELL Maplin Sands tests
   (3), and with data which will be produced in the up-
   coming Thorney Island Heavy Gas Trials to be conducted by the British Health and Safety Executive
   (4) and the U.S. Fertilizer Institute-sponsored ammonia
   spill tests to be conducted at China Lake, California
   (5).

2. A laboratory experimental program to provide accurate
   measurements of gravity spreading and associated
   air entrainment into heavy gas released in the absence
   of wind. It is hoped that these data can be used to
   aid in the assessment of currently available air
   entrainment models as well as to develop a better
   understanding of the mixing processes present in
   gravity-spreading heavy gas clouds.

S. Hartwig (ed.), Heavy Gas and Risk Assessment - II, 181–210.
Copyright © 1983 by Battelle-Institut e.V., Frankfurt am Main, Germany.

3.   Intercomparison of models, for a range of spill
     scenarios of interest to the U.S. Coast Guard, as
     a means for identifying the important differences
     in methodology used.

Work on all three of these phases is in progress, but we consider
here only some comparisons of predictions of the dispersion of a
heavy gas cloud which might be formed following a catastrophic
release of LNG in a marine vessel collision.

One of the authors (6) previously described the wide range of predictions, obtained using mathematical models in current use in 1977, of
the maximum downwind travel of a flammable gas-air mixture following
a postulated instantaneous reléase of 25,000 m^3 LNG (one marine carrier
tank) onto water. Although the author concluded that some of the
differences in predicted downwind distances were associated with several representations of atmospheric conditions such as atmospheric stability category and wind speed, important variations were associated
with the modeling methods used to describe the gravity spreading phase
of the dispersion process and the modeling of energy exchange between
the heavy, cold LNG vapor cloud and its surroundings.

Some of the models reported in the 1977 report have been recognized
as inadequate for prediction of catastrophic heavy gas release-
dispersion scenarios (7). Other models which appear to offer more
realistic methodology have since been introduced. We consider here
predictions of dispersion of a vapor cloud formed following the
instantaneous release of 25,000 m^3 LNG onto water using models
recently proposed by Eidsvik (8,9,10) and Colenbrander (11).

The Eidsvik model is a top-hat model; i.e. the gas-air cloud is
represented as a spatially uniform cylindrical cloud the size and properties (including concentration) of which, change with time due to dilution by air entrainment and gravity spreading. The prediction of air
entrainment and energy exchange between the cloud and its surroundings
and the downwind advection velocity of the cloud determines the maximum downwind distance to a specified lower concentration limit and
the area coverage of the cloud prior to its dilution below that limit.

The olenbrander model assumes a modified Gaussian distribution
of concentration in the heavy gas cloud dispersing downwind of a gas
release and predicts the rate of atmospheric takeup of gas from the
source into the wind field and its subsequent gravity spreading and
dispersion downwind.

Although there are many important differences in these two
modeling approaches, they both use a similar data base for predicting
air entrainment into the heavy gas cloud. They represent, in our
opinion, application of more realistic methodology than any of the

four "simpler" models described in Havens' earlier review (6). We believe it is worthwhile to extend the analysis of the catastrophic LNG release scenario using these models and suggest that although such predictions are necessarily based at present on an inadequate data base and on inadequately tested theory, some improvements in modeling approaches have been made which should give increased confidence in such predictions for risk assessment and emergency response use.

EIDSVIK MODEL DESCRIPTION

The description here is based on the reports and papers by Eidsvik (8,9,10). Primary emphasis is given to modeling a quasi-instantaneous LNG vapor release; details for modeling continuous releases are described in the reference documents.

The LNG vapor cloud is represented to form over a liquid spill onto water. The liquid spill is assumed instantaneous, with a specified liquid pool radius and evaporation rate. Figure 1a represents the cloud at time t_1 (liquid spill at $t = 0$) prior to the end of the liquid pool evaporation, and Figure 1b represents the vapor cloud at a time t_2 after the end of the liquid pool evaporation.

The cloud is assumed to retain a cylindrical shape over the liquid release point during the evaporation period, and to be translated downwind (as a spatially uniform property, cylindrical volume) with an average velocity $u_c(t)$. We wish to predict the time-varying properties of the gas cloud and its position and size until the cloud concentration decreases to the time average 5% molar or volume concentration which represents the lower flammability limit for methane (whose properties are used herein to represent LNG vapor).

Material and energy balances over the cloud volume are, respectively,

$$\frac{dM}{dt} = \dot{M}_g + \dot{M}_a$$

$$\frac{d(MCT)}{dt} = \dot{M}_g \, C_g \, T_g + \dot{M}_a \, C_a \, T_a + \dot{Q}_w + \dot{Q}_v .$$

(1)

Gravity Spreading

The radius of the cloud is computed using a gravity intrusion relation (16,17)

$$\frac{dR_c}{dt} = 1.3 \sqrt{g\left(\frac{\rho - \rho_a}{\rho}\right)H} .$$

(2)

Air Entrainment

Designating air entrainment velocities into the side and top surfaces

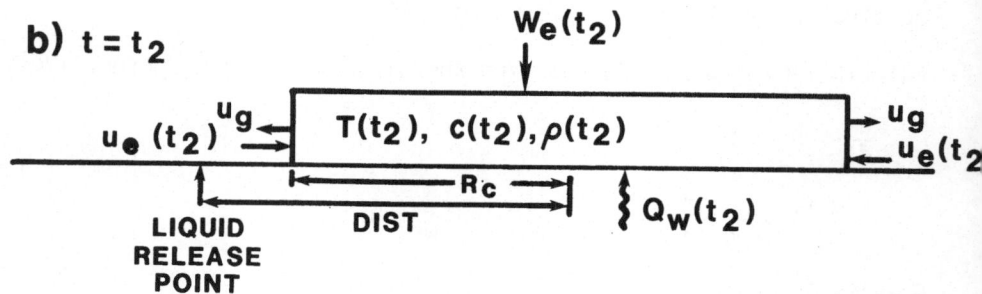

a) $t = t_1$

AIR ENTRAINMENT
VELOCITY,
$W_e(t_1)$ AIR ENTRAINMENT
VELOCITY,
$u_e(t_1)$

$u(z)$

u_g

$T(t_1)$, $c(t_1)$, $\rho(t_1)$ u_g $H(t_1)$

$Q_w(t_1)$

GAS CLOUD   LIQUID
RADIUS   RADIUS
$R_c(t_1)$ $R(t_1)$

b) $t = t_2$

$W_e(t_2)$

$u_e(t_2)$ u_g $T(t_2)$, $c(t_2)$, $\rho(t_2)$ u_g
$u_e(t_2)$

R_c

DIST $Q_w(t_2)$

LIQUID
RELEASE
POINT

Figure 1. Schematic Representation of Heavy Gas
Cloud Formation and Dispersion

of the cylindrical cloud by u_e and w_e, respectively,

$$\dot{M}_a = (w_e + 2 H u_e/R_c) \pi R_c^2 \rho_a . \tag{3}$$

The vertical entrainment velocity is computed as follows:

$$w_e = 3.5 w/(11.7 + Ri)$$

with a bulk Richardson number defined by

$$Ri = g H(\rho - \rho_A)/\rho w^2$$

and w representing a turbulence velocity scale calculated as a weighted vector sum of mechanically-induced turbulence velocity w_M and thermally-induced turbulence velocity w_T

$$w = ((0.7 w_T)^2 + (1.3 w_M)^2)^{1/2} . \tag{4}$$

The turbulence velocities w_T and w_M are computed using the relations

$$w_T = (\theta_{w_0} g H/T)^{1/3} \tag{5}$$

and

$$w_M = u_* = (1/2 C_F)^{1/2} u \tag{6}$$

with the "free stream velocity" for the cloud represented as

$$u^2 = (2/3 \frac{dR_c}{dt})^2 + u_a^2(H) . \tag{7}$$

The two-thirds factor in Equation (7) determines the average spreading velocity over the top of the cloud whose radius is increasing as dR_c/dt. We have represented the expected variation of u_a with height as a power law function

$$u_a(z) = (z/z_0)^\alpha u_0 \tag{8}$$

with α chosen to fit Equation (8) to a logarithmic profile determined from specification of a surface roughness and Monin-Obukhov length scale.

The horizontal or "edge" entrainment velocity is computed as

$$u_e = 0.5(dR_c/dt)^2/(dR_c/dt)_0 \tag{9}$$

where the normalization velocity $(dR_c/dt)_0$ is taken as the maximum value of the front spreading velocity.

Energy Balance Calculations

Heat transfer from the earth surface to the overlying cloud is modeled with the relation

$$\dot{Q}_w = h(T_w - T) \; \pi \; R_c^2 \tag{10}$$

with the heat transfer coefficient determined (assuming Reynolds' analogy between heat and momentum transfer) from

$$h = N_{ST} \, C \, G = 1/2 \; C_F \, C \, G \tag{11}$$

where C_F is the surface drag coefficient. Eidsvik's suggested value of $C_F = 0.002$ for flow over relatively smooth sea has been used.

Heat addition to the gas-air mixture due to condensation of moisture is calculated as

$$\begin{aligned} \dot{Q}_v &= \dot{M}_a \; X \; L_w &&\text{if } T < T_{DA} \\ &= 0 &&\text{if } T > T_{DA} \end{aligned} \tag{12}$$

where X represents the mass of water vapor condensed per unit mass of air entrained into the cloud. X is estimated using the Clausius-Clapeyron equation and assuming that water vapor is condensed until the dew point of the entering air is reduced to the temperature of the cloud:

$$X_a = 3.74 \times 10^{-3} \, \exp \left[\frac{L_w}{R_w} \left[\frac{1}{273} - \frac{1}{T_{DA}} \right] \right] \tag{13}$$

$$X_c = 18/29 \; A \, \exp \, (-a/T) \tag{14}$$

$$\text{and } X = X_a - X_c \, . \tag{15}$$

Cloud Translation with Wind

The center of mass of the cloud is translated downwind with a velocity given by

$$u_c = u_0 (\beta H/z_0)^\alpha \tag{16}$$

where the constant β determines the height at which the cloud translation velocity is computed.

Model Solution

The density of the cloud was calculated using the ideal gas equation of state

$$\rho = P/R \, T \tag{17}$$

$$\text{with } R = (M_a \, R_a + M_g \, R_g)/(M_a + M_g)$$

and the heat capacity in Equation (1) was computed using the relation

$$C = (M_a C_a + M_g C_g)/(M_a + M_g) \qquad (18)$$

Equations (1) - (18) were solved using the IBM Continuous Systems Modeling Program (12).

COLENBRANDER MODEL DESCRIPTION

The description herein follows generally the outline of Colenbrander's paper (11). However, there were instances in which some "interpretation" of Colenbrander's intent was necessary, and in a few instances, the authors have incorporated methods to deal with questions not fully treated by Colenbrander. The steady state form of Colenbrander's model is described first, followed by a description of the simulation of a transient heavy gas release as a series of quasisteady releases.

The Steady State Model

The model treats dispersion of gas released from an idealized rectangular-shaped source of half width $y = B$ and length L (in the downwind direction), with the gas source center at the origin of a Cartesian coordinate system as shown in Figure 2. Similarity forms for the concentration profiles are assumed which represent the plume as being composed of a horizontally homogeneous section in which dispersion only occurs vertically, with Gaussian (concentration profile) edges. A power law velocity profile is assumed. With the assumed similarity forms for the concentration and velocity profiles, the variables c_A (ground level, centerline gas concentration), b (width of the horizontally homogeneous center section of the plume), and s_y and s_z (scale factors in the similarity forms for concentration) are constrained by ordinary differential equations.

$s_z(x)$ is determined by requiring that it satisfy the diffusion equation

$$u_x \frac{\partial c}{\partial x} = \frac{\partial}{\partial z} \left[K_z \frac{\partial c}{\partial z} \right] \qquad (19)$$

with a vertical turbulent diffusivity given by

$$K_z = \frac{k \, u_* \, z}{\phi \, (Ri_*)} \qquad (20)$$

The function $\phi(Ri_*) = 0.74 + 0.25 \, Ri_*^{0.7} + 1.2 \times 10^{-7} \, Ri_*^3$, proposed by Colenbrander, is a curve fit of laboratory scale data for vertical mixing in density-stratified fluids reported by McQuaid (13) and Kantha et al. (14). The friction velocity in Equation (20) is

$$C(x,y,z) = C_A(x) \exp\left[-\left(\frac{|y|-b(x)}{S_y(x)}\right)^2 - \left(\frac{z}{S_z(x)}\right)^{1+a} \right], |y| > b$$

$$C(x,y,z) = C_A(x) \exp\left[-\left(\frac{z}{S_z(x)}\right)^{1+a} \right], |y| \leq b$$

$$u_x = u_0 \left(\frac{Z}{Z_0}\right)^a$$

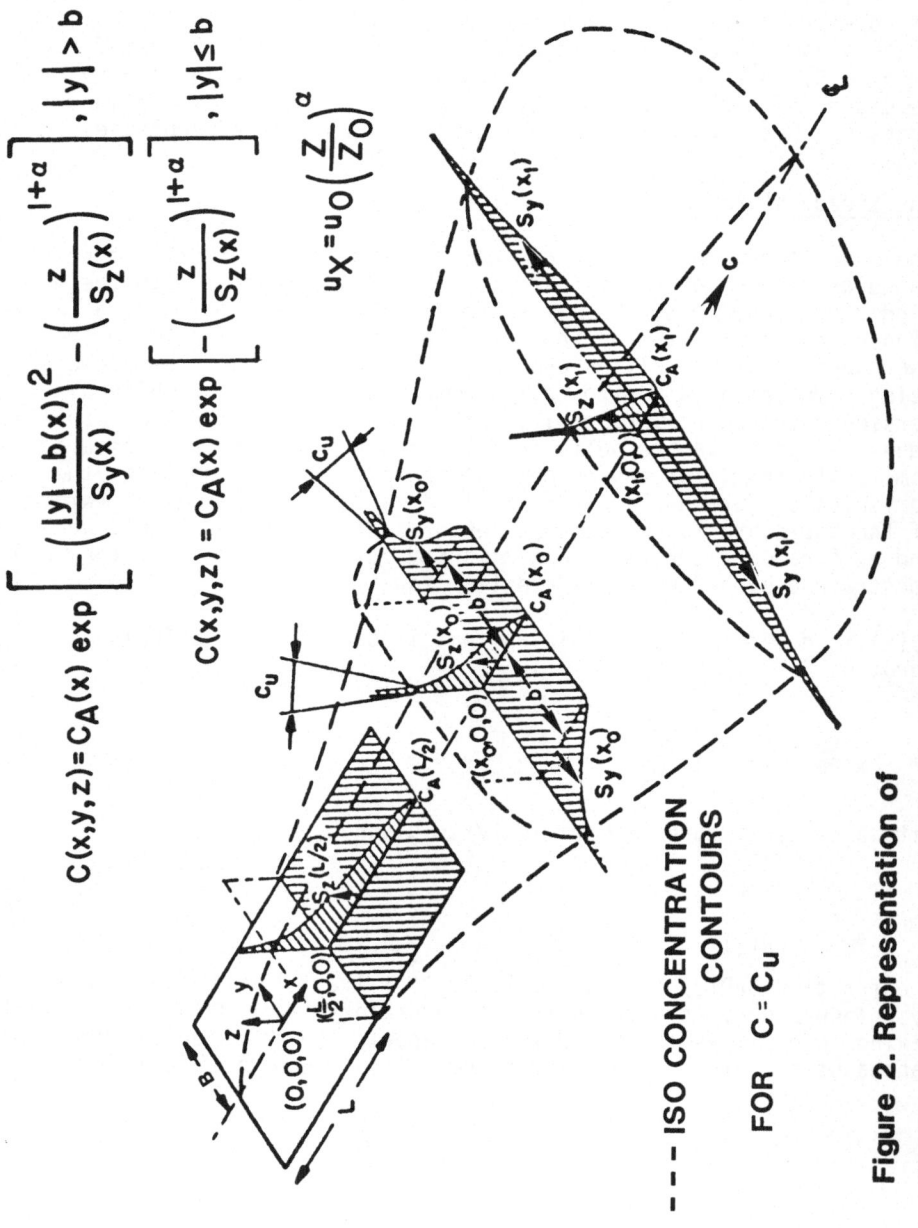

– – – ISO CONCENTRATION
CONTOURS

FOR $C = C_u$

Figure 2. Representation of
Concentration and Velocity Profiles Downwind of Area Source—Colenbrander

calculated assuming a logarithmic wind velocity profile (15)

$$u_* = k\, u_0 \left[\ln \frac{z_0 + z_r}{z_r} - \psi_1 \left(\frac{z_0}{\lambda} \right) \right]^{-1} \tag{21}$$

Combining the assumed similarity forms for concentration and wind velocity with Equations (19) - (21) gives

$$\frac{d(s_z/z_0)^{1+\alpha}}{dx} = \frac{k}{z_0} \frac{u_*}{u_0} \frac{(1+\alpha)^2}{\phi(Ri_*)} \tag{22}$$

where the Richardson number is computed as

$$Ri_* = g\, \frac{\rho(c_A) - \rho_a(H_{EFF})}{\rho_a(z=0)} \frac{H_{EFF}}{u_*^2} \tag{23}$$

with the effective cloud depth defined as

$$H_{EFF} = \frac{1}{c_A} \int_0^\infty c\, dz = \frac{\Gamma(1/(1+\alpha))}{1+\alpha}\, s_z \tag{24}$$

Colenbrander has "generalized" Equation (22), which is derived for two-dimensional dispersion, to the following form for application to a heavier-than-air gas plume which will spread laterally as a density intrusion.

$$\frac{d}{dx} \left[B_{EFF} \left(\frac{s_z}{z_0} \right)^{1+\alpha} \right] = \frac{k}{z_0} \frac{u_*}{u_0} \frac{(1+\alpha)^2}{\phi(Ri_*)} B_{EFF} \tag{25}$$

where the plume "effective half width" is defined by

$$B_{EFF} = b + \frac{\sqrt{\pi}}{2}\, s_y \tag{26}$$

and determined using the gravity intrusion relation (16,17)

$$\frac{d\, B_{EFF}}{dt} = \sqrt{g \left(\frac{\rho(c_A) - \rho_A(z=0)}{\rho(c_A)} \right) H_{EFF}}. \tag{27}$$

The average transfer velocity in the plume can be shown to be

$$u_{EFF} = u_0(s_z/z_0)^\alpha \; (\Gamma(1/(1 + \alpha)))^{-1} \tag{28}$$

and, it follows that

$$\frac{d \; B_{EFF}}{dx} = C' \sqrt{1 - \frac{\rho_A(z = 0)}{\rho(c_A)} \left(\frac{s_z}{z_0}\right)^{\left(\frac{1}{2} - \alpha\right)}} \tag{29}$$

$$\text{with } C' = \left[\frac{g \; z_0 \; \Gamma^3(1/(1 + \alpha))}{u_0^2 \; (1 + \alpha)}\right]^{1/2} .$$

The crosswind similarity parameter $s_y(x)$ is also determined by requiring that it satisfy the diffusion equation

$$u_x \frac{\partial c}{\partial x} = \frac{\partial}{\partial y} K_y \frac{\partial c}{\partial y} \tag{30}$$

with the horizontal turbulent diffusivity given by

$$K_y = K_0 \; u_x \; B_{EFF}^\gamma. \tag{31}$$

When $b = 0$, $s_y = \sqrt{2} \; \sigma_y$, where σ_y is the similarity parameter corre-lated by Pasquill-Gifford (18) as $\sigma_y = \delta x^\beta$ for use in the relation

$$c = c_A \exp\left[-\frac{1}{2} \left(y/\sigma_y\right)^2\right] \tag{32}$$

Further, when $b = 0$ it can be seen that Equations (30) - (32) require that

$$\sigma_y \frac{d\sigma_y}{dx} = K_0 \; B_{EFF}^\gamma \tag{33}$$

with $\gamma = 2 - 1/\beta$ and $K_0 = \frac{2\beta}{\pi} \; (\delta\sqrt{\pi/2})^{1/\beta}$,

then

$$s_y \frac{ds_y}{dx} = \frac{4\beta}{\pi} \; B_{EFF}^2 \left\{\frac{\delta\sqrt{\pi/2}}{B_{EFF}}\right\}^{1/\beta} \tag{34}$$

Colenbrander "generalizes" Equation (34) by assuming it applicable when determining s_y even when b is not zero.

At a downwind distance x_t where b has decreased to zero, the cross-wind concentration profile is assumed Gaussian with s_y given by

$$s_y = \sqrt{2} \ \delta(x + x_v)^\beta \tag{35}$$

where x_v is a virtual source distance given by

$$s_y(x_t) = \sqrt{2} \ \delta(x_t + x_v)^\beta . \tag{36}$$

It is assumed that gravity spreading is terminated for $x > x_t$.

For a steady gas/air plume, the center line concentration c_A is determined from the material balance

$$E = \int_0^\infty \int_{-\infty}^\infty c \ u_x \ dy \ dz = \frac{2 \ c_A \ u_0 \ s_z^{1+\alpha} \ B_{EFF}}{(1 + \alpha) \ z_0^\alpha} \tag{37}$$

where E is the plume gas source strength.

Equations (21) - (26), (29), and (34) - (37) are combined with an equation of state relating cloud density to gas concentration and are solved simultaneously to predict s_z, s_y, c_A, and B_{EFF} as functions of downwind distance beginning at the downwind edge of the gas source.

Initial Condition (Gas Source) Specification. Neglecting lateral dispersion over the gas source, s_y = 0 is specified at the downwind edge of the source. The initial values of c_A, s_z, and B are source-dependent. Letting Q^* be the gas take-up rate (flux) into the atmosphere from the source, it follows from a material balance that

$$\int_{-L/2}^{L/2} Q^*(x) \ dx = \int_0^\infty u_x(z) \ c_A(x = L/2) \ \exp\left[-\frac{z}{s_z(x)}\right]^{1+\alpha} dz. \tag{38}$$

If the gas source flux Q is greater than the atmospheric take-up flux Q^*, a pure gas layer with concentration c_A equal to the emitted gas density ρ_E will form over the source. The maximum take-up flux over the entire source of length L would then be:

$$Q^*_{max} (L) \equiv \frac{\rho_E \ u_0 \ (s_{z0})^{1+\alpha}}{(1 + \alpha) \ z_0^\alpha \ L} . \tag{39}$$

Two possibilities exist. The source flux Q is either greater than Q^*_{max} or it is less than Q^*_{max}. If $Q > Q^*_{max}$, a pure heavy gas layer forms, which at steady state conditions would be characterized by length and width constrained to satisfy the relation

$$Q^*_{max} \ L \ B = W \ L_s \ B_s \tag{40}$$

where L_S and B_S represent the gas source dimensions (such as an LNG pool). For non-rectangular gas sources, the source is represented as a square of equal area. For example, a circular pool of LNG with radius R_S is represented as square with length $L_S = \sqrt{\pi}\, R_S$.

If $Q < Q^*_{max}$, no pure gas layer accumulates over the gas source. The source length and half width are then $L = L_S$ and $B = B_S$, and for an average value of the ground level concentration $\tilde{c}_A$,

$$Q^* = \frac{\tilde{c}_A\, u_0 (s_{z0})^{1+\alpha}}{(1+\alpha)\, z_0^\alpha\, L}\ . \tag{41}$$

In this case, Equation (41) must be solved by iteration, since the values of $\tilde{c}_A$ and s_z are related through the concentration dependence of the Richardson number. s_{z0}, the value of s_z at the downwind edge of the heavy gas source, is computed using Equation (22) over the source, starting with $s_z = 0$ at the upwind edge of the source, and neglecting lateral dispersion over the source.

The values of c_A, B_{EFF}, s_y and s_z at the downwind edge of the source ($x = L/2$) are then used as initial conditions for the integration.

Quasi-Steady Approximate Treatment of Transient Gas Releases

Colenbrander has treated transient gas releases, i.e., releases in which the area of the gas source and/or the gas emission rate vary with time. The dispersion downwind of such releases is computed for a series of source configurations representing the time-varying source. Information from this series of steady state predictions is then extracted to provide a quasi-steady state description of the transient dispersion downwind of the source.

Following Colenbrander's description, consider a series of "observers" who travel (in the wind field) over the transient gas source which is assumed circular with radius $R_S(t)$ and vertical emission (flux) $Q(t)$. The vertical vapor flux is assumed spatially uniform over the source.

The transient gas source is represented as an equivalent square shape with length $L_S(t) = \sqrt{\pi}\, R_S(t)$. If $Q(t) < Q^*_{max}(L_S(t))$, $B(t) = B_S(t)$ and $Q^*(t) = Q(t)$. The ground level concentration at the source downwind edge is determined as indicated in Equation (41). If $Q(t) > Q^*_{max}(L_S(t))$, a pure heavy gas layer accumulates over the gas source. The pure heavy gas layer is assumed cylindrical in shape with radius $R(t)$ and height $H(t)$. Material balance requirements dictate that

$$\frac{d}{dt}\left[\pi\,\rho_E\,R^2(t)\,H(t)\right] = \pi\,R_S^2(t)\,Q(t) - \pi\,R^2(t)\,Q^*_{max}\ (L)\ . \tag{42}$$

It is assumed that the center of the pure heavy gas layer remains at
x = 0. The gravity intrusion model for dense gas spreading is used
to relate R(t) and H(t);

$$\frac{d}{dt}\,R(t) = \sqrt{g\left[1 - \frac{\rho_A\,(z = 0)}{\rho_E}\right]}\,H(t)\ . \tag{43}$$

Equations (22), (39), (42) and (43) are solved simultaneously to
obtain values of R(t) and Q*(t).

Let the subscript i refer to quantities associated with "observer" i.
Consider that the observers originate from the point which corresponds to the maximum upwind extent of the gas layer (x = -R_{max}).

The desired observer velocity is the average transport velocity of
the gas u_{EFF}; however, the value of u_{EFF} will differ from observer
to observer with the unfortunate consequence that some observers
may be overtaken by others. If one examines a neutrally buoyant
cloud, the value of u_{EFF} becomes a function of downwind distance alone.
With this functionality and the value of s_{z_0} when the averaged source
rate ($\pi\,R^2\,Q^*$) is a maximum (denoted with ^{z_0}the subscript m),
Colenbrander models the observer velocity to be:

$$u_i(x) = \frac{u_0}{\Gamma\left(\frac{1}{1 + \alpha}\right)}\left[\frac{s_{z_{om}}}{z_0}\right]^{\alpha}\left\{\frac{x + R_{max}}{\frac{\sqrt{\pi}}{2}\,R_m + R_{max}}\right\}^{\alpha/(1+\alpha)} \tag{44}$$

noting that $u_i(x) = \frac{dx}{dt}$, observer position and velocity as functions
of time are determined.

Colenbrander computes an averaged source seen by each observer.
If t_{up_i} and t_{dn_i} denote the times when observer i encounters the
upwind and downwind edges of the source, respectively, then the
source length seen by observer i is:

$$L_i = x_i(t_{dn_i}) - x_i(t_{up_i}).$$ (45)

The half width of the source, $B_i'(t)$, is:

$$B_i'(t) = \sqrt{R^2(t) - x_i^2(t)}$$ (46)

Consequently, the gas source "half area" A_i seen by the observer is given by

$$A_i = \int_{t_{up_i}}^{t_{dn_1}} u_i(t) \, B_i'(t) \, dt \; .$$ (47)

An average observed half width (required for representing the source as rectangular) is given by

$$B_i = A_i/L_i$$ (48)

and the average take-up flux from the gas source during the passage of observer i is given by

$$Q_i^* = \frac{\displaystyle\int_{t_{up_i}}^{t_{dn_i}} Q^*(t) \, B_i'(t) \, u_i(t) \, dt}{B_i \, L_i}$$ (49)

In order to close the integral material balance (Equation (37)) over the source, an averaged value of s_{z0_i} is calculated:

$$s_{z0_i} = \int_{t_{up_i}}^{t_{dn_i}} \left(\frac{ds_z}{dx}\right) u_i(t) \, dz$$ (50)

where $\left(\dfrac{ds_z}{dx}\right)$ is substituted from Equation (22). Unfortunately, this method of determining s_{z0} can result in values of c_A computed from the material balance (Equation (37)) which exceed ρ_E. When this occurs, the value of s_{z_0} is calculated with the integral material balance and ρ_E.

For each of several observers, released successively from $x = -R_{max}$, the observed dimensions L_i and B_i, the upwind and downwind edges of the observed source, the average vertical dispersion coefficient, s_{z0_i}, and the average take-up flux Q_i^* applicable during the observer's passage can be determined. For each such observer, a steady state calculation of $c_{A_i}(x)$, $s_{y_i}(x)$, $s_{z_i}(x)$ and $B_i(x)$ is made.

The concentration distribution at any specified time t_s is then determined by locating the position of the series of observers at time t_s, i.e. $x_i(t_s)$, and noting that quasi-steady values of the concentration, s_y, s_z, and b at times t_s are $c_A(x_i(t_s))$, $s_y(x_i(t_s))$, $s_z(x_i(t_s))$ and $b(\bar{x}_i(t_s))$. The corresponding concentration distribution is then computed from the assumed profiles given in Figure 2.

Allowance for Dispersion Along the Wind Direction

Colenbrander applies a "correction" or adjustment to the values of c_A calculated, as described above. The calculated ground level concentration $c_A(x)$ is considered to have resulted from the release of successive puffs of gas without any dispersion in the x-direction. It is then assumed that such puffs, which would be planar, would diffuse in the x-direction as the puff moves downwind with an x-direction concentration dependence given by the relation

$$c'(x; x_p) = \frac{c_A(x_p)}{\sigma_x(x_p)\sqrt{2\pi}} \exp\left\{-\frac{(x - x_p)^2}{2\,\sigma_x^2\,(x_p)}\right\} \tag{51}$$

where x_p denotes the position of the puff center.

This relation follows if it is assumed that each planar puff diffuses in the x-direction independently of any other puff and the dispersion is one-dimensional and Gaussian. The concentration at x is then determined by superposition, i.e., the contribution to c_A at a given x from neighboring puffs is added to give an x-direction corrected value of c_A:

$$c_A'(x) = \frac{1}{\sqrt{2\pi}} \int_0^\infty \frac{c_A(\zeta)}{\sigma_x(\zeta)} \exp\left\{-\frac{1}{2}\left\{\frac{x - \zeta}{\sigma_x(\zeta)}\right\}^2\right\} d\zeta. \tag{52}$$

Colenbrander suggests

$$\sigma_x(\zeta) = 0.13\,\zeta. \tag{53}$$

Resulting values of c_A^i can exceed ρ_E. We take the reference for σ_x to be $-R_{max}$.

The system of equations described above have been solved using the IBM Continuous Systems Modeling Program (12) and a DEC MINC-23 with 64 k bytes memory.

SIMULATION OF GAS DISPERSION FOLLOWING 25,000 m^3 LNG RELEASE ON WATER

An instantaneous release of 25,000 m^3 LNG onto water was assumed with a predicted liquid pool radius and pool evaporation rate as shown in Figure 3. The pool size and evaporation rate predictions were taken from the report on the SIGMET model application to catastrophic LNG releases by Havens (19). This pool spread and evaporation rate appear to be as justifiable now as when the SIGMET calculations were reported. It was desired to compare the results with those obtained from SIGMET with the same gas source description.

Releases under the following atmospheric and sea conditions have been simulated.

Atmospheric Stability	Pasquill D	Pasquill F
Monin-Obukhov length, λ $\psi(z/\lambda)$ in Eq. (21)	∞	5.62 m
Wind Velocity		
u at z = 10 m	2.24, 4.48, 8.96 m/sec	2.24 m/s
z_R	10^{-4} m	10^{-4} m
Temperature		
Water	288 K	
Air	293 K	
Air Dew Point	287,273 K	(68% - 20% relative humidity)

The exponent in the power law wind profile was calculated to give a weighted least squares fit of the logarithmic wind profile given in Equation (21). The power law exponents used were 0.1017 for D stability and 0.2997 for F stability.

Properties of LNG vapor were assumed to be those of methane.

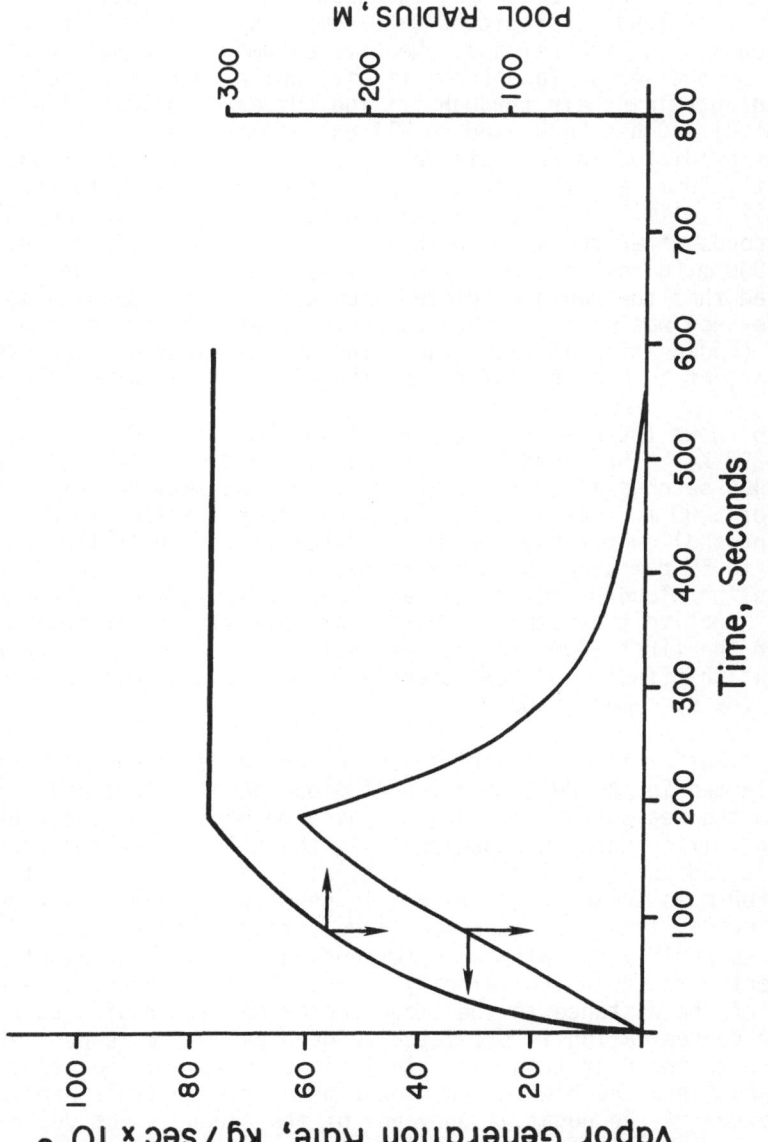

Figure 3. Liquid Pool Source Rate and Radius as Functions of Time

Eidsvik Model Predictions

Figure 4 shows predicted cloud radius, concentration, density, and
downwind distance to the cloud center for a neutral atmospheric flow
characterized by u_0 = 2.24 m/s, T_A = 293 K at 68% relative humidity
(R.H.). The cloud is predicted to become positively buoyant about
500 seconds after the release. We have assumed the cloud subsequently
remains on the sea surface (no lift-off) while being translated downwind and entraining air through its top surface. A lower limiting
value of Ri = 0 has been used in all entrainment relations when the
cloud is predicted to be positively buoyant. The maximum cloud radius
is about 1400 m, and the cloud is predicted to decrease to the lower
flammability limit (LFL) concentration (assumed 5% by volume) about
1600 seconds after the spill with a maximum downwind LFL extent of
about 4900 m (downwind distance + radius). Analysis of the simulation
indicated that the rapid predicted decrease of cloud density to positive buoyancy was primarily due to addition of heat to the cloud
through condensation of water vapor in the entrained air and, to a
lesser extent, to heat transfer from the sea surface to the cloud.

Figure 5 shows the predicted effect of low humidity air (20% R.H.
at T = 293 K). The cloud is now predicted to drop below the LFL at
about 1400 seconds after the release, with a maximum downwind LFL
extent of 5200 m. However, in this case the predicted cloud is
wider and shallower (about 18 m depth when the concentration has
dropped to 5% compared to 53 m for the case shown in Figure 4). With
20% R.H. the cloud radius continues to increase, since the cloud
remains negatively buoyant. Analysis of this simulation indicated
that for the first 1200 seconds the heat addition to the cloud by
condensation effects and heat transfer from the sea surface area
were of the same magnitude.

If heat transfer from the surface and water condensation heat effects
are neglected in the determination of cloud density (but heat transfer from the sea surface is included as a mechanism for inducing
turbulent mixing through Equation (5)), the cloud is predicted to
grow to a radius of 3650 m (H $\approx$ 7.5 m) at a maximum LFL extent of
about 6600 m as shown in Figure 6. If negative buoyancy is reduced
through energy transfer or phase change effects, the cloud radius
grows less rapidly but with a corresponding increase in height.
Consequently, the downwind velocity of the cloud is increased and
the sum of the distance to the cloud center and its radius at the
time the concentration is predicted to drop below LFL is not very
sensitive to the heat transfer calculation. However, the cloud
radius and, hence, the hazard zone downwind of the spill is predicted
to be larger if the negative buoyancy of the cloud is not decreased
by energy exchange effects. Finally, these predictions indicate that
there might be a cloud lift-off for conditions of high humidity,
although we have assumed that lift-off does not occur.

The value of β used in Equation (16) was varied from 0.3 to 0.6 to
determine the sensitivity of the downwind translation velocity

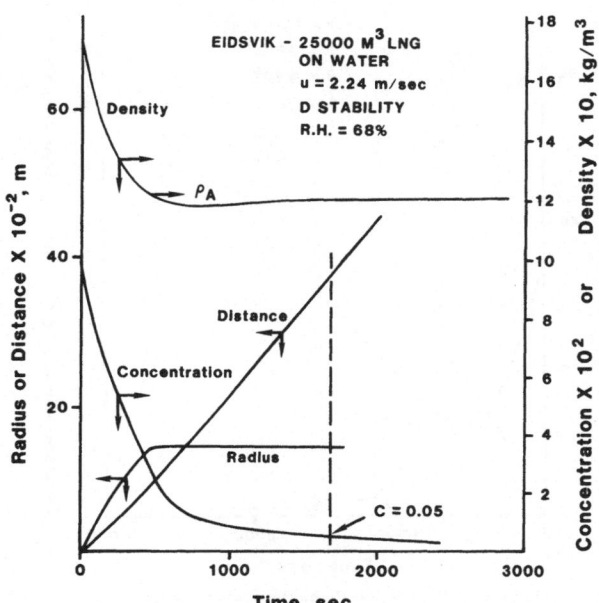

Figure 4. Eidsvik Model Predictions, High Humidity

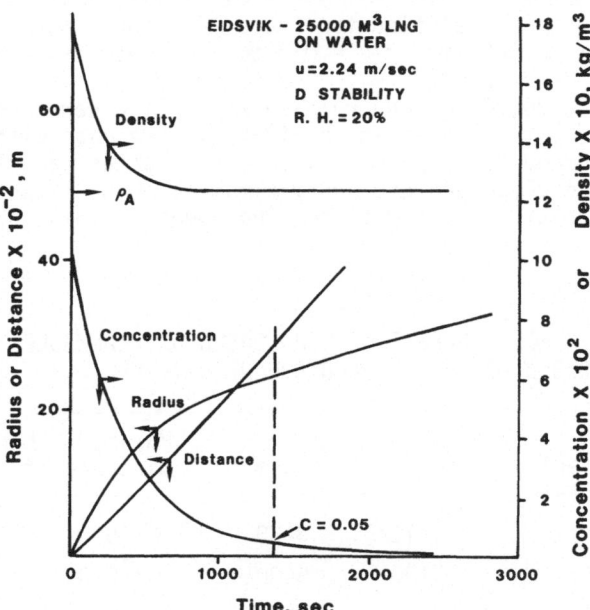

Figure 5. Eidsvik Model Predictions, Low Humidity

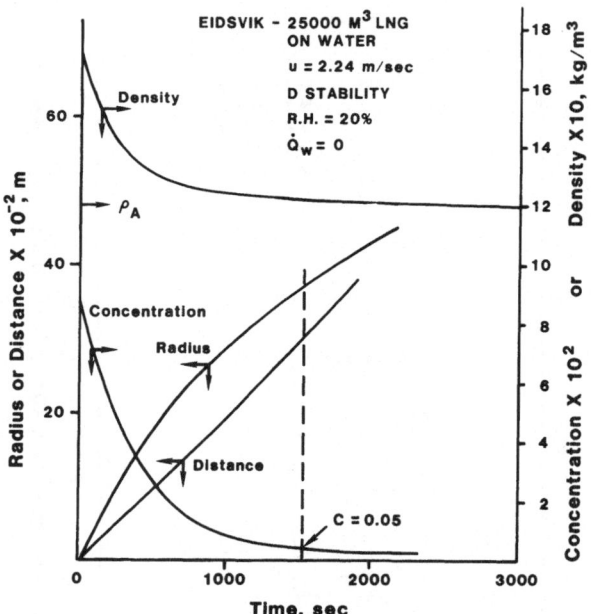

Figure 6. Eidsvik Model Predictions, Low Humidity, Adiabatic Energy Balance

its specification. The maximum downwind extent of the LFL varied (for other conditions as specified for the simulation shown in Figure 4) from 4700 m to 5000 m for β = 0.3 to 0.6, respectively.

Simulation of a 25,000 m^3 LNG release with all other conditions specified as for the run shown in Figure 4 were made with wind velocities of 4.48 and 8.96 m/sec. Table 1 indicates the cloud radius, distance to the cloud center, and maximum downwind extent of cloud at LFL concentration for the three wind speeds simulated.

TABLE 1

EIDSVIK MODEL EFFECT OF WIND SPEED ON FLAMMABLE
CLOUD DIMENSIONS AND DOWNWIND TRAVEL

u_0(m/sec)	R, m	X_c, m	Maximum LFL Distance, m
2.24	1410	3480	4900
4.48	1190	5400	6600
8.96	900	9900	10800

Colenbrander Model Predictions

LFL (5% by volume) contours for Pasquill stability classes D and F at a wind velocity (u_0) of 2.24 m/s are shown for several times after the spill beginning (t = 0) in Figures 7 and 8, respectively. LFL contours for Pasquill D stability at a wind velocity (u_0) of 8.96 m/s are presented in Figure 9; Figure 10 summarizes the effect of wind velocity (u_0) on the maximum LFL extent, the time range when the cloud is at the maximum LFL extent, and the maximum half width. For all of these simulations, cloud density as a function of concentration was calculated assuming adiabatic mixing of the emitted gas with humid ambient air. As a result, the density of the gas/air mixture is less than the ambient air for some values of concentration. The function ϕ (Equation (20)), however, is valid only for positive Richardson numbers, and the limiting value of $Ri_* = 0$ has been used.

Figures 7 through 9 also show the gas blanket radius for various times, the outermost of which indicates the maximum extent of the gas blanket in each case. The vertical dispersion over the source controls the atmospheric take-up flux and the corresponding pure gas blanket. As would be expected by the specification for the vertical dispersion coefficient K_z in Equation (20), the vertical dispersion is enhanced by higher values of u_* (caused by either higher wind speeds or decreased atmospheric stability), and s_{z_0} (which follows the same dependence as K_z) reaches a limiting value for large sources. Following from Equation (39), the maximum take-up flux (Q^*_{max}) is increased for the same conditions which increase vertical dispersion; then for these conditions, the quasi-steady state sources are narrower (B_{EFF} is less) and deeper (s_{z_0} is greater).

For dispersion downwind of the area source, enhanced vertical dispersion decreases the maximum LFL distance calculated by the model. In the case of Pasquill F stability, however, the maximum LFL distance is limited due to the increased half width; otherwise, the maximum LFL distance would be expected to be greater than the D stability calculation. For all of these cases, the differences between the horizontal dispersion represented by s_y seem to be unimportant.

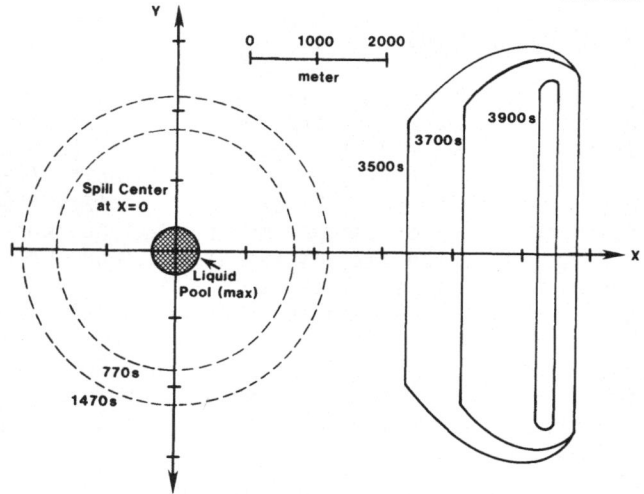

**Figure 7. Gas Source Radius and LFL Contours (at Z=0.5m)
for U_0= 2.24 m/s, D Stability**

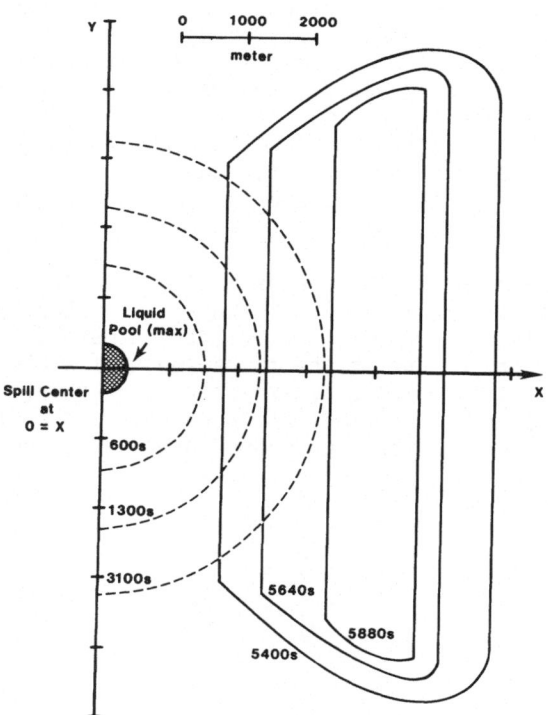

**Figure 8. Gas Source Radius and LFL Contour (at Z=0.5m)
for U_0= 2.24 m/s, F Stability**

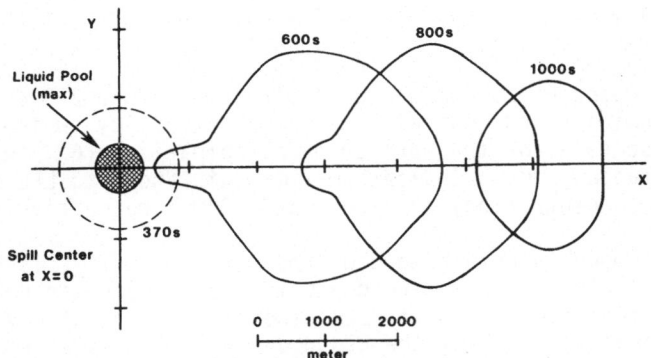

**Figure 9. Gas Source Radius and LFL Contours (at Z=0.5m)
for U_0= 8.96m/s, D Stability**

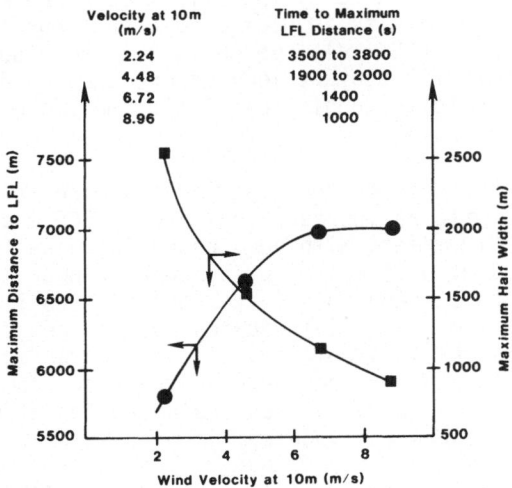

Figure 10. Summary of Calculations for D Stability

CONCLUSIONS

Predictions of the downwind atmospheric dispersion of liquid methane vapor following an instantaneous release of 25,000 m^3 LNG onto water have been made with the Eidsvik and Colenbrander models for comparison with previously published predictions of this accident scenario.

For a wind speed of 2.24 m/sec with air temperature 293 K and 68% relative humidity and neutral atmospheric stability, which were used as base case for the earlier model comparison (6), the Eidsvik and Colenbrander models indicate maximum distances to LFL (time average 5% concentration) of 4900 m and 5800 m, respectively.

The Eidsvik model's predicted downwind distance to LFL is highly dependent on wind speed if the cloud is assumed (as was done herein) to be advected with a velocity corresponding to the wind velocity at a height equal to one-half the cloud depth. If this assumption is made, which amounts to decoupling the gravity spreading and downwind advection processes and neglecting cloud acceleration, the maximum downwind LFL extent is predicted to increase from 4900 m for a wind velocity of 2.24 m/sec (at 10 m) to 10,800 m for a wind velocity of 8.96 m/sec. The Eidsvik model predictions of cloud size are sensitive to air humidity specification since moisture condensation in the cloud results in decreased cloud density. For the high humidity condition (68% relative humidity) the cloud is predicted to become neutrally buoyant long before the cloud is diluted to the LFL; in low humidity conditions (20%), the cloud is predicted to remain negatively buoyant.

The Colenbrander model's predicted downwind distance is less dependent on wind speed, with an increase from about 5800 m at 2.24 m/sec to about 7000 m at 6.0 m/sec, with no further increase (up to the maximum value tested of 8.96 m/sec). The Colenbrander model shows little effect, for this scenario, of atmospheric stability on the maximum distance to the LFL, but indicates much larger cloud widths with increasing atmospheric stability. The Colenbrander model used herein does not include provision for heat transfer from the sea surface to the cloud, and we have not yet evaluated the effect of humidity with the Colenbrander model.

The predictions described herein of a range of 4900-5800 m for the maximum downwind extent for the same spill conditions (2.24 m/sec wind and D stability atmosphere) used in previously published comparisons (6) show rough agreement with similar estimates made with the SIGMET model when the range of uncertainty in vertical diffusivity suggested by Havens (19) is incorporated. All three models, which represent three difference approaches to the dispersion prediction problem, indicate increased downwind distances to the LFL with increased wind speed for this scenario.

LITERATURE CITED

1. U.S. Coast Guard Contract DTCG23-80-C-20029, "Development of Vapor Dispersion Models for Non-Neutrally Buoyant Gas Mixtures," with the University of Arkansas, J. A. Havens, Principal Investigator

2. Koopman, R. P., L. M. Kamppinen, W. J. Hogan, and C. D. Lind, Burro Series Data Report (1981), LLNL/NWC 1980 LNG Spill Tests, Lawrence Livermore Laboratories, Livermore, California.

3. Puttock, J. S., D. R. Blackmore, and G. W. Colenbrander, Field Experiments on Dense Gas Dispersion, Journal of Hazardous Materials, 6, No. 1/2, 1982.

4. McQuaid, J., Future Directions of Dense-Gas Dispersion Research, Journal of Hazardous Materials, 6, No. 1/2, 1982.

5. Ammonia Spill Study, sponsored by the Fertilizer Institute, 1015 18th St. N.W., Washington, D.C., to be conducted at China Lake Naval Weapons Center in 1982.

6. Havens, J. A., An Assessment of Predictability of LNG Vapor Dispersion from Catastrophic Spills onto Water, Journal of Hazardous Materials, 3, 1980, pp 267-278.

7. Blackmore, D. R., M. N. Herman, and J. L. Woodward, Heavy Gas Dispersion Models, Journal of Hazardous Materials, 6, No. 1/2, 1982.

8. Eidsvik, K. J., A Model for Heavy Gas Dispersion in the Atmosphere, Report No. 18/79, Norwegian Institute for Air Research, P.O. Box 130, N-2001, Lillestrom, Norway, June 1979.

9. Eidsvik, K. J., A Model for Heavy Gas Dispersion in the Atmosphere, Atmospheric Environment, 14, 1980, pp 769-777.

10. Eidsvik, K. J., Heavy Gas Dispersion Model with Liquefied Release, Atmospheric Environment, 15, No. 7, 1981, pp 1163-1164.

11. Colenbrander, G. W., A Mathematical Model for the Transient Behavior of Dense Vapor Clouds, 3rd International Symposium on Loss Prevention and Safety Promotion in the Process Industries, Basel, Switzerland, 1980.

12. Continuous System Modeling Program III (CSMP III) Program Reference Manual, IBM Publication No. SH19-7001, 1975.

13. McQuaid, J., Some Experiments on the Structure of Stably Stratified Shear Flows, Technical Paper P21, Safety in Mines Research Establishment, Sheffield, UK, 1976.

14. Kantha, L. H., O. M. Phillips, and R. S. Azad, On Turbulent Entrainment at a Stable Density Interface, Journal of Fluid Mechanics, 79, 1977, pp 753-768.

15. Businger, J. A., Turbulent Transfer in the Atmosphere-Surface Layer, in Workshop on Micrometeorology, American Meteorological Society, Boston, MA (1973).

16. Yih, C. S., Dynamics of Non-Homogeneous Fluids, Macmillan, New York (1965).

17. Turner, J. S., Buoyancy Effects in Fluids, Cambridge University Press (1973).

18. Pasquill, F., Atmospheric Diffusion, Halsted Press Division of John Wiley and Sons, Inc., New York (1974).

19. Havens, J. A., A Description and Assessment of the SIGMET LNG Vapor Dispersion Model, U.S. Coast Guard Report No. CG-M-3-79, February 1979.

NOMENCLATURE

A	constant in vapor pressure correlation
A_i	source area seen by observer i (m^2)
a	constant in vapor pressure correlation
B	source half width (m), Figure 2
B_i'	local half width of source seen by observer i (m)
B_s	liquid source half width (m)
B_{EFF}	effective width of gas plume (m)
b	half width of horizontally homogeneous central section of gas plume (m)
C	constant pressure heat capacity of the gas-air mixture (J/kg^oK)
C_F	surface-friction drag coefficient
c	concentration (kg/m^3)
c_A	centerline, ground level concentration (kg/m^3)
c_A'	centerline concentration corrected for x-direction dispersion (kg/m^3)
$\tilde{c}_A$	average ground level concentration (kg/m^3)
C_a	constant pressure heat capacity of the ambient air (J/kg^oK)
C_g	constant pressure heat capacity of the gas (J/kg^oK)
E	source rate (kg/s)
G	mass flux $(kg/m^2 \, s)$
g	acceleration of gravity (m^2/s)
H	height or depth of density intrusion or cloud (m)
H_{EFF}	effective cloud depth, Equation (24) (m)
h	heat transfer coefficient $(J/m^2s \, ^oK)$
K_0	constant in Equation (31)

K_y horizontal turbulent diffusivity (m^2/s)

K_z vertical turbulent diffusivity (m^2/s)

k von Karman's constant, 0.35

L source length (m), Figure 2

L_s liquid source length (m)

L_w latent heat of freezing of water (J/kg)

M total cloud mass (kg)

M_a total air in the cloud (kg)

M_g total emitted gas in the cloud (kg)

$\dot{M}_a$ mass rate of air entrainment into the cloud (kg/s)

$\dot{M}_g$ mass rate gas enters the cloud (kg/s)

N_{ST} Stanton number

P ambient pressure (N/m^2)

Q gas source flux from liquid pool $(kg/m^2 \ s)$

Q^* atmospheric take-up flux $(kg/m^2 \ s)$

Q^*_{max} maximum atmospheric take-up flux $(kg/m^2 \ s)$

$\dot{Q}_v$ rate of heat added to the cloud due to condensation of water from the air (J/s)

$\dot{Q}_w$ rate of heat transfer from the earth's surface to overlying cloud (J/s)

R mass average gas constant (J/kg K) or gas source radius (m)

R_a mass gas constant-air (J/kg K)

R_c cloud radius, Eidsvik model (m)

R_g mass gas constant--emitted gas (J/kg K)

R_m value of R when $(\pi \ R^2 \ Q^*)$ is a maximum (m)

R_{max} maximum radius of the cloud

R_s liquid source radius (m)

R_w mass gas constant--water (J/kg K)

Ri   Richardson number, Eidsvik model

Ri_* Richardson number, Colenbrander model

s_y horizontal concentration scaling parameter (m)

s_z vertical concentration scaling parameter (m)

s_{z0} s_z at the downwind edge of the source (x = L/2) (m)

s_{z0_m} value of s_{z0} when ($\pi R^2 Q*$) is a maximum (m)

t   time (s)

T   cloud temperature (K)

T_{DA} dew point temperature (K)

T_w average water temperature (K)

T_a ambient air temperature (K)

t_{dn_i} time when observer i encounters downwind edge (s)

T_g pure gas saturation temperature (K)

t_{up_i} time when observer i encounters upwind edge (s)

u   average flow velocity (m/s)

u_a characteristic wind velocity, such as velocity at a specified height (m/s)

u_c horizontal cloud velocity (m/s)

u_e horizontal entrainment velocity (m/s)

u_i velocity of observer i (m/s)

u_x wind velocity, along x-direction (m/s)

u_0 wind velocity measured at $z = z_0$ (m/s)

u_{EFF} effective cloud advection velocity (m/s)

u_* friction velocity (m/s)

w characteristic turbulent velocity (m/s)

w_e vertical entrainment velocity (m/s)

w_m mechanically-induced turbulence velocity (m/s)

w_T thermally-induced turbulence velocity (m/s)

X mass fraction of water in air (kg/kg)

x,y,z Cartesian coordinates (m)

X_a mass fraction of water in ambient air (kg/kg)

X_c dew point mass fraction of water in gas at T (kg/kg)

x_t downwind distance where gravity spreading terminates (m)

x_v virtual point source distance (m)

z_0 reference height in wind velocity profile specification (m)

z_R surface roughness (m)

α constant in power law wind profile

β height for computation of u_c in Equation (16), or
            constant in σ_y correlation in Equation (32)

γ constant in Equation (31)

δ constant in σ_y correlation in Equation (33) (m)

θ_{w0} vertical temperature flux at earth's surface (K m/s)

ρ density of gas-air mixture (kg/m^3)

ρ_a density of air (kg/m^3)

ρ_E density of emitted gas (kg/m^3)

σ_x x-direction dispersion coefficient (m)

σ_y Pasquill-Gifford lateral dispersion coefficient (m)

$\phi(Ri_*)$ function describing influence of density stratification on
            vertical diffusion

MAPLIN SANDS EXPERIMENTS 1980: COMBUSTION OF LARGE LNG AND REFRIGERATED
LIQUID PROPANE SPILLS ON THE SEA

W.J.S. Hirst and J.A. Eyre
Shell Research Ltd., Thornton Research Centre, P.O. Box 1,
Chester CH1 3SH

This paper presents the results of experiments performed to investigate
the combustion of large gas clouds evolved from LNG and refrigerated
liquid propane spills on the sea. Extensive results for flamespeed
and flame-generated pressure are reported as well as detailed
observations of combustion behaviour. These experiments represent the
largest and most fully instrumented trials of this type to date.

1. INTRODUCTION

When LNG or refrigerated liquid propane is spilled onto water
boiling of the fuel produces a dense, low-lying cloud of vapour which
spreads under the influence of gravity, mixes with air and is advected
away from the spill point. Some portion of the resulting cloud will
be flammable and if ignited will give rise to a propagating flame
within this region. Speculation exists as to whether such an event
can develop into an unconfined vapour cloud explosion (U.V.C.E.) with
combustion taking place at a sufficiently high rate to generate a
damaging blast wave extending the hazard boundary beyond the perimeter
of the cloud. The extreme case of fast combustion is detonation, in
which the combustion zone is travelling supersonically relative to the
unburnt gas. The overpressure wave associated with a detonation wave
can be greater than 18 bars and would produce damage over a large area.
Extensive work[1,2] has shown that direct initiation of detonation
in methane/air or propane/air clouds is extremely difficult even under
ideal experimental conditions of homogeneous stoichiometric mixtures;
hence its accidental occurrence is very unlikely. No accidents producing detonation of unconfined clouds of methane have been reported.
Although some doubt exists detonation may have taken place in one case
involving confinement of propane.[3]

Deflagration is the other mode of combustion. In this case the
flame travels subsonically through the flammable region. The speed of
deflagration varies widely, depending on a variety of interrelated
factors, and cannot be predicted reliably for large clouds. Some
limited data are available for flamespeeds from early field trials

211

S. Hartwig (ed.), Heavy Gas and Risk Assessment - II, 211–224.

carried out by the U.S. Coast Guard and Gaz de France.[5]   These produced
flamespeeds in natural gas and propane clouds of up to 20 m s^{-1}.   In
smaller scale experiments flamespeeds considerably in excess of this
value have been induced by means of partial confinement and obstacle-
generated turbulence.   A fast deflagration with a flamespeed of
150 m s^{-1} or greater through a large cloud would produce a blast wave of
significant damage potential.   It is a matter of considerable practical
importance to understand if or how such behaviour might come about.
An accurate appreciation of the behaviour of large-scale deflagration
would facilitate the development of more effective preventative measures
and deployment of safety resources to optimum effect in reducing the
actual hazard.   The motivation for such research is therefore high.
Since neither analytical nor simulation techniques can predict the
behaviour of large-scale combustion events, direct observation of
representative tests is the most reliable method of anticipating actual
behaviour.   In order to gather such data, a series of large-scale
combustion experiment has been performed by Shell Research Ltd.

2.   OBJECTIVES

The overall objective of the Maplin Sands test programme was to
provide reliable information on the dispersion and combustion of LNG
and refrigerated liquid propane spills on the sea.   Results of the
dispersion investigation are presented in ref. 4.   In the combustion
experiments measurements were made of heat radiation and flame-generated
pressure, the two mechanisms potentially capable of extending the
danger zone beyond the boundary of the flammable cloud.   The results
of this thermal radiation work will be presented elsewhere.

The objectives of the flame-generated pressure investigation were:
(1)   To ignite large volumes of essentially pre-mixed gas within
      dispersing clouds of natural gas and propane.
(2)   To measure the flame propagation throughout the cloud, thereby
      obtaining flamespeeds and accelerations.
(3)   To measure the air pressure waves generated by the deflagration of
      such clouds.
(4)   To perform such experiments from both instantaneous and continuous
      spills of LNG and refrigerated liquid propane under a variety of
      meteorological conditions.

3.   EXPERIMENTAL PROCEDURE

The experiments were performed on extensive tidal mudflats at
Maplin Sands in the Thames estuary.   Seventy instrument stations were
positioned on arcs concentric with the spill point at radii ranging
from 40 to 650 m.   The spill point was also surrounded by a 300 m
diameter dyke to retain the tide.   Pontoons were used to carry the
instrumentation required to measure the dispersion and subsequent
combustion of the clouds.   Figure 1 shows a typical deployment of the

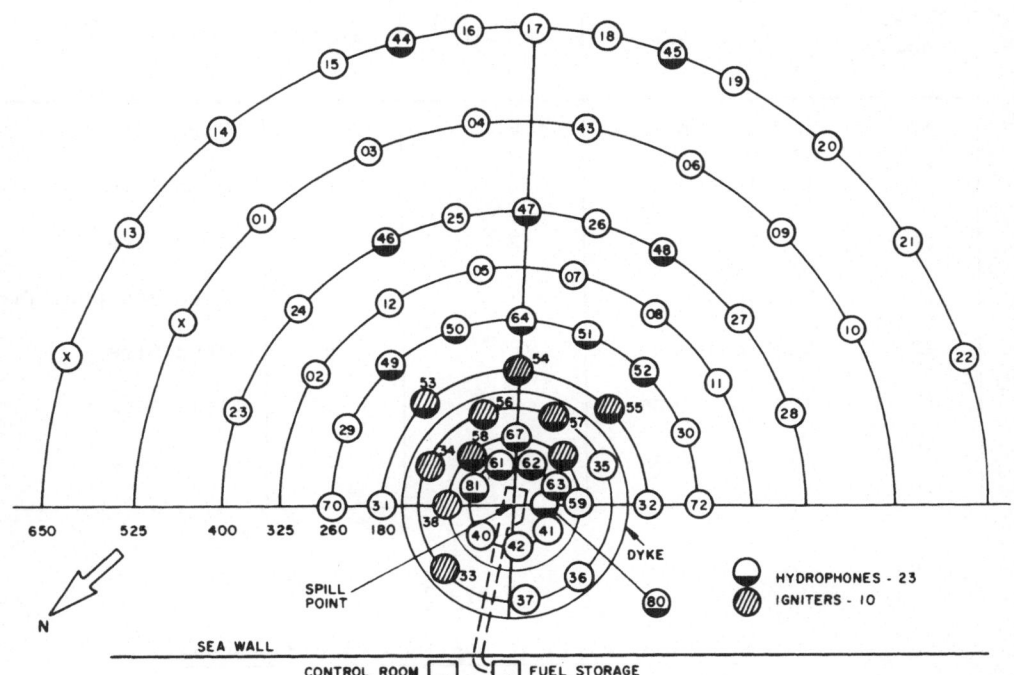

FIG.1 - Pontoon deployment for the tests showing hydrophone
and igniter positions

pontoons, though this varied slightly during the tests.   The
specifically combustion-related instrumentation comprised 10 electronic
spark igniters and 24 hydrophones mounted in air to measure pressure
fluctuations generated by the deflagration of the cloud.   Twenty-six
radiometers were deployed to measure the heat radiated from the cloud
and pool fires.

Each test began with pre-cooling of the insulated fuel discharge
pipe used to carry the LNG or refrigerated liquid propane from the
onshore cryogenic plant to the spill point located 300 m offshore.
For instantaneous spills the insulated submersible barge was also
pre-cooled.   The liquefied gases could be pumped out to the spill
point at rates of up to 5.8 m^3 per minute, for continuous spills, or
loaded into the spill barge for instantaneous spills of up to 20 m^3.

Table 1

Combustion trials at Maplin - LNG

Trial no.	Spill type	Volume $m^3 min^{-1}$ or m^3	Wind speed, $m s^{-1}$	Comments
17	Continuous	2.8	8	Flame failed
27	Continuous	3.2	6	Cloud fire
38	Continuous	5.8	5	Cloud fire
39	Continuous	4.7	4	Cloud and pool fire
22	Instantaneous	12.0	5	Cloud fire
23	Instantaneous	8.5	5	Flame failed
24	Instantaneous	12.0	5	Cloud fire

Table 2

Combustion trials at Maplin - Liquid propane

Trial no.	Spill type	Volume $m^3 min^{-1}$ or m^3	Wind speed, $m s^{-1}$	Comments
49	Continuous	2.1	6	Cloud fire
50	Continuous	4.3	8	Cloud and pool fire
51	Continuous	5.6	7	Cloud and pool fire
68	Instantaneous	(5-10)	6	Cloud fire

4. RESULTS

Eleven combustion tests were performed comprising instantaneous and continuous spills of both LNG and refrigerated liquid propane.   Tables 1 and 2 give the details of fuel and spill type as well as rate or quantity spilled.   For each of these tests the 24 channels of pressure data have been examined and rapid sequence stills analysed to yield plan views of the pre-mixed flame development.   The results from just two of the tests will be presented here, one LNG, one refrigerated liquid propane.   These tests illustrate the more important combustion characteristics of the programme as a whole.   The conclusions presented at the end of the paper are drawn from all the tests.

4.1 AN LNG EXPERIMENT

Test 27 was a continuous spill of LNG at $3.2 \text{ m}^3 \text{ min}^{-1}$ in a wind speed of 6 m s^{-1}.   The visible plume extended out to a radius of 300 m at the time of ignition which occurred on the edge of the cloud at a radius of 90 m.   The flame propagated downwind to a radius of 130 m before extinguishing.   The upwind flame burnt along one side of the plume towards the spill point and only at a radius of 60 m did the flame burn across the plume whereupon it burnt down the other side out to a radius of 110 m.   The remaining unburnt plume was clearly bifurcated with the richer yellow flame forming a 'V' pointed at the spill point, indicating that the concentration in the central region of the plume was lower than at the edges.   Plate I shows the development of the cloud fire and the transition to the richer yellow flame; the bifurcation is particularly clear in frames 5 and 6.

The major part of the pre-mixed flame history is shown in Figure 2. The perimeters shown were obtained from the disappearance of the water fog; allowance has been made for the effects of perspective.   In the case of LNG under the conditions of temperature and humidity prevailing at Maplin the gas concentration outside the water fog was insufficient to support a flame, thus the cloud always encompassed the flame and delineated its boundary.   Figure 3 shows that the maximum flamespeed observed was 10 m s^{-1} relative to the ground, the average flamespeed was about 4 m s^{-1}.   Some flame-generated pressure was measured and Figure 4 shows the largest pressure obtained in this test.

4.2 REFRIGERATED LIQUID PROPANE EXPERIMENT

Test 51 was a continuous spill of refrigerated liquid propane at $5.6 \text{ m}^3 \text{ min}^{-1}$ initially in a wind of 7 m s^{-1}.   A low-lying wide plume developed which was visible out to a radius of 190 m at ignition, which occurred on the edge of the cloud at r = 130 m.   The flame burnt downwind to a radius of 260 m where it extinguished. This agrees well with the LFL distance derived from the gas concentration measurements in this region.   Plate II shows the development of the pre-mixed cloud fire and the transition to a horseshoe-shaped diffusion flame burning back towards the spill point, indicating a rich core to the cloud.   Figure 5

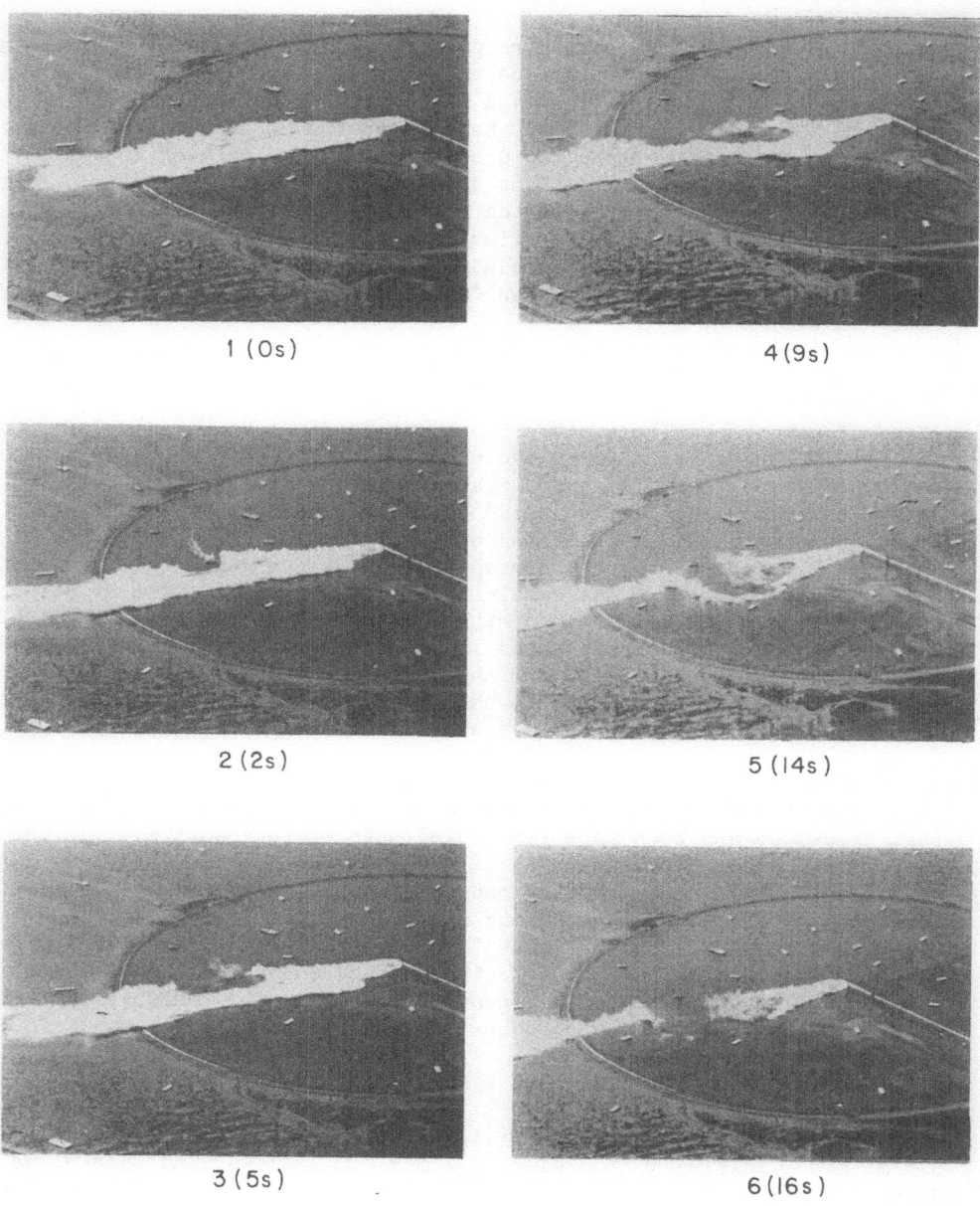

1 (0s)

4(9s)

2 (2s)

5 (14s)

3 (5s)

6 (16s)

PLATE I – The helicopter view of test 27 showing the development of the pre-mixed
cloud fire and transition to richer slower burning, bifurcation in the plume
shows particularly clearly in the last two frames

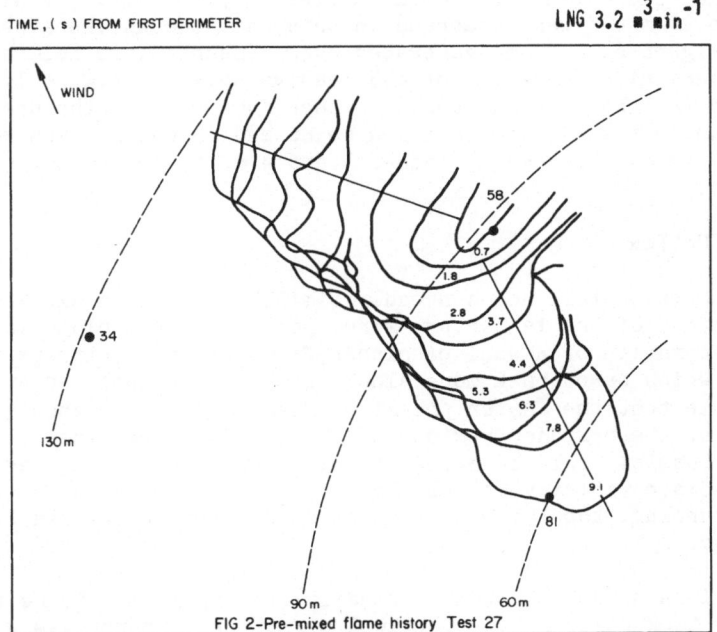

TIME , (s) FROM FIRST PERIMETER

LNG 3.2 $\mathrm{m}^3\mathrm{min}^{-1}$

FIG 2-Pre-mixed flame history Test 27

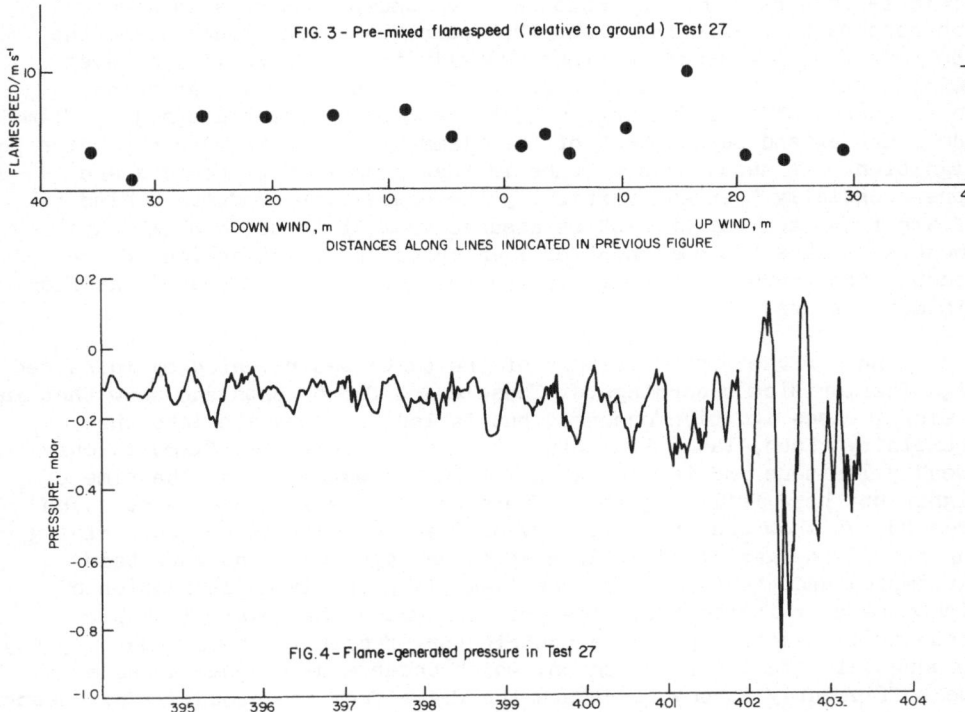

FIG. 3 - Pre-mixed flamespeed (relative to ground) Test 27

DISTANCES ALONG LINES INDICATED IN PREVIOUS FIGURE

FIG. 4 - Flame-generated pressure in Test 27

shows the development of the cloud fire in plan view.   Flamespeeds
of up to 20 m s^{-1} were observed in this test though the average was 12
m s^{-1}, Figure 6.   Flame-generated overpressures were recorded on 15
transducers with a maximum of 0.3 mbar as shown in Figure 7.   As
observed in other propane tests, richer fuel beneath the pre-mixed flame
was drawn in towards the ignition point and formed a transient tall
flame; this can be seen in frames 3 and 4 of the Plate II.

5. DISCUSSION

     Combustion followed a broadly similar pattern in all the tests.
Deflagration of the leaner pre-mixed portion of the cloud was followed
by the formation of a wall of richer flame (less continuous in the case
of LNG) which burnt back more slowly towards the spill point.   It was
noticeable that the faster pre-mixed flame did not advance quickly over
the top of these richer regions of the cloud but remained as a wall of
fire.   Possibly this behaviour resulted from the strong convection in
the immediate vicinity of the flame boundary drawing in the leaner,
faster burning, top layer of gas faster than the pre-mixed flame could
burn away.

     It is also notable that virtually all expansion of the burnt
products took place in the vertical direction; unburnt gas was not
pushed significantly ahead of the advancing fast pre-mixed flame.
This is in direct contradiction of the assumptions made in some
approaches to predicting blast from pancake-shaped clouds where the
burning gas is assumed to expand equally in all directions or even
exclusively in the horizontal direction.   An important practical
consequence of this finding is that the ground area traversed by flame
does not extend beyond that of the flammable cloud immediately prior to
ignition.   A second consequence is that because hot products are
preferentially "vented" vertically the combustion products behind the
flame front are not at rest as assumed in most acoustic blast wave
models.   Thus the component of flamespeed due to expansion of the
combustion products pushing forward the unburnt gas ahead of the flame
is much reduced.

     Another unexpected feature of the tests was revealed by infra-red
Aga Thermovision recordings.   These exist for 6 tests and show that in
several cases ignition occurred but failed to propagate into the
remaining cloud, Table 3.   Since the leaner pre-mixed flame is only
weakly luminous and is concealed within the water fog at the time of
ignition these failed ignitions were not otherwise detectable.   The
recordings show the strongly buoyant flame/combustion products rising
up out of the gas cloud.   In most cases edge ignitions were being
attempted and the failure of the flame is probably an indication of
inhomogeneity in the gas concentration around the lower flammable
boundary.   Another possible contributory factor is the strong buoyancy
produced in the ignition region, which under some circumstances might
be sufficiently strong to detach the flame from the cloud.   This seems

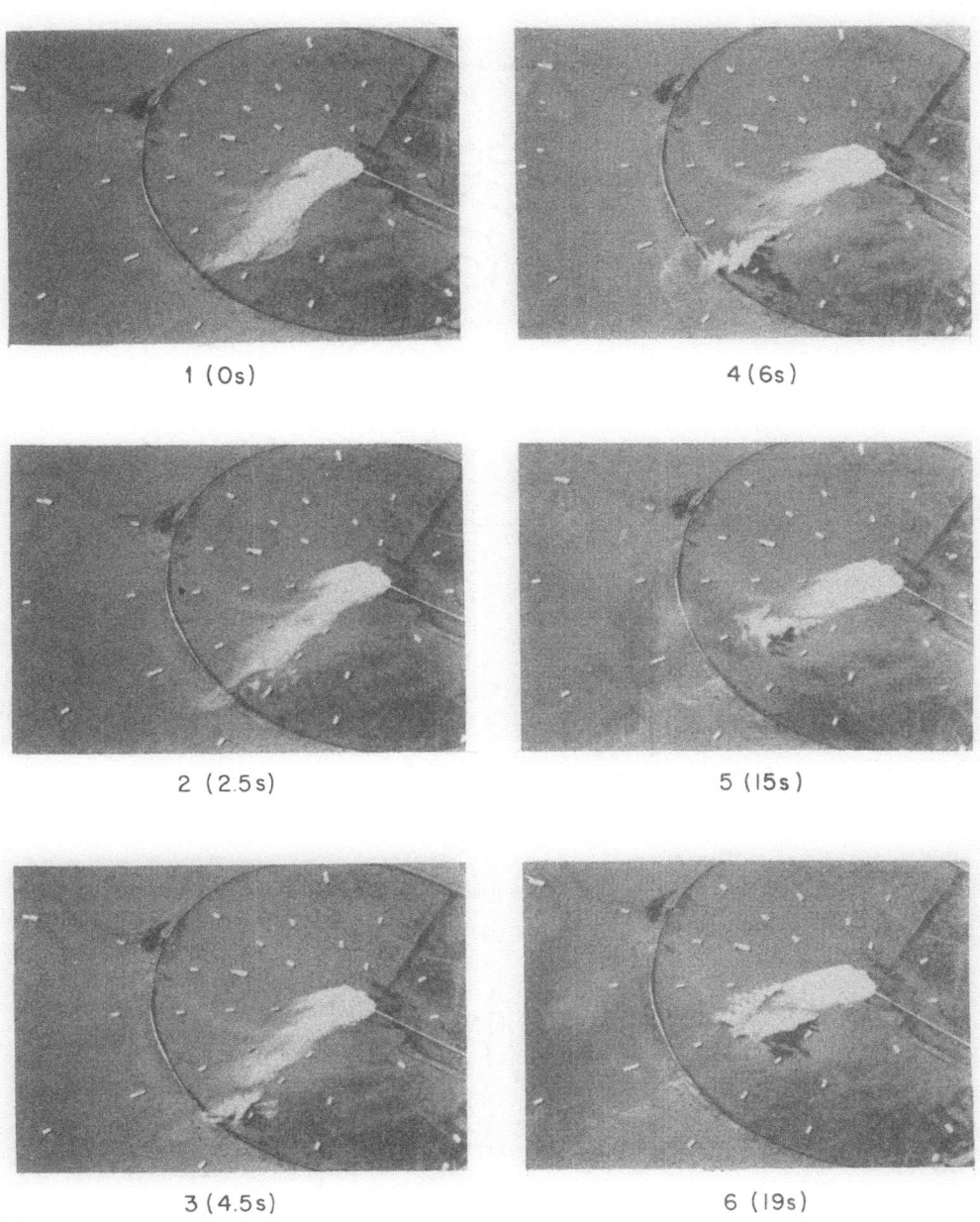

1 (0s)

4 (6s)

2 (2.5s)

5 (15s)

3 (4.5s)

6 (19s)

PLATE Ⅱ—The helicopter view of test 51 showing the pre-mixed flame development and transition to richer burning nearer the spill point

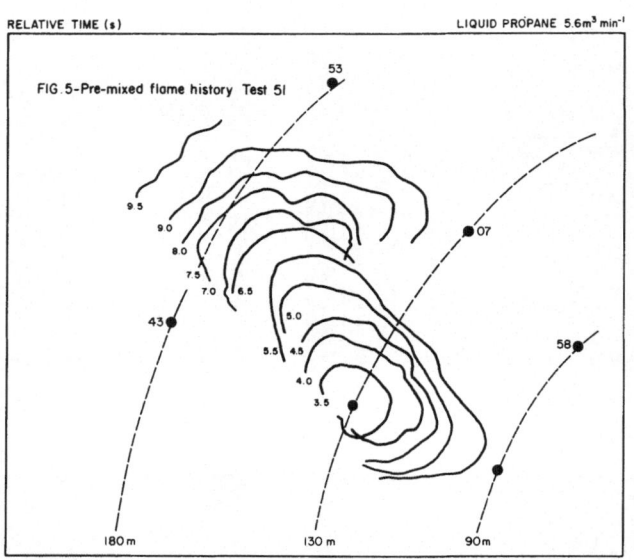

RELATIVE TIME (s)   LIQUID PROPANE 5.6m³ min⁻¹

FIG.5-Pre-mixed flame history Test 51

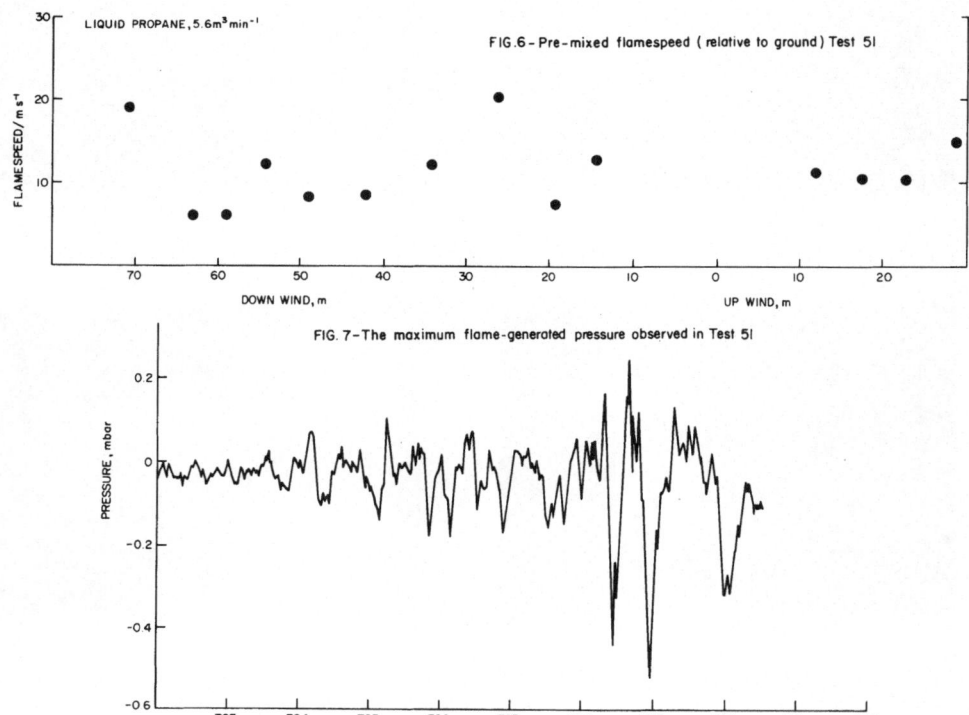

FIG.6 - Pre-mixed flamespeed (relative to ground) Test 51

FIG. 7-The maximum flame-generated pressure observed in Test 51

Table 3

Ignition behaviour of the tests derived from
the Aga Thermovision infra-red video recordings

Test	No. of ignitions	No. of failed ignitions	Fuel	Comments
17	1	1	N. Gas	Visual evidence
22	3	2	N. Gas	Aga
23	1	1	N. Gas	Aga
24	3	2	N. Gas	Aga
39	2	1	N. Gas	Aga
50	2	1	Propane	Aga
68	1	0	Propane	Aga, flare igniter
TOTAL	13	8		

to have happened in at least one test, where the burnt region did not
appear to be an isolated pocket of gas.   This type of behaviour would
be more likely for slower flames, such as those in natural gas, where
buoyancy has longer to act.

The pre-mixed flamespeeds observed in the tests did not exhibit any
sustained acceleration such as might lead to higher flamespeeds in
larger similar clouds.   The maximum flamespeeds observed were 28 m s^{-1}
in propane and 10 m s^{-1} in natural gas.   It is initially surprising
that there was such a marked difference between the flamespeeds - a
factor of 2.8.   A similar difference was observed between the average
flamespeeds, 12 and 5 m s^{-1}, respectively.   However, at 290 K the laminar
burning velocities of propane and methane are similar at 0.41 and
0.34 m s^{-1}, ref. 6.   Several factors may contribute to the differences
observed at Maplin, two of which are temperature and the presence of
dense water fog.   The effect of the first of these is negligible for
the differences in cloud temperature observed, 10°C.   The differences
in liquid water content of the clouds are considerable.   Whereas all
pre-mixed combustion of natural gas clouds occurred within dense water
fogs, for propane, not only was there less liquid water content in the
fogs (as a result of their higher temperatures) but also about half the
pre-mixed cloud contained no water fog at all.   The extent of the
effect this will have in slowing the methane flame is not clear, though
by reducing the flame temperature, the rate of conduction, and hence the
flamespeed, must drop.

The flamespeeds observed at Maplin are slow in relation to the
speeds necessary to produce damaging blast waves, consequently the
maximum flame-generated overpressures were very low, 0.8 mbar in propane
and 0.4 mbar in natural gas.   Generally the rarefactions in the pressure
traces were more pronounced than the overpressures.   This is believed
to be due to the sudden decrease in the rate of combustion occurring at
the extinction of the flame.   Since the flame-generated pressure is
proportional to the rate of change of the rate of fuel consumption, it
follows that the synchronous extinction of a large area of flame, as
would occur at the LFL boundary, will produce a relatively large
rarefaction.

Examination of the pressure data for one particular test provides
confirmation that the flame-generated pressure decays as 1/R in the far
field; R is the distance from the flame.   This is shown for test 51 in
Figure 8 where peak to peak pressure fluctuation has been plotted as a
function of distance from the region of flame responsible for the
pressure measured (as deduced from time of arrival).

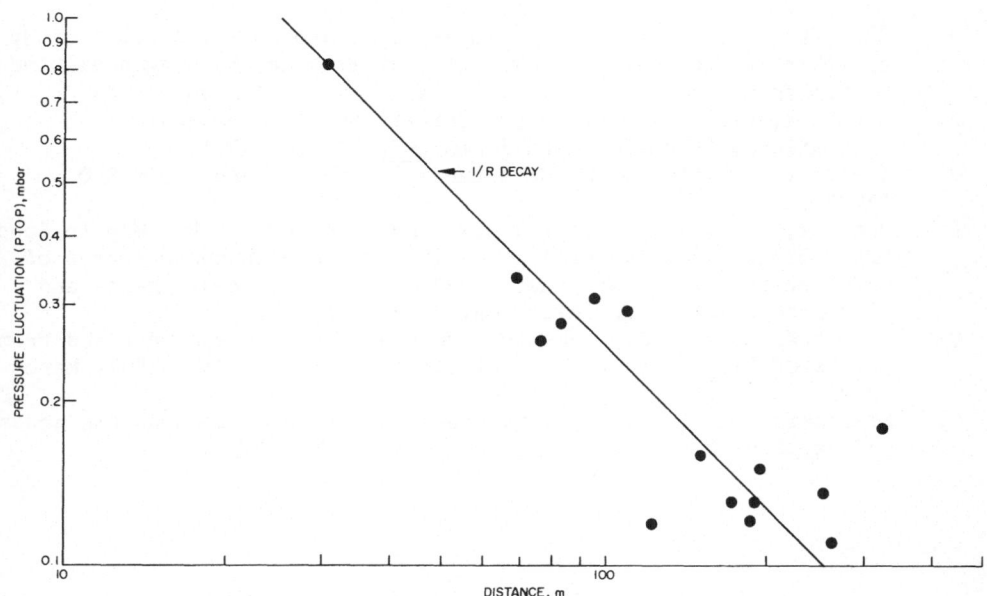

FIG.8-Plot of flame-generated pressure v distance from flame for test 51 showing the I/R decay
of pressure with distance in the far-field

6. CONCLUSIONS

1. On the evidence of the trials performed at Maplin Sands the deflagration of truly unconfined flat clouds of natural gas or propane does not constitute a blast hazard.

2. The maximum flamespeeds observed were 28 m s^{-1} for propane and 10 m s^{-1} for natural gas, average pre-mixed flamespeeds were 12 and 5 m s^{-1} respectively; all speeds are quoted relative to the ground.

3. Flame-generated overpressures were <0.8 mbar for propane and <0.4 mbar for natural gas, both measured in the immediate vicinity of the flame.

4. Flame-generated pressure decayed as 1/R.

5. No sustained flame acceleration, which might lead to faster flames developing in similar larger clouds, was observed.

6. Expansion of the combustion products took place exclusively in the vertical direction, thus the area traversed by the flame did not extend beyond that within the LFL boundary at the time of ignition.

REFERENCES

1. Bull, D.C., Elsworth, J.E., Hooper, J.E. and Quinn, C.P., A study
    of spherical detonation in mixtures of methane and oxygen diluted
    by nitrogen.   J. Phys. D. Appl. Phys., 9, 1991-2000 (1976).
2. Benedick, W.B., High explosive initiation of methane-air
    detonations, Combustion and Flame, 35, 89-93 (1979).
3. Advisory Committee on major hazards.   Second report H.M.S.O.
    (1979).
4. Puttock, J.S., Colenbrander, G.W. and Blackmore, D.R., Maplin Sands
    experiments 1980: Dispersion results from continuous releases of
    refrigerated liquid propane.   2nd Symposium on heavy gases and
    risk assessment.   Frankfurt May 25/26 1982.
5. RAJ, PPK, et al., Experiments involving pool and vapour fires from
    spills of LNG on water, A.D. Little inc., Report No. 82404, March
    1979.
6. Zabetakis, M.G., Flammability characteristics of combustible gases
    and vapours, U.S. Dept. of Commerce AD-701576 (1965).

PRESENT UNDERSTANDING OF THE EXPLOSION PROPERTIES OF FLAT

VAPOUR CLOUDS

W. Geiger
Battelle-Institut e.V.
Frankfurt am Main

SUMMARY

An outline is given of the present knowledge of the explosion properties of flat vapour clouds. The models which are available for detonation and for deflagration of flat vapour clouds in ideal situations
are briefly discussed. For deflagration, which represents the common
mode of explosion, it is shown that deviations from the ideal situation may significantly enhance or attenuate the pressure buildup. The
effects of typical features of real situations on the explosion pressure are discussed. It is concluded that severely enhanced overpressures
may have to be taken into account locally within the flat cloud, particularly in regions of partial confinement or repeated obstacles.

1. INTRODUCTION

After the release of large amounts of liquefied hydrocarbons
large pancake-shaped flammable clouds,i.e. of low height but of large
extension in downwind and crosswind directions, may be formed. The
probability that such a cloud will be ignited by some ignition source
in the environment is,in the case of land spills,rather high. This is
shown e.g. by a survey of unconfined vapour cloud explosions from
81 rail-car spills of flammable liquids /1/. If ignition occurs at an
early stage of cloud formation a large fire will result. However, if
ignition is delayed until intense mixing with air has been achieved,
instead of or in addition to fire an explosion will develop in that
part of the vapour cloud where concentrations are between the flammability limits.

The pressure wave due to the unconfined explosion of a flat
vapour cloud differs considerably from the pressure wave due to the
unconfined explosion of a hemispherical cloud of the same amount of
flammable mixture. In this paper an outline of the present knowledge

225

S. Hartwig (ed.), Heavy Gas and Risk Assessment - II, 225–236.
Copyright © 1983 by Battelle-Institut e.V., Frankfurt am Main, Germany.

of the explosion properties of flat vapour clouds is given. First, the models that are available for the two distinct cases of detonation and deflagration of flat vapour clouds in ideal situations will be briefly described. For deflagration, which represents the common mode of explosion, the deviations from ideal behavior caused by typical features of real situations will then be discussed.

2. DETONATION OF A FLAT VAPOUR CLOUD (IDEAL)

An ideal situation is characterized by the following properties:
- Quiescent pancake-shaped cloud (constant thickness, circular)
- Homogeneous stoichiometric mixture
- Flat, unobstructed terrain
- Central ignition.

Detonation characterized by supersonic propagation of the reaction front and by overpressures of several bar can occur only if a high-energy source (e.g. high explosive) for shock initiation of the mixture is present and if, in addition, the thickness of the cloud is larger than the critical size for self-supported propagation of detonation. The initiation energy required for detonation and the critical thickness depend on the reactivity of the mixture. In general, these figures are such that detonation is highly improbable for most conceivable situations. Nevertheless, detonation is often taken into consideration in view of its simplicity and because it represents an upper limit for the hazard of vapour cloud explosion.

In Fig. 1 the wave front contours during the propagation of detonation are shown for a flat cloud and, for comparison , for a hemispherical cloud with the same volume of explosive mixture /2/.

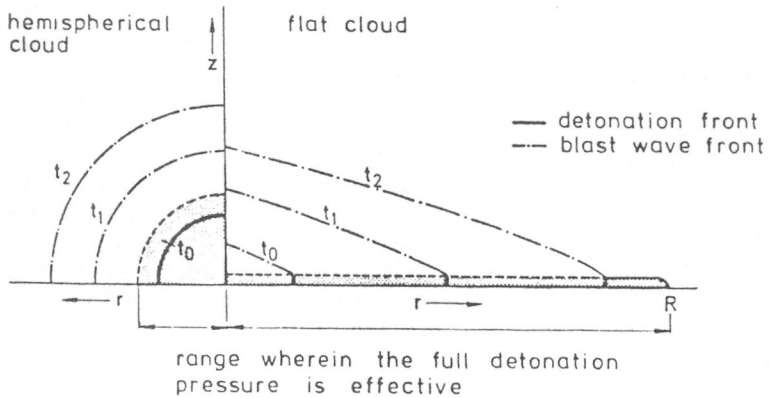

Fig. 1: Wave front contours at successive times during detonation propagation

The detonation front propagates to the edge of the cloud without attenuation. Most of the released combustion energy is used up for the blast wave which is emitted upwards during detonation propagation to the edge (vertical expansion of the reaction products). Beyond the cloud edge, i.e. after the end of detonation, there remains only the induced air blast wave which decays rapidly with horizontal distance.

The spatial distribution of overpressure at the ground can be calculated easily using a simple analytical model /2/ /3/, see for example Fig. 2. The suitability of this model is confirmed by the results of a 2-D hydro-code /4/. As is shown by Fig. 2, the area of high overpressure is, in the case of detonation, restricted essentially to the area covered by the cloud at the time of ignition. This area is, however, considerably larger than the area of high overpressure predicted by hemispherical cloud models.

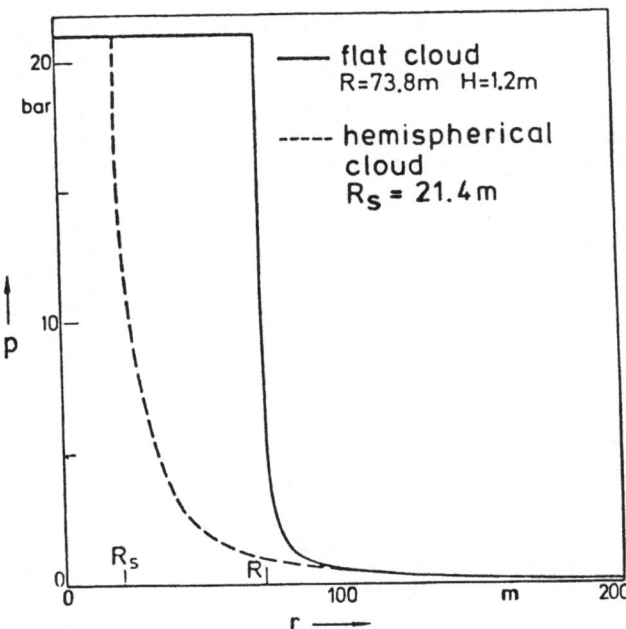

Fig. 2: Peak overpressure vs. horizontal distance for detonation of a flat cloud with H/R = 0.016 and of a hemispherical cloud of equal volume (R,H original radius and thickness of the flat cloud, R_s original radius of the hemispherical cloud)

3. DEFLAGRATION OF A FLAT VAPOUR CLOUD (IDEAL)

For the hazard analysis of unconfined vapour cloud explosions the case of deflagration is much more significant. The velocity of the reaction front (flame front) is subsonic, the explosion pressure is accordingly much smaller than in the case of detonation.

The explosion process for spherical clouds is usually described by the expanding piston model according to which the expansion of the hot reaction products (which generate a pressure wave and a flow velocity ahead of the flame front) corresponds to that of an expanding spherical piston /5/. For a spherical cloud, centrally ignited, a value of 0.1 bar may be assumed as an upper limit for the deflagration overpressure; this corresponds to a flame speed of about 85 m/s.

The piston model for spherical clouds has been used at Battelle as a starting point to develop a simple analytical model for the deflagration of flat clouds /6/ /2/ /3/. This model takes into account the essential feature of flat cloud explosion that during horizontal propagation of the flame front towards the edge of the cloud a blast wave is induced at the interface between reaction products and air, similar to the case of detonation. This, in turn, implies that a rarefaction wave propagates into the high pressure zone immediately behind the flame front which relieves the pressure there. Accordingly, the high pressure is restricted to a narrow zone (width of the order of the cloud thickness) travelling with the flame front like a horizontally expanding torus. Beyond the cloud edge the overpressure drops rapidly analogous to the case of detonation. In Fig. 3 an example for the spatial distribution of peak overpressure for a deflagrating flat cloud is given. Obviously, as in the case of detonation, the area over which the original explosion pressure is effective is much larger than the area derived from hemispherical models.

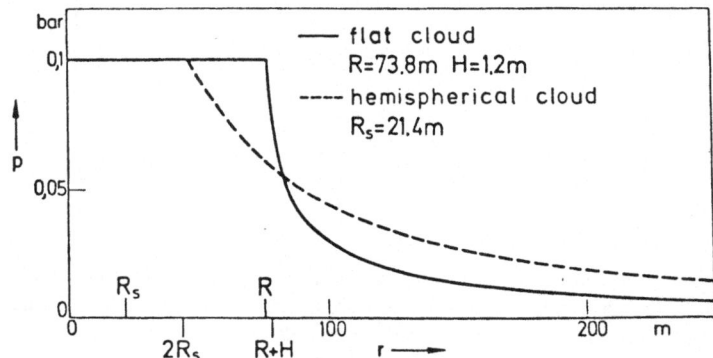

Fig. 3: Peak overpressure vs. horizontal distance for deflagration of a flat cloud with H/R = 0.016 and of a hemispherical cloud of equal volume (cloud parameters as in Fig. 2)

Because the pressure zone in a flat cloud is very narrow the
deflagration pressure at a given flame speed will be lower than in the
case of hemispherical cloud deflagration. In the Battelle model this
reduction of explosion pressure has not been taken into account; for
ideal situations the model is, hence, to be regarded as conservative.
The pressure reduction is implicitly allowed for in so-called acoustic
models by Auton and Pickles /7/ and by Strehlow /8/ which, however, do
not adequately represent the situation in the near field, apart from
other deficiencies.

In Fig. 4 the overpressure distribution for a flat cloud (cloud
parameters as in Fig. 2) using the different models is shown. According to Auton and Pickles the peak overpressure is reduced by a factor
of 7.5 with regard to the value 0.1 bar chosen in the Battelle model;
according to Strehlow the reduction even amounts to a factor of 100.

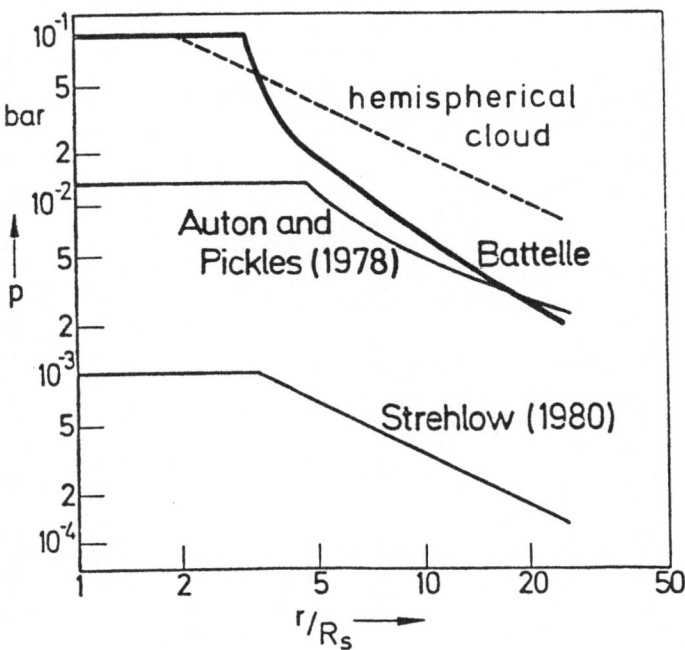

Fig. 4: Peak overpressure vs. horizontal distance for the deflagration
of a flat cloud: comparison of different models

4. FLAT CLOUD DEFLAGRATION IN NON-IDEAL SITUATIONS

Experiments with flat shape of the explosive volume have been
performed on a small scale using soap bubbles to enclose the explosive
mixture /9/. These experiments demonstrate that under completely ideal
conditions the overpressure during deflagration of a flat volume is

indeed much smaller than in the case of a centrally ignited hemispherical volume, see <u>Fig. 5</u> (time interval 20 to 50 ms).

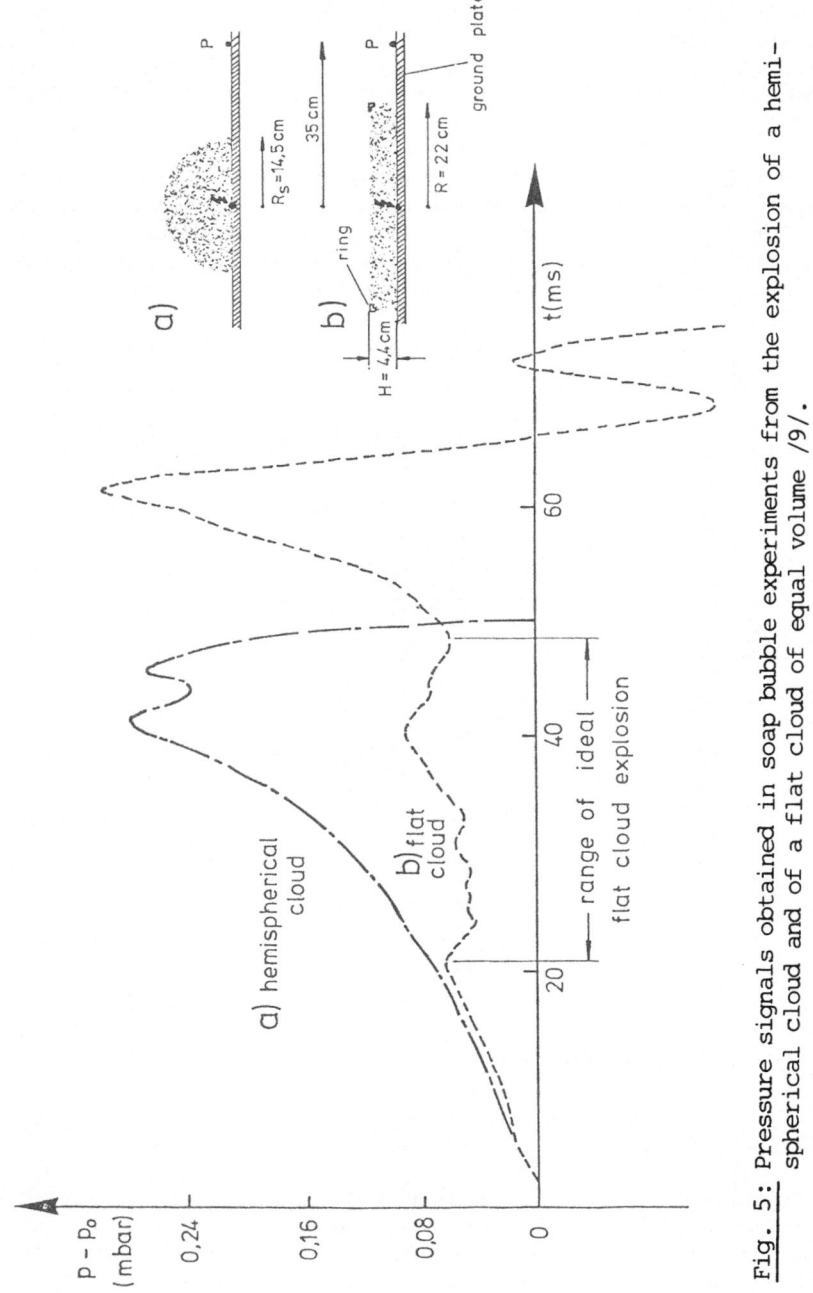

<u>Fig. 5:</u> Pressure signals obtained in soap bubble experiments from the explosion of a hemi-
spherical cloud and of a flat cloud of equal volume /9/.

However, as can also be seen from Fig. 5, even very small deviations from the ideal situation (here the presence of a thin supporting ring at the circumference of the explosive volume) may cause a strong increase of overpressure so that, with respect to hazard assessment, the figure for ideal flat cloud explosion becomes irrelevant. Similar experiences have been made in large-scale balloon experiments in which the envelope foil is removed immediately prior to ignition; the small disturbance remaining at the interface between explosive mixture and air causes flame acceleration when the flame propagates through this zone /10/.

Involuntary experiments involving large flat clouds are supplied by actual vapour cloud explosion accidents. A particular example is the accident in Port Hudson, USA, in 1970 /11/. Following a propane pipeline break a flat flammable propane-air cloud with an estimated extension in downwind direction of about 500 m, several meters high, had been formed. The accident area was essentially farmland with a few houses and trees scattered over a large area. In spite of these mild, seemingly near-ideal conditions considerable structural damage was done to the houses and trees indicating a pressure level of the order of 0.1 bar over a large range.

From these findings the conclusion can be drawn that in the explosion hazard analysis of flat vapour clouds considerably higher overpressures than the small figures predicted by the acoustic models will have to be taken into account.

What are the main factors or parameters which we do have to consider in a real situation with regard to the possible level of overpressure? In Table 1 several parameters which are of significance for real situations are listed together with the possible effect on pressure buildup. Depending upon the particular parameters which have to be considered in a given situation the explosion pressure may be enhanced or reduced, as compared with the ideal situation. An example for pressure reduction and even failure of cloud ignition is supplied by the Maplin Sands experiments which have been reported at this symposium /12/.

In most cases of pressure enhancement turbulence plays a prominent role. A turbulent flow field may be present already before ignition due to atmospheric turbulence or to the conditions of release and of cloud formation. Additional turbulence will be included during the combustion process itself when the flow ahead of the flame front, which is generated by the expansion of the hot combustion products, interacts with obstacles. The effect of turbulence on flame propagation is twofold. The large-scale eddies will stretch and distort the flame leading to an increase in flame surface area. Fine-scale turbulence increases the local transport properties and hence promotes the mixing of the hot combustion products with the unburned mixture resulting in an increase of the local burning velocity.

Tab. 1: Parameters responsible for enhancement or attenuation of pressure buildup in non-ideal situations

Parameter	Consequence	Effect on pressure buildup
Weather conditions (wind, atmospheric stability)	Non-quiescent, turbulent mixture	Enhancement
Edge ignition Non-circular shape	Early relief of pressure zone	Attenuation
Multi-point ignition	Increased burning rate	Enhancement
Buoyancy of burnt gas	Early dispersion below LFL / Generation of turbulence and instabilities	Attenuation / Enhancement } ambiguous
Non-stoichiometric composition	Reduced reaction energy	Attenuation
Fluctuations in concentration (eddy patterns) resulting from release or cloud formation conditions	Trapped (fuel-rich) pockets which will be partially enclosed by the flame and rapidly burn behind the flame front	Enhancement
Dissemination of dust within the cloud	Heating of the unburned part of the cloud due to absorption of heat radiation, multi-point ignition	Enhancement
Terrain roughness	Generation of turbulence	Enhancement
Obstacles in the flame path (e.g... buildings, trees, installations), in particular repeated obstacles	Formation of large-scale eddies and trapped pockets, generation of turbulence, creation of preferred semi-confined flame .paths possibly leading to multi-point ignition	Enhancement
Local confinement by assembled structures or within buildings	Generation of turbulence by the expansion flow out of the partially confined volume, jet ignition of the mixture outside	Enhancement

Both large-scale and small-scale turbulence lead to an increase
in the overall burning rate, hence to flame acceleration and increased explosion pressure. The significance of turbulence is stressed
by the fact that in special situations, particularly in the case of
repeated obstacles, a positive feedback between the burning rate and
the turbulence generated ahead of the flame is effective.

        As an example for pressure buildup attenuation compared to the
ideal situation the reduction of overpressure for noncentral ignition
is shown in Fig. 6. (This is for a hemispherical volume; for a flat
cloud the effect of pressure reduction when moving the point of ignition towards the cloud edge is expected to be less pronounced.)

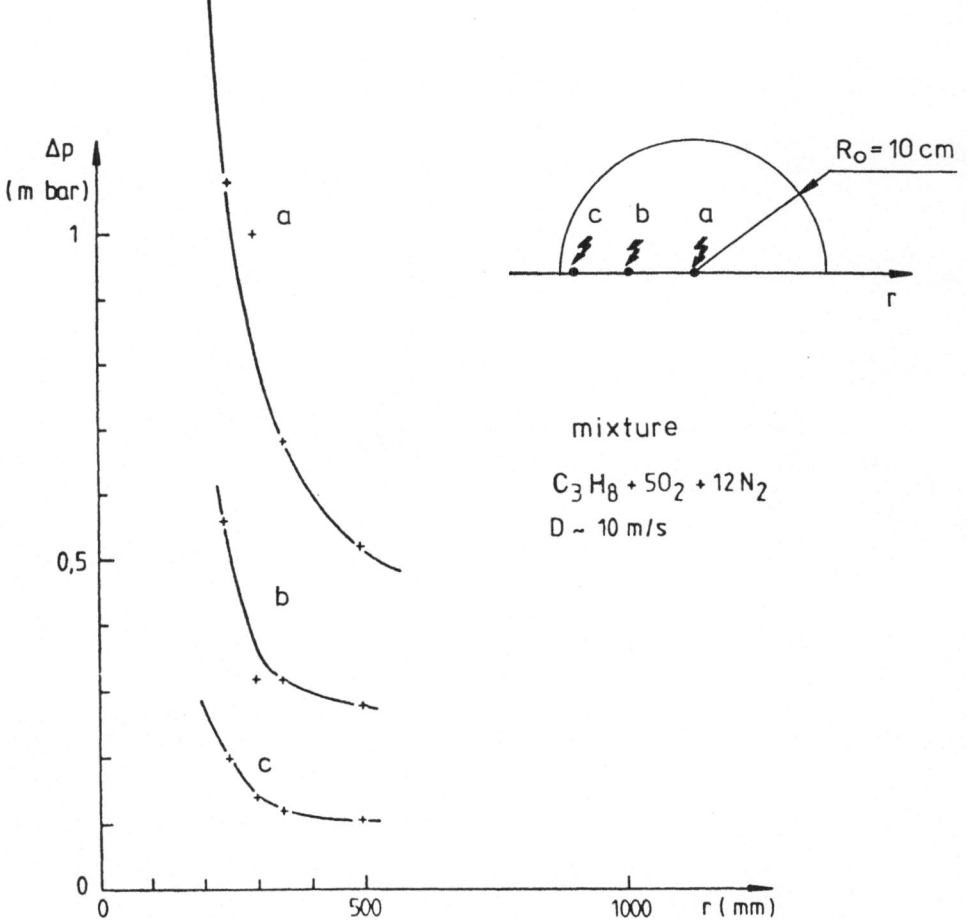

Fig. 6: Variation of peak overpressure with distance for different
        points of ignition (according to Charuel and Leyer /13/)

   An example of pressure enhancement will also be given here. The
expansion of the combustion products out of a partially confined volume
may occur in the form of free jets. A free jet acts as a powerful ignition source which, compared with the weak ignition case, triggers
more violent combustion in the surrounding unconfined part of the
cloud. Fig. 7 presents a striking example of this event: For jet
ignition, the peak overpressure at point P of the unconfined mixture
is larger by a factor of 200 than for spark ignition. The increase of
flame speed and explosion pressure due to jet ignition is, however,
limited to a region near the partial confinement, i.e. it is not expected to extend over the whole cloud.

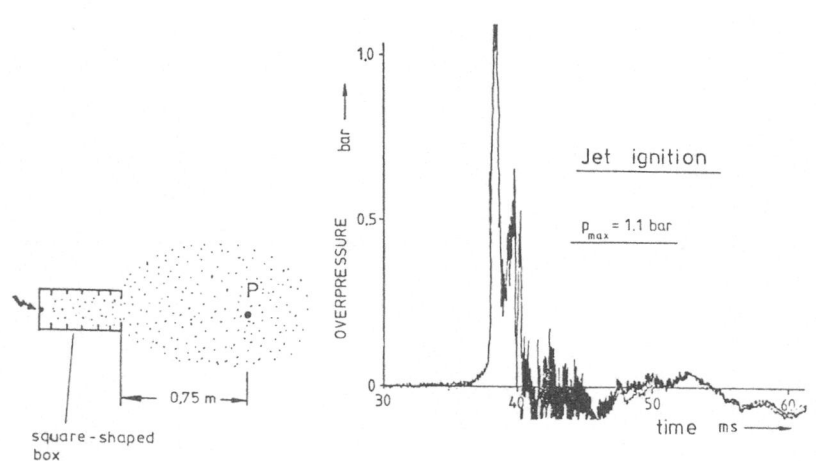

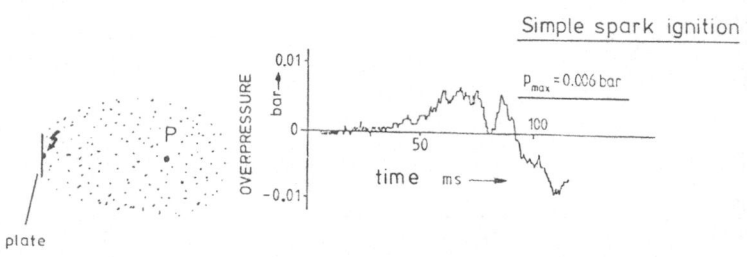

Fig. 7: Pressure signals at point P of an unconfined ethylene-air
        mixture for strong and weak ignition

5. CONCLUDING REMARKS

In view of the present knowledge of flame propagation and pressure generation in ideal and non-ideal situations we favour the following concept of the explosion properties of flat vapour clouds:

An ideal situation as defined in section 2 will never occur in reality. A spill onto calm sea /12/ represents a case which approximates an ideal situation. For this case the level of overpressure as derived from acoustic models may be a realistic prediction. Turbulence generators, such as trees, shrubs, installations, buildings, will always - to a smaller or larger extent - be present in the cloud, in addition to atmospheric turbulence. These turbulence generators will counteract the tendency in flat vapour clouds to reduce flame speed and explosion pressure due to the early relief of the pressure zone from above. For a global assessment of flat cloud explosion we believe that an overpressure value of 0.1 bar as adopted in the Battelle flat cloud model can be assumed to be a reasonable, possibly conservative, average figure for the overpressure within the cloud. The same argument is valid for the predicted decay of overpressure in the surrounding area outside the cloud.

As has been discussed, much larger overpressures than 0.1 bar can occur locally within the cloud, particularly in regions of partial confinement or repeated obstacles. In order to obtain a comprehensive picture, it will be necessary, supplementary to the global analysis, to scrutinize the potential area of flat cloud explosion with regard to the possible presence of configurations involving locally enhanced overpressure. The special case when such configurations are distributed rather densely over the area of flat cloud explosion must be treated differently; then not only local enhancement of pressure but also an increase of the average pressure level is to be expected.

REFERENCES

/1/   D.R. Stull, Fundamentals of Fire and Explosion, p. 72, American Institute of Chemical Engineers Monograph, No. 10, Vol. 73, 1977

/2/   W. Geiger und R. Synofzik, Explosion flacher Gaswolken: ein einfaches Modell zur Berechnung der Druckwelle, Chemische Rundschau Nr. 7/1981, Verlag Vogt-Schild AG, Solothurn und Zürich

/3/   W. Geiger and R. Synofzik, A Simple Model for the Explosion of Pancake-Shaped Vapour Clouds, 3rd Int. Symp. on Loss Prevention and Safety Promotion in the Process Industry, Basle, September 15-19, 1980

/4/   A.H. Wiedermann, T.V. Eichler and C.A. Kot, Air Blast Effects
        on Nuclear Power Plants from Vapour Cloud Explosions, 6th Int.
        Conf. on Structural Mechanics in Reactor Technology, Paris,
        17-21 August 1981

/5/   A.L. Kunl, M.M. Kamel and A.K. Oppenheim, Pressure Waves
        Generated by Steady Flames, University of California
        Scientific Report, Berkeley, 1972

/6/   W. Geiger and R. Synofzik, Influence of Dispersion Behavior of
        Dense Explosive Gases on the Possible Strentgth of Explosion,
        Proc. Symp. on Heavy Gas and Risk Assessment (Frankfurt,
        September 3-4, 1979), Reidel Publishing Comp., Dordrecht/Holland,
        1980

/7/   T.R. Auton and J.H. Pickles, The Calculation of Blast Waves
        from the Explosion of Pancake-Shaped Vapour Clouds, Note No.
        RD/L/N 210/78, Central Electricity Research Laboratories, UK

/8/   R.A. Strehlow, The Blast Wave from Deflagration Explosions, an
        Acoustic Approach, 13th Loss Prevention Symposium AIChE,
        Philadelphia, June 1980

/9/   J.C. Leyer, An Experimental Study of Pressure Field by Ex-
        ploding Cylindrical Clouds, Euromech Colloquium 139 on Un-
        controlled Blasts and Explosions in Industry, Aberystwyth, UK,
        12-15 April 1981

/10/   J.C. Leyer, Rapport d'Interprétation des Essais "Charles"
        (Déflagration Lente), Progress Report to the Commision of the
        European Communities for Contract 005 SRF, December 1981

/11/   D.S. Burgess and M.G. Zabetakis, Detonation of a Flammable
        Cloua Following a Propane Pipeline Break, (The December 9,
        1970, Explosion in Port Hudson, Mo.), Bureau of Mines Report
        of Investigations 7752, 1973

/12/   W.J.S. Hirst, Combustion Experiments with Large LNG and Propane
        Spills on Sea, Second Symposium on Heavy Gases and Risk
        Assessment, Frankfurt, May 25-26, 1982

/13/   P. Charuel and J.C. Leyer, Caractéristiques du Champ de Pression
        Engendré par une Flamme Accelerée en Expace Libre, Final
        Report to the Commission of the European Communities for Study
        Contract ECI 699-80-F, November 1981

AN ANALYSIS OF VAPOUR CLOUD EXPLOSIONS BASED ON ACCIDENTS

Ir. B.J. Wiekema
Department of Industrial Safety TNO
p.o. box 342
7300 AH Apeldoorn
The Netherlands

ABSTRACT

The accidental release of a combustible gas or liquid may result
in an explosive vapour cloud which upon ignition will form a threat to
the surrounding area. Models have been developed in the past in order
to quantify these effects, but the reliability and accuracy of such
models is nevertheless questioned. As research shows that this topic
is very complicated an alternative approach is worked out in this paper.
This approach is based on accidents that have taken place in the past.
The analysis leads to the derivation of trends under which the accidents
took place and to a comparison of the actual effects with a theoretical
vapour cloud explosion model.

INTRODUCTION

The release of a combustible gas or liquid in the open may result
in the formation of a vapour cloud. An ignition of the explosive part
of this vapour cloud leads to the creation of a flame front that will
propagate with a varying velocity through the entire explosive region of
the cloud. The flame speed as a function of time determines the overpressure level in the blast wave, that results from the vapour cloud
explosion. In fact, this flame speed (and the related flame front
surface) determines the rate of release of combustible energy as a
function of time and the blast wave, with a velocity almost equal to
the speed of sound, transports the energy in the surrounding area. As
the flame speed can vary over two orders of magnitude and the speed of
sound is almost constant for this case, it is easily understood that
the higher the flame speed, the higher the overpressure level in the
blast wave, and in principle, the more damage will be caused.

The flame speed as a function of time is the most useful variable,
known to occur in actual situations, for estimating possible damage. The

237

S. Hartwig (ed.), Heavy Gas and Risk Assessment - II, 237–248.

a priori estimation of the flame speed is very complicated if not
impossible. All models that have been developed for such a vapour
cloud explosion incorporate a way to by-pass this problem. This is
either done by defining a yield in so-called TNT-equivalence methods
or by defining flame speeds. All models have been checked with some
accidental explosions and therefore they all predict damage circles
in the same order of magnitude. However, preference should be given
to models using flame speeds as it is known that the detonation process in TNT takes place at unrealistically high speeds in comparison
with flame speeds in a vapour cloud. This has an important influence
on the pressure time profile in a blast wave.

As the determination of flame speeds for actual situations is
still under investigation and progress is only made slowly, an alternative approach has been adopted. This approach is based on the information that is available after accidents with vapour clouds have occurred.
A detailed analysis of damage patterns of several objects should lead
to quantification of the explosion process. As a first step in the
analysis of accidents that have happened in the past shows that quantification is difficult. The approach adopted in this study on the analysis
of accidents leads to two types of results, i.e. the derivation of
patterns in which accidents took place and the comparison of accidents
with a vapour cloud explosion model.

CASUISTRY OF VAPOUR CLOUDS

A number of publications and reports have surveyed accidents and
made descriptions of single accidents. Together with our own information, some accurate and some less accurate descriptions have been obtained of 165 vapour clouds that were ignited. Clouds that were not
ignited were excluded from the analysis. Figure 1 shows the distribution of discovered accidents during the period 1920 - 1980.

An analysis of accidents can be based only on the available information. Two boundaries therefore limit the value of such accidents,
namely incomplete descriptions of accidents and incomplete registra-
tion of accidents. The conclusion of such an analysis should therefore be read with those boundaries in mind.

For the purpose of the analysis a number of characteristic
features of ignited vapour clouds were selected. The choice was
based on the availability of data in accident reports, on the importance for vapour cloud explosion modelling and on aspects directly
related to safety studies and risk analyses. Each selected feature was
divided into categories in order to assess relationships. The available
reports, documents and descriptions were analysed in terms of the
chosen features in order to classify each accident accurately as
possible.

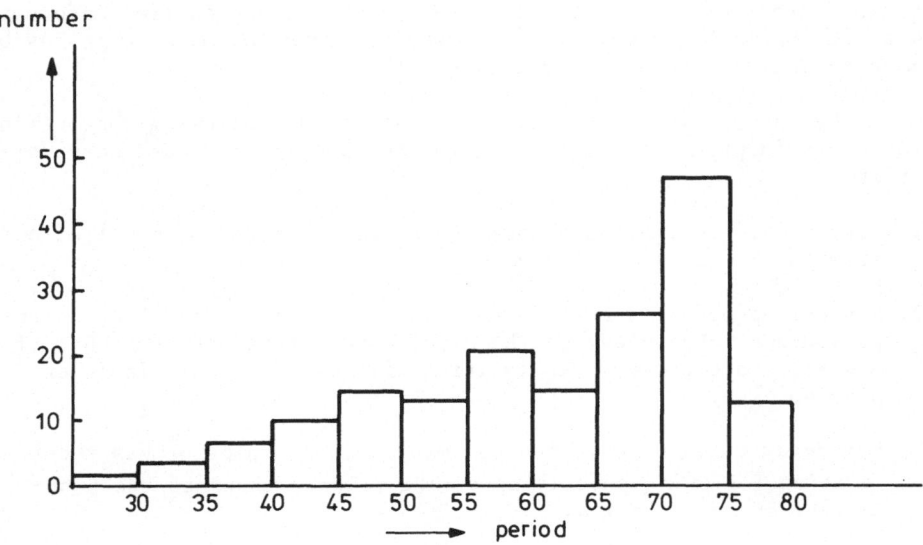

Figure 1. Number of registered ignited vapour clouds per
5-year period.

RELEVANT FEATURES OF VAPOUR CLOUDS

With regard to the possibility of damage to the surrounding area
one of the most important questions is: "What happens after ignition
of a combustible vapour cloud?" The resulting flame speed(s) can be
such that either a pressure wave, which is capable of causing damage
to the surrounding area, is created, or no pressure wave of any significance with respect to damage is created. Thefirst case is called
an explosion, the latter a flash fire.

Some characteristic features were selected to learn more about
factors that influence or determine the flame propagation behaviour.
These features were: the amount of mass released in the accident, the
distance that the cloud drifted before ignition, the time delay between release and ignition, the location (were semi-confinement and
obstacles present), the number of fatalities and the number of wounded
people.

Table 1 presents the relative percentages of occurrence of the different features in two groups, namely flash fires and explosions. Only gases that are considered to be medium reactive, like propane and butane, are included (1, 2). In cases in which the required information was not available, the feature was read as unknown. No attempt has been made to derive a result by combining other data.

The interpretation of table 1 leads to the following conclusions, which are in principle only valid for the population under consideration:

1. Nearly all the ignited vapour clouds were ignited within 1 kilometre from the location of the accident.

2. Explosions occurred only in semi-confined situations, i.e. in the presence of obstacles. On the other hand, it is not true that if obstacles are present, an ignition will always result in an explosion.

3. For delay times of more than 30 minutes, no explosion was recorded. Ignition following shortly after the release enhances the probability of an explosion.

4. In nearly half the accidents there were no fatalities. The number of fatalities is relatively higher for explosions than for flash fires.

5. The number of wounded people is relatively higher for explosions than for flash fires.

COMPARISON WITH A VAPOUR CLOUD EXPLOSION MODEL

Some of the vapour cloud explosions on which detailed information was available have been used to analyse the validity of a vapour cloud explosion model (1). This model, presented in figure 2, shows horizontally the distance R to the explosion centre in a dimensionless form R/L. L stands for the characteristic explosion length which is defined as the cube root of the quotient of the total available combustion energy (in Joules) in the cloud and the ambient pressure p_o (in N/m^2). The upper half of the graph shows vertically the overpressure Δp of the expected blast wave, whereas the lower half shows the duration t_r of the overpressure in the blast wave. In order to compare accidents, some damage criteria have to be formulated. The criteria used in this study were limited to the peak overpressure of the blast wave and they read as follows:
- $\Delta p = 3.10^4$ N/m^2: extensive structural damage
- $\Delta p = 10^4$ N/m^2: boundary of structural damage to houses
- $\Delta p = 3.10^3$ N/m^2: extensive window pane damage
- $\Delta p = 10^3$ N/m^2: boundary of window pane damage

		Explosion	Flash fire
Mass (kg)	$< 10^2$	3%	0%
	$10^2 - 10^3$	14%	15%
	$10^3 - 10^4$	43%	38%
	$10^4 - 10^5$	31%	27%
	$> 10^5$	9%	19%
Drift (m)	$< 10^2$	67%	73%
	$10^2 - 10^3$	33%	25%
	$> 10^3$	0%	2%
Location	semi-confined	100%	62%
	unconfined	0%	38%
Delay (min)	$<$ 1	25%	14%
	1 $-$ 5	35%	48%
	6 $-$ 15	25%	14%
	16 $-$ 30	15%	0%
	$>$ 30	0%	24%
Fatalities	0	39%	50%
	1 $-$ 5	33%	36%
	6 $-$ 15	17%	10%
	16 $-$ 50	9%	4%
	$>$ 50	2%	0%
Wounded	0	14%	40%
	1 $-$ 5	22%	40%
	6 $-$ 15	17%	10%
	16 $-$ 50	26%	8%
	$>$ 50	20%	2%

Table 1. Relative percentages of occurrences of important
features with respect to vapour cloud explosions
and flash fires.

Also peak overpressures that were explicitly mentioned in the reports
in relation to damage were used. The following vapour cloud explosions
have been used in the presented comparison:

1. FRG (1943). A railroad tankcar with 16,5 to $19,5.10^3$ kg of butadieen
   exploded. The resulting cloud was ignited within 10 to 25 seconds and
   caused damage to the surroundings [3]. A detailed damage analysis is
   presented in [3] and resulted, together with accidents 2 and 9, in
   a peak overpressure versus distance relation outside the vapour cloud
   which reads as follows in the dimensionless variables used in this
   paper:

$$\frac{\Delta p}{p_0} = 0,2 \times \frac{L}{R}$$

2. Ludwigshafen, FRG (1948). A railroad tankcar containing 30.10^3 kg
   of dimethylether suddenly exploded. The vapour cloud ignited shortly
   thereafter. The overpressure distance relation derived was similar
   to accident 1 and 9 [3].

3. Portland, USA (1954). A vapour cloud explosion resulting from an
   LPG release caused damage to storage tanks in or very close to the
   cloud. The overpressure at the tanks was estimated to be 2,4 to
   $4,8.10^4$ N/m^2 [4].

4. Freeport, USA (1961). About 23 m^3 of cyclohexane was released. The
   subsequent explosion caused damage to a control building that was
   close to the vapour cloud. The overpressure was estimated to be
   $1,4.10^4$ N/m^2 [5, 6, 7].

5. Raunheim, FRG (1966). An explosion of 500 kg methane/ethane caused
   extensive glass damage up to a distance of 400 metres minor damage
   up to 1.200 m   [5, 7]. The characteristic explosion length for
   this accident is 63 metres, based on 500 kg combustible gas.

6. Pernis, NL (1968). A vapour cloud of about 50 to 100 tons of higher
   hydrocarbons exploded 13 minutes after the release. The explosion
   caused extensive damage up to a distance of 170 metres, damage to
   buildings and installations up to 1.000 metres and window pane damage
   up to 7 kilometres [8]. With an average combustion energy of
   4.10^7 J/kg the characteristic explosion length is estimated to be
   270 to 340 metres.

7. Port Hudson, USA (1970). About 24 minutes after the rupture of an
   LPG pipeline an explosion took place [8, 9]. Calculations
   based on [3] indicate that with the appropriate weather conditions
   and terrain about 9.600 kg of propane was within the explosive region
   at the time of the explosion. This determines the characteristic
   explosion length to be 164 metres. Table 6 of [8] relates the observed
   damage to different amounts of TNT. These data were transferred to
   the variables used in this study. Further, table 7 of (9) states

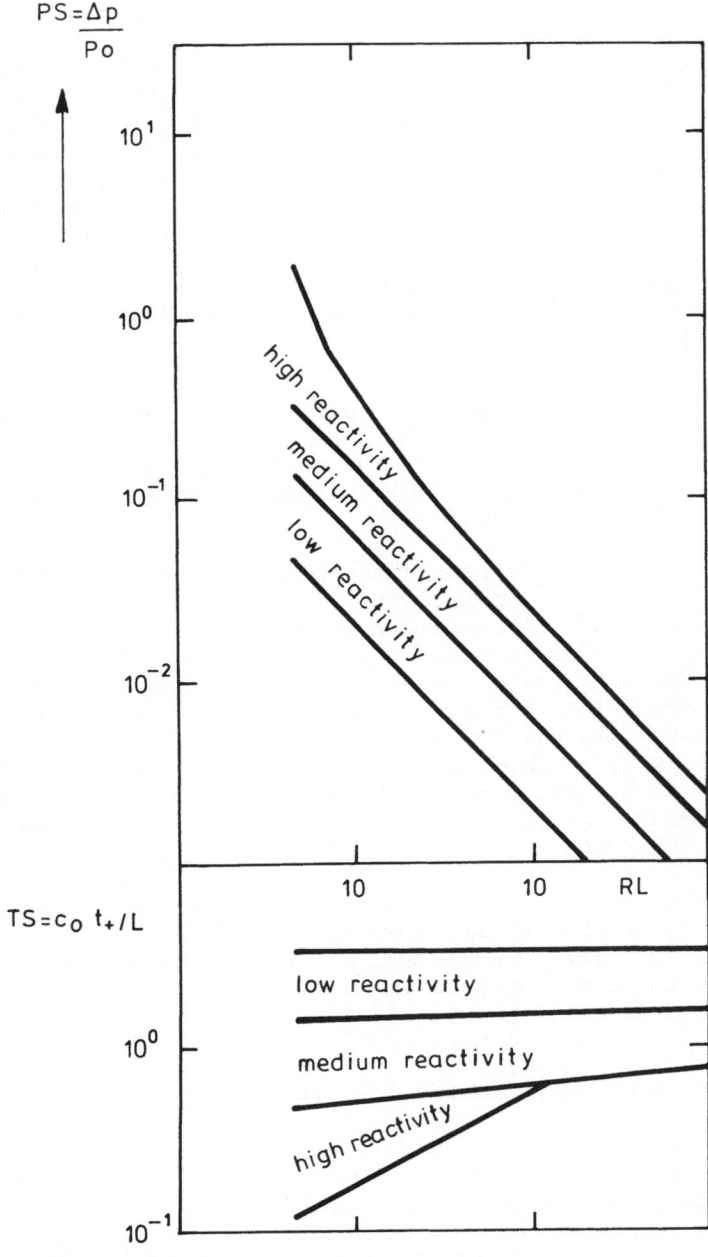

Figure 2. Vapour cloud explosion model.

that extensive window pane damage occurred, up to 1 mile from the
explosion centre and the boundary for window pane damage was found
to be at 2 1/2 miles.

8. East St. Louis, USA (1972). A railroad tankcar carrying propylene
   was punctured. As the tankcar was moving the release took place over
   some distance. The vapour cloud exploded after 8 to 10 minutes [11,
   12]. From the accident data it is deduced that the cloud had a
   length/width ratio of 10 with a total area of 2.10^4 m^2. It is also
   mentioned that the cloud was of low height. This leads to an estimation of the characteristic explosion length of about 100 to 160
   metres. The damage pattern shows a circular but not concentric
   trend. The centre of the circles is diverted in the wind direction
   whereas the spill took place over some distance perpendicular to the
   wind direction.

9. Flixborough, UK (1974). About 50 to 60.10^3 kg of cyclohexane was
   released, creating a cloud of spherical shape. The overpressure
   distance relation was found to be the same as for accidents 1 and
   2 [7]. It is also interesting to note that the damage patterns were
   shifted in the wind direction[13].

10. Decatur, USA (1974). A railroad tankcar was punctured causing the
    release of about 69.10^3 kg of butane. The explosion took place after
    8 to 10 minutes [14]. If the source is considered to be instantaneous
    about 70% of the released butane could maximally be within the ex-
    plosion limits [10]. Because the cloud was of low height and covered
    an area of 800x1200 m^2 a similar value for the amount in the explosive
    region is found. The characteristic explosion length in this accident
    is then deduced to be about 280 metres. The damage data show that the
    boundary for structural damage to houses is 750 metres from the
    explosion centre. The damage circles are in this case also shifted
    in the wind direction.

The data derived from the 10 vapour cloud explosions are compared
with the values predicted by the vapour cloud explosion model in figure
3. It is shown that, given the assumption made, the predicted effects
for a hemispherical cloud are in agreement with the real effects of
vapour cloud explosions.

INFLUENCE OF THE WIND DIRECTION AND WIND VELOCITY

The damage boundaries as found in accidents 8, 9 and 10 show
that their contours are shifted in the wind direction. This effect is,
in number, more pronounced for the longer distances, i.e. minor damage,
as is to be expected on the basis of a well-known theory[15]. From the
descriptions of these accidents the relative effect for several damage

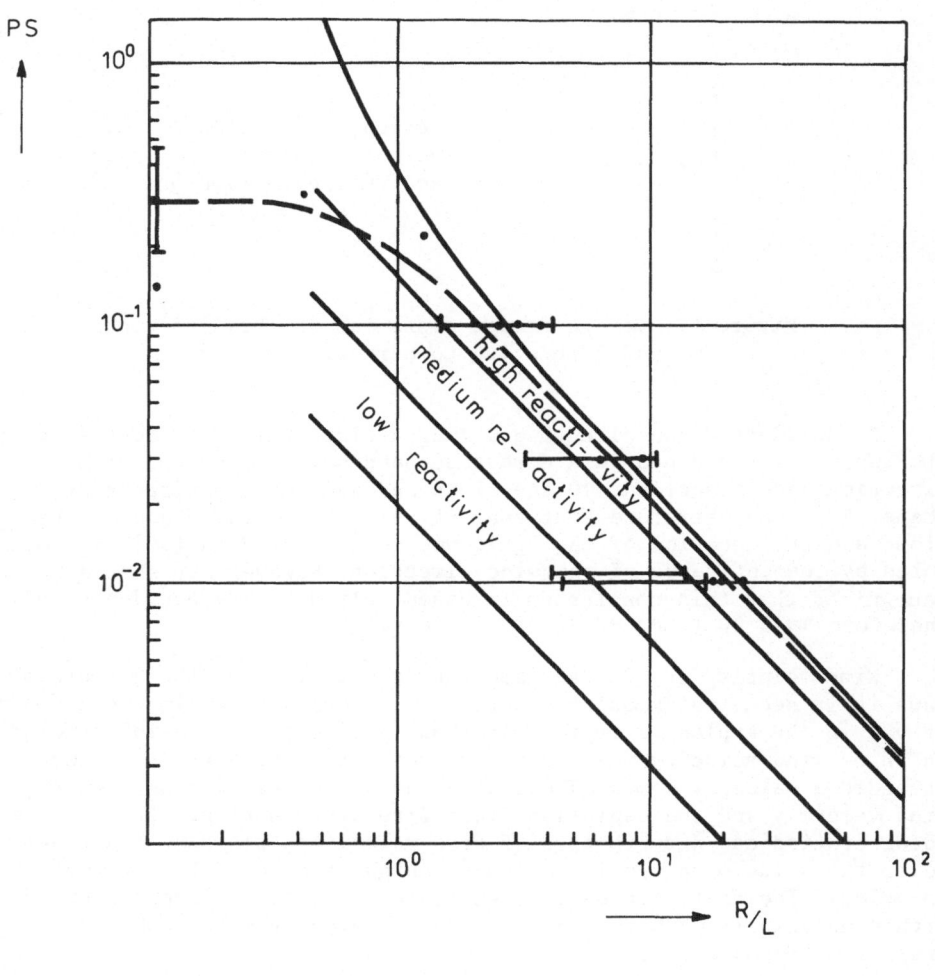

Figure 3. Comparison of accidents with the vapour cloud explosion model.

levels is deduced (Figure 4).

wind direction

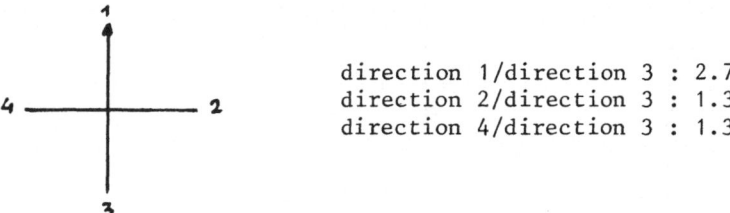

direction 1/direction 3 : 2.7
direction 2/direction 3 : 1.3
direction 4/direction 3 : 1.3

Figure 4. Average damage distances in the directions 1, 2
and 4 relative to direction 3.

In accident 8 the cloud had a length/width ratio of about 10 with
the length of the cloud in a direction perpendicular to the wind
direction and in accident 10 the cloud had an almost hemispherical
shape. It seems therefore that the effects of the real shape of the
cloud and the location of the ignition, as estimated in [16], are over-
ruled by the influence of the wind direction. However, it should be
recognized that this conclusion is based only on three accidents and
therefore must be regarded as preliminary.

Wind velocity is also an important variable. From the dispersion
laws it is seen that amongst others wind velocity determines the amount
of gas in the explosive region. In those accidents of the 165 studied
in which wind velocity was mentioned, the actual wind velocity never
exceeded a value of 6 m/s. There was also no correlation between the
wind velocity and the explosion/flash fire distribution. Although the
local statistical distribution of wind velocities plays an important
role, the value of 6 m/s is not such a high wind velocity for most
locations. The fact that neither explosions nor flash fires occurred
with wind velocities higher than 6 m/s is relevant with respect to
dispersion methodology.

EVALUATION

Several models have been developed in order to quantify the effects
of a vapour cloud explosion. As research shows that the topic is very
complicated, an alternative approach based on accidents has been
adopted. It was the purpose of this study to determine whether such an
approach could lead to relevant results.

This alternative approach covered an analysis of 165 accidents
involving an ignition of a vapour cloud and an analysis of 10 vapour

cloud explosions. The analysis of vapour clouds that were ignited showed
that for those accidents it was a necessary condition that obstacles or
other forms of semi-confinement were present within the vapour cloud in
order to generate an explosion. However, the reverse is not true because
in the absence of **obstacles** or semi-confinement explosions were not recorded. But flash fires have been observed.

It has also been shown that the real shape of the cloud is probably not an important factor with respect to the damage caused by vapour
cloud explosions. In general the same seems to be true for the location
of the ignition source.

The vapour cloud explosion model, based on a hemispherical cloud
seems to predict reasonably the explosion effects of clouds of limited
height and width, given the available accuracy of damage patterns and
criteria.

This all leads to the following description of a vapour cloud
explosion. When a vapour cloud encounters an ignition source a flame
front is generated that propagates through the combustible part of the
cloud. If obstacles or semi-confinements are present within the cloud,
the flame front velocity can be enhanced significantly at those locations.
The subsequent explosive behaviour(s) will have a nearly spherical
characteristic(s). Most of the combustible material will contribute to
the blast wave.

REFERENCES

1. Wiekema, B.J. *Vapour cloud explosion model*. Journal of Hazardous
   Materials 3 (1980), pp 221-232.
2. Zeeuwen, J.P., and Wiekema, B.J. *The measurement of relative
   reactivities of combustible gases*. Conference of mechanisms of
   Explosions in Dispersed Energetic Materials. Dover, USA (1978).
3. Giesbrecht, H. et al. *Analyse des potentiellen Explosionswirkung
   von kurzzeitig in die Atmosphäre freigesetzte Brenngasmengen*.
   Teil I: Chemie Ingenieur Technik 52 (1980), pp 114-122.
   Teil II: Chemie Ingenieur Technik 53 (1981), pp 1-10.
4. *Railroad- and ship fires*. NFPA Quarterly 48 (1955) 3.
5. Gugan, K. *Unconfined vapour cloud explosions*. Institution of
   Chemical Engineers in Association with George Godwin Ltd, Rugby,
   UK (1978).
6. Lewis, D.J. *Unconfined vapour cloud explosions, historical perspective and predictive method based on incident records*. Progress
   in Energy and Combustion Science, 6 (1980).
7. Davenport, J.A. *A study of vapour cloud incidents*. 83rd. National
   Meeting of the AIChE, Houston (1977).
8. Burgess, D.S. et al. *Detonation of a flammable cloud following a
   propane pipeline break*. Report of Investigations 7752. US Bureau
   of Mines, Pittsburgh (1973).

9. Eisenberg, N.H. et al. *Vulnerability Model. Appendix F.* Report
   CG-D-136-75. US Coast Guard Washington (1975).
10. *Methods for the calculation of the physical effects of the escape
    of dangerous material (liquids and gases).* Report Directorate-
    General of Labour, Voorburg, The Netherlands (1979).
11. *Hazardous materials railroad accident in the Alton and Southern
    Gateway Yard in East St. Louis, Illinois, January 22, 1972.*
    RAR-73-1, NTSB, Washington (1973).
12. Strehlow, R.A. *Equivalent explosive yield of the explosion in the
    Alton and Southern Gateway Yard in East St. Louis, January 22, 1972.*
    Report AAE-TR-73-3. University of Illinois (1973).
13. Marshall, V.C. *The Decatur Anomaly.* The Chemical Engineer, February
    1980.
14. *Hazardous materials accident in the railroad yard of the Norfolk
    and Western Railway at Decatur, Illinois, July 19, 1974.* RAR-75-4,
    NTSB, Washington (1975).
15. Cox, E.F. et al. *Meteorology directs where blast will strike.*
    Bulletin of the American Meteorological Society, vol. 35, 3 (1954).
16. Strehlow, R.A. *The blast wave from deflagrative explosions, an
    acoustic approach.* 13th Loss Prevention Symposium AIChE, Philadelphia
    1980.

MODELING OF FIRE FOR RISK ASSESSMENT IN PETROCHEMICAL INDUSTRIES

J. Hofmann
Battelle-Institut e.V.
Am Römerhof 35
D 6ooo Frankfurt/Main

INTRODUCTION

The assessment of potential hazards from chemical plants plays a relevant role for the establishment of safety measures to protect the plant personel and the public from the consequences of an accidental release of toxic or inflammable materials. Fire is a major hazard in process plants. Hence risk assessment has to focus on the probabilities and the consequences of large fires in chemical and petrochemical installations.

The objective of this paper is to review the present status of fire modeling. It will mainly focus on large scale pool fires.

Special emphasis is given to analytical models which allow fast calculations for practical purposes, i.e. safety analyses in the chemical industries.

The effect of a flame on an observer is determined by various factors
- processes in the flame (fuel burning rate, surface emissive power, flame geometry)
- external influences (e.g. wind resulting into flame tilt, flame drag and a change of the flame length or the status of the atmosphere between flame and observer)
- status of the observer (orientation with respect to the flame, absorptivity etc.).

Most of these topics are discussed briefly below. Other review discussions can be found in the work of Hall (1973) on pool burning, de Ris (1978) on fire radiation and Brötz and Schönbucher (1978a) on transport phenomena in pool flames. Lees (1980) gives an excellent discussion of all hazard assessment aspects including a general presentation of fire hazards. The most recent review on thermal radiation hazards from large pool fires has been presented recently

S. Hartwig (ed.), Heavy Gas and Risk Assessment - II, 249–260.
Copyright © 1983 by Battelle-Institut e.V., Frankfurt am Main, Germany.

by Moorhouse and Pritchard (1982).

POOL FIRES

     Flames are generally grouped according to their fluid dynamic
behaviour into

- laminar flames
- turbulent flames

or according to the status of the combustion system into

- premixed flames
- diffusion flames.

     In pool fire modelling we are faced with diffusion flames, which
may be turbulent but more generally are buoyancy-dominated.

FLAME LENGTH

     Correlations for the flame length of a buoyant diffusion flame have
been given by various authors. The equation used most frequently in
hazard assessment is that of Thomas (1963)

$$\frac{L}{d_o} = 42 \ (\frac{m''}{\rho_a \sqrt{g d_o}})^{0.61}$$

(1)

which was derived from experiments with wooden crib fires under calm
conditions. For wind blown flames Thomas (1965) gives

$$\frac{L}{d_o} = 55 \ (\frac{m''}{\rho_a \sqrt{g d_o}})^{0.67} \ (u^*)^{-0.21}$$

(2)

with $u^* = u/u_c$ (for $u < u_c$: $u^* = 1$). The range of applicability of
eq. (1) and (2) was given by Thomas as $3 < L/d_o < 10$. Putnam and Speich
(1961) converted their data on large city gas flames into

$$\frac{L}{d_o} = 29 \ (\frac{m''}{\rho_a \sqrt{g d_o}})$$

(3)

which applies to $100 < L/d_o < 200$.

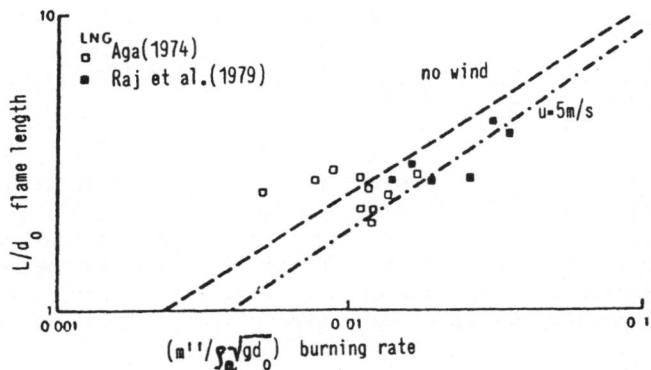

Fig. 1: Dimensionless flame length as function of dimensionless
burning rate. Experimental data on LNG spills on land (AGA)
and on water (Raj). Curves are Thomas' eq. (1) and (2)
without wind and u = 5 m/s (Moorhouse and Pritchard, 1982)

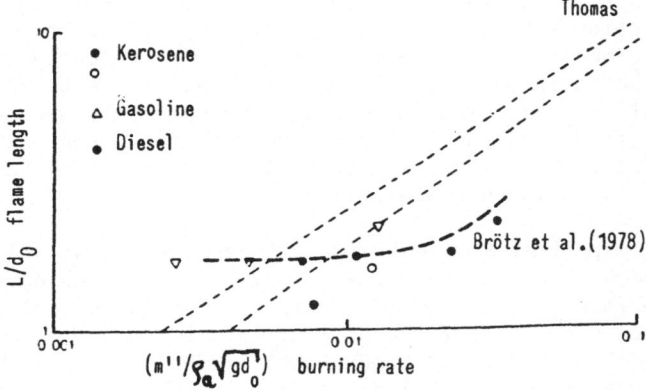

Fig. 2: Dimensionless flame length as function of dimensionless
burning rate. Experimental data are on kerosine, gasoline
and diesel. Upper curves are Thomas' eq. (1) and (2)
(with and without wind). Lower curve is according to
Brötz and Schönbucher (1978 a/b) (Moorhouse and Pritchard,
1982)

Fig. 1 shows LNG data on land (AGA, 1974) and on water (Raj et al., 1979) compared to Thomas' predictions (without wind and at u = 5 m/s) as given by Moorhouse and Pritchard (1982). Other LNG data from the AGA project showed the expected increase of the flame length with increasing pool diameter, but a least square fit to the data gave (AGA, 1974)

$$\frac{L}{d_o} = (\frac{m''}{\rho_a \sqrt{g d_o}})^{-0.19} (u^*)^{0.06} \tag{4}$$

which reflects the rather unphysical result of a decreasing flame length with increasing mass burning rate. In addition eq. (4) shows that as the wind speed increases so does the flame length in contrast to Thomas' equation. In recent experiments with LNG pit fires by Moorhouse (1982) the visible flame envelope was described as a cylinder or a cone. The resulting flame lengths were correlated numerically yielding for the cylindrical flame

$$\frac{L}{d_e} = 6.2 (\frac{m''}{\rho_a \sqrt{g d_o}})^{0.254} (u^*_{10})^{-0.044} \tag{5}$$

where d_e is the equivalent radius of a circular pool and the index 10 refers to the wind speed at 10 m height.

Eq. (5) shows the same physical behaviour as Thomas' correlation. It has been fitted to data in the range of $1 < L/d_e < 3$.

From that we conclude that the general behaviour can be described by a correlation like eq. (2), i.e. with proportionality to the mass burning rate and inverse dependence on the wind velocity. The actual numerical factor in eq. (2) depends on the substance burned and on the chemical reaction kinetics whereas the exponents depend on the size of the fires and the status of the wind field and the degree of turbulence in the flame. A more unified treatment of basic principles may be found in Sunavala's (1967) study.

Recent experiments with kerosene, gasoline and diesel show rather different behaviour (see Moorhouse et al., 1982) which deviates considerably from Thomas' equation (see fig. 2). For large pool diameters L/d_p tends to a constant value of 1.7. This behaviour is consistent with the model of Brötz and Schönbucher (1978b) which assumes that the turbulent transport in the flame can be described as a diffusion process, i.e. a Gaussian profile. The predictions of both approaches are shown in fig. 2.

The knownledge of the mass burning rate m'' is very crucial for the determination of the flame length. m'' is usually expressed as $m'' = v \cdot \rho$ with the burning velocity v, which is usually measured.

An equation for m'' has been given by Burgess and Hertzberg (1978):

$$m'' = C_1 \frac{H_c}{H_v} \cdot 10^{-3} \ (kg/m^2/s) \tag{6}$$

with $C_1 = \dfrac{H_v}{C_p(T_b - T_a) + H_v}$ $(T_b \geq T_a)$

$C_1 = 1$ $(T_b \leq T_a)$

which results into $m'' = 0.12 \ (0.10) \ kg/m^2/s$ for LPG (LNG). These values agree reasonably well with recent experimental data by Mizner and Eyre (1982) ($m''_{exp} = 0.13 \ (0.106) \ kg/m^2/s$). It is well known that the burning rate depends on the pool size and the wind velocity but the actual dependence is known only in narrow ranges. Whereas eq. (6) fits the LNG, LPG and methanol data reasonably well, measured burning rates for e.g. gasoline, hexane and kerosene lie by a factor of 3 or even more below the calculated value (Moorhouse and Pritchard, 1982), which may change the dimensionless flame length by a factor of 2.

FLAME TILT

Wind blown fires are modified in two ways; the influence of wind on the flame length is described by equations like (2); additionally the wind tilts the flame from the horizontal by an angle ϑ. Based on AGA (1974) tests Atallah et al. derived

$$\cos \vartheta = 1 \quad , \quad u_v^* < 1$$

$$\cos \vartheta = \frac{1}{\sqrt{u_v^*}} \quad , \quad u_v^* > 1 \tag{7}$$

where in $u_v^* \rho_a$ has to be replaced by ρ_v, the vapour density. Another equation given by Welker and Sliepcevich (1966), which correlates the AGA data equally well, is

$$\frac{\tan \vartheta}{\cos \vartheta} = 3.2 \ (Re)^{0.07} \ (\frac{\rho_v}{\rho_a})^{-0.6} \ (Fr)^{0.8} \tag{8}$$

where the Froude and Reynolds numbers are

$$Fr = \frac{u^2}{g\,d_o} \qquad , \quad Re = \frac{d_o\,u}{\nu} \qquad (9)$$

Experimental data on large LNG and LPG fires (Mizner and Eyre, 1982) show excellent agreement with both correlations. Moorhouse (1982) fitted large LNG pool data by a cylindrical flame approach and obtained

$$\cos\vartheta = 0.86\ (u_v^*\ (10\ m))^{-0.25} \qquad (10)$$

or

$$\frac{\tan\vartheta}{\cos\vartheta} = 1.9\ (Re_{10})^{0.05}\ (Fr_{10})^{0.399} \qquad (11)$$

(the index 10 relates to the wind velocity in 10 m height). The fit obtained has no reasonable advantage compared to eq. (8) or (9), i.e. those equations represent the flame tilt to a sufficient accuracy for practical calculations.

FLAME DRAG

The phenomenon that the flame base of a pool fire is elongated in the wind direction has been discussed by Welker and Sliepcevich (1966). They give:

$$\frac{d_o'}{d_o} = 2.1\ (\frac{\rho_v}{\rho_a})^{0.48}\ (Fr)^{0.21} \qquad (12)$$

with d_o', the elongated flame base. This elongation may have some influence on the heat radiation effects which, up to now, has not been studied.

HEAT RADIATION

To calculate the consequences of heat radiation on an observer two simplifying approaches have been used which describe the flame as point source or as a solid flame with a regular shape, e.g. a cylinder or a cone. The point source model may be used when the object is sufficiently far away from the flame (more than five pool diameters).

The heat radiated to the environment Q is

$$Q = f \; \frac{Q_c}{4 \pi r^2} \qquad\qquad (13)$$

where Q_c is the heat developed by the fire. f gives the fraction of the heat produced which is radiated from the flame surface. $Q_c = m'H_c$ and f can be calculated from the fact that the total heat radiated from the pool surface, i.e. $fm''H \frac{\pi}{4} d_o^2$ equals the heat radiated from the cylindrical flame surface. i.e. $(\frac{\pi}{4} d_o^2 + \pi d_o L)P$ (top + mantle of cylinder) and therefore

$$f = \frac{P}{m''H_c} \quad (1 + 4 \frac{L}{d_o}). \qquad\qquad (14)$$

In most applications values for f in the range of $0.2 < f < 0.4$ are used. In the range $d_o > 10$ m a value of f = 0.15 is appropriate (TNO, 1979).

The point source approach assumes that the target surface is always receiving the maximal possible heat flux. This assumption is too conservative for practical purposes.

To overcome the shortcomings of the point source model a solid extended flame of cylindrical or conical form is assumed. In this case the radiation incident on the target is given by

$$Q = \tau \; FP \qquad\qquad (15)$$

The surface emissive power P may be written as

$$P = \epsilon \sigma \; T_f^4 \qquad\qquad (16)$$

The emissivity of the flame is $\epsilon \approx 1$ for pool fires exceeding $d_o \sim 1$ m (Raj and Kalelkar, 1974). The average flame temperature of small scale hydrocarbon fires (kerosene, petrol) has been determined by Rasbash et al. (1956) to be in the range of 1200-1300 K.

Usually in hazard assessment the emissive power is used instead of flame temperatures. Experimental data give (Mizner and Eyre, 1982)

LNG	150-200 KW/m^2
LPG	50- 60 KW/m^2
Kerosine	35 KW/m^2

The decrease of the emissive power signals an increase in the soot content of the flames (Mizner and Eyre, 1982) according to the increase in molecular weight.

Generally the emissive power or the flame temperature profile can be calculated from basic principles (e.g. the Navier-Stokes equation and the chemical reaction kinetics) (see Cox and Chitty, 1980). These calculations are usually very computer-time-intensive and not suitable for practical purposes. Hence it seems appropriate to use rough estimates from the point source model e.g. calculate P from eq. (14) with the assumption $f \sim 0.15$ (valid for $d_0 \geqslant 10$ m).

The geometrical view factor F accounts for the orientation of the radiation target element with respect to the more-or-less idealized flame (cylinder or cone with constant average temperature). It can be calculated using standard techniques from

$$ F = \int_{A_2} \frac{\cos\phi_1 \, \cos\phi_2}{\pi r^2} \, dA_2 \tag{17} $$

where the integral over the flame surface can be performed only for simple geometrical flame-target arrangements (see eg. VDI-Wärmeatlas, 1977). For practical purposes it is more convenient to subdivide the flame into a large number of small surface elements and perform the integral as a sum of the flame segments.

The atmospheric transmission coefficient τ takes into account the absorption of the flame infrared radiation by infrared absorbing tracer gases in the atmosphere, i.e. water vapour and carbon dioxide. Accurate calculations are not possible because emission spectra of large pool fires in the region of the H_2O and CO_2 bands are not rigourosly known. Raj (1977) assumes a blackbody emission spectrum (for LNG at 1150 K) and calculates τ at T_a = 27 °C and 50 % relative humidity. The calculates values of τ range from τ = 0.75 at 30 m down to τ = 0.67 at 110 m. Hence in the range of interest for fire hazard assessment a value of τ = 0.7 may be used without loss of accuracy. To obtain the heat absorbed by a target Q (eq. (15)) has to be multiplied by the absorptivity of the target. Most hazard assessment models (e.g. Eisenberg et al., 1975) use α = 1, i.e. all incident radiation is absorbed. This may be a very conservative estimate. Actual values for e.g. painted steel surface are about $\alpha \sim 0.9$, i.e. the use of realistic values may avoid unnecessary conservatisms.

CONCLUSIONS AND RECOMMENDATIONS

The previous sections summarize the methods most frequently used for pool fire hazard assessment. It is obvious, that at every step there is a number of uncertainties resulting from either the lack of suitable modeling according to physical and chemical principles or the lack of experimental data in the required scale size.

Most theoretical investigations rely on small scale experiments. Only recently data on moderatly large size pool fires became available (Mizner and Eyre, 1982). Dimensional analysis results into correlations for flame length and flame tilt depending on Reynolds and Froude numbers. Exponents were adjusted for various pool sizes. Obviously different functional dependences resulted for different scales (eq. (1) and (3)) and substances. Hence the general concept may be rather doubtful. An alternative more promising approach may be based on the diffusion concept (Brötz and Schönbucher, 1978).

The final calculation of heat radiation effects on an observer is usually performed under extremely conservative assumptions (τ, α, ϵ are frequently set unity). Additionally the estimate of the average flame temperature or the emissive power may lead to an uncertainty of a factor of 2. To avoid those shortcomings an experimental clarification of the role of the parameters entering the calculation is required.

Modeling efforts should utilize sophisticated computer codes based on first principles to derive practical correlations which can be used by safety engineers without extensive numerical calculations, i.e. what we need in the near future to overcome conservatisms and overestimates is the bridge guided by experimental data from basic fire research to applied engineering.

REFERENCES

AGA, (1974), Consequences of LNG Spills on Land, LNG Safety Program, Project IS-3-1, Battelle Columbus Laboratories

Brötz, W.; Schönbucher, A.; (1978a); Wärme- und Stofftransport in Tankflammen, Chem. Ing. Tech. 50, 573-585

Brötz, W.; Schönbucher, A.; (1978b); Der konvektive Stofftransport als geschwindigkeitsbestimmender Schritt in Tankflammen und seine Analogie zur Ausbreitung luftfremder Stoffe in der Atmosphäre, Bericht Bunsengesellschaft Phys. Chem. 82, 1202-1217

Burgess, D.; Hertzberg, M.; (1978); Radiation from Pool Flames, Chap. 27 from "Heat Transfer in Flames" ed. by N.H. Afgan and J.M. Beèr

Cox, G.; Chitty, R.; (1980); A Study of the Deterministic Properties of Unbounded Fire Plumes, Combustion and Flame, 39, 191

de Ris, J.; (1978); Fire Radiation - A Review, 17th Symp. on Combustion, Univ. Leeds, England

Eisenberg, N.A.; Lynch, C.J.; Breeding, R.J.; (1975); Vulnerability model: A Simulation System for Assessing Damage Resulting from Marine Spills, US Coast Guard Report CG-D-136-75

Hall, A.R.; (1973); Pool Burning, Oxidation and Combustion Reviews 6 169-225

Lees, F.P.; (1980); Loss Prevention in the Process Industries, Vol. I+II, Butterworths, London

Mizner, G.A.; Eyre, J.A.; (1982); Large Scale LNG and LPG Pool Fires, I.Chem. E. Symposium Series No. 71, 147, Rugby, England

Moorhouse, J.; (1982); Scaling Criteria for Pool Fires Derived from Large Scale Experiments, I.Chem. E. Symposium Series No. 71, 165, Rugby, England

Moorhouse, J.; Pritchard, M.J.; (1982); Thermal Radiation Hazards from Large Pool Fires and Fireballs - A Literature Review, I.Chem. E. Symposium Series No. 71

Putnam, A.A.; Speich, C.P.; (1961); A Model Study of the Interacting Effects on Mass Fires, Battelle Memorial Institute, Columbus, Ohio, Research Report

Raj, P.P.K.; (1977); Calculation of the Thermal Radiation Hazards from LNG Fires, AGA Transmission Conference, St. Louis, Missouri, USA

Raj, P.P.K.; Kalelkar, A.S.; (1974); Assessment Models in Support of the Hazard Assessment Handbook, Arthur D. Little Inc., Cambridge, Mass., US Coast Guard Report CG-D-65-74

Raj, P.P.K.; Moussa, A.M.; Aravamudan, K.; (1979); US Coast Guard Report CG-D-55-79

Rasbash, D.J.; Rogowski, Z.W.; Stark, G.W.P.; (1956); Fuel 35 (1), 94

Sunavala, P.D.; (1967); Dynamics of the Buoyant Diffusion Flame, Journal of the Inst. of Fuel, 40, 492

Thomas, P.H.; (1963); The size of Flames from Natural Fires, 9th Int. Symp. on Combustion, Academic Press Inc., London, p. 844-859

Thomas, P.H.; (1965); Fire Research Note No. 600, Fire Research Station, Borehamwood, England

TNO; (1979); Methoden voor het berekenen van de fysische effekten van het incidental vrijkomen van gevaarlijke stoffen, uitgave van het Directoraat Generaal van de Arbeid, Voorburg, Nederland

VDI-Wärmeatlas (1977); Enlarged Version of the 1974 Edition, VDI-Verlag

Welker, J.R.; Sliepcevich, C.M.; (1966); Fire Technology 2, 127

Symbols Used

C_p	Specific heat at constant pressure (J/kg/K)
d_o	pool radius (m)
F	geometrical view factor
g	acceleration of gravity (m/s^2)
H_c	heat of combustion (J/kg)
H_v	heat of vaporization (J/kg)
L	flame length (m)
m'	mass burning per unit time (kg/s)
m"	mass burning rate $(kg/s/m^2)$
P	emissive power of flame (kW/m^2)
Q	heat flux (kW/m^2)
r	distance of flame center to target (m)
T_a	ambient temperature (K)
T_b	boiling temperature (K)
T_f	average flame temperature (K)
u	wind velocity (m/s
u_c	characteristic wind velocity: $u_c = (gm''d_o/\rho_a)^{1/3}$
α	absorptivity of the target
ϵ	emissivity of the flame
ν	kinematic viscosity of air $(m^2/s$
ρ_v	vapour density (kg/m^3)
ρ_a	density of ambient air $(1.29\ kg/m^3)$
σ	Stefan-Boltzmann constant $(kW/m^2/K^4)$
τ	atmospheric transmissivity

PROCEDURES FOR THE USE OF RISK
ANALYSIS IN DECISION MAKING

Paul Baybutt
Risk Assessment Group
Battelle's Columbus Division
Columbus, Ohio 43201

ABSTRACT

Risk analysis is used to determine the probabilities and/or consequences of certain undesirable events. The consequences may be
related to the health and safety of workers or the public, or they may
involve economic losses such as property damage. Usually one is
concerned with undesirable events such as fires, explosions, and the
release of hazardous materials in engineered systems such as chemical
process plants. It is not sufficient merely to calculate such risks.
Once determined, they must be compared with safety goals or risk
criteria to judge their acceptability and then, if necessary, a
suitable plan for remedial action must be developed.

Often it is not a simple matter to judge the acceptability of the
calculated risks. First, risk must be expressed in a form in which it
can be compared with the acceptability criterion. Then uncertainties
in the risk estimates must be considered in the decision on acceptability. When remedial action is needed, corrective actions must be
prioritized, and this involves use of risk/cost/benefit analysis in
order to ensure that the most cost efficient approaches to risk
reduction and control are used. This paper discusses approaches for
handling these issues and describes methods which can be used to make
decisions on the basis of risk estimates.

1. INTRODUCTION

Risk is defined as "the chance of bad consequences" by the Oxford
English Dictionary and as "the chance of injury, damage, or loss" by
Webster's Dictionary. Risk is present in all human activity. Many
everyday risks, such as those from driving, are routinely accepted,
perhaps because the risks are well known. By contrast, risks from
industrial facilities, such as chemical process plants, are often not
well understood. These risks can be health and safety related (for
example, both immediate and long-term health effects of exposure to

261

toxic chemicals) or economic (for example, destruction of equipment
and lost production due to fires, explosions, or other accidents).

Risks arising from modern technologies frequently can be reduced by
the use of risk assessment which provides the decision maker with an
evaluation of both the probabilities and consequences of undesirable
occurrences that could result from the use of a technology. Risk
assessment is a useful tool in facilitating objective decisions on the
acceptability of risk, identifying health and safety problems and
approaches to their solution, and helping to prevent economic losses
due to accidents.

Modern risk assessment represents a synthesis of such techniques as
safety, hazards and reliability analysis, and is used for the evaluation and control of risks associated with the operation of industrial
plants. A typical example of such a risk assessment is provided by an
analysis, performed by the Health and Safety Commission in the United
Kingdom, of a petrochemical facility located on Canvey Island in the
River Thames, east of London.[1] Studies of this type provide
estimates of the risks associated with the operation of engineered
systems. In order to control and manage these risks it is necessary
that the risk estimates be used in a decision-making process. This is
not necessarily straightforward. Some acceptable level of risk must
be selected and often it is difficult to reach a consensus on a suitable value. Indeed, even the form of the criterion for acceptable
risk can be subject to debate. Additionally, the uncertainties which
are usually present in the risk estimates must be considered in the
decision-making process. Furthermore, when remedial actions are
required to reduce risk, corrective actions must be prioritized, and
this involves the use of risk/cost/benefit analysis in order to ensure
that the most cost-efficient approaches to risk reduction and control
are used. This paper discusses these issues and describes methods
which can be used to make decisions on the basis of risk estimates.

2. RISK AND DECISION MAKING

Risk assessment may be conducted for a variety of reasons. First,
there may be concern about the health and safety of plant workers or
of members of the public who work or live in the vicinity of the
plant. Risk assessments for these purposes may be motivated by a
desire of the plant operator to minimize such risks, by government
regulations, or by public concern. Estimates of risk can be used to
compare the risks of plant operation with those of other technologies,
to demonstrate compliance with government regulations, and to help
allay public concern.

A second reason for conducting risk assessments is concern which may
exist owing to the financial risks posed by plant operation. Accidents can result in damaged or destroyed equipment, and process
downtime which causes a loss of product and possibly the interruption

of supply to other processes if the product is a key intermediate.
Furthermore, loss of product for a significant period of time can
result in losing ground to competitors. All these risks add up to
financial losses to the company operating the plant. Risk assessment
provides the information needed for rational decision-making on the
control of these risks. Risk analysis can be used to help decide
appropriate levels of self-insurance, insurance deductibles, and
insured sums. This requires not only risk estimates but also
estimates of the costs that would be incurred should the risks be
realized. Risk-benefit and cost-benefit analysis can then be used to
facilitate decision making.

Ideally, risk assessment is conducted at the design stage of a new
facility. This permits iteration between design and risk analysis,
and modifications to the design can be made at minimal cost. A close
collaboration between designers and risk analysts at the pilot plant
stage is also important. Risk analysis can be used here to guide
cost- and risk-effective modifications to the design based on actual
operating experience. For large-scale facilities risk analysis can be
used to ensure that modifications which are made over the years, such
as those to improve productivity, do not increase the risk. Often
this can be done by simply revising a risk analysis that was performed
at the design or pilot plant stage. Risk analysis can also play an
important role for plants which have been in existence for some time
and for which a detailed risk analysis has not been performed
previously. In these cases a risk analysis will help ensure that
plant risk does not increase unacceptably as the plant ages and
wearout becomes significant.

Risk analysis can be conducted at several levels depending on specific
needs. The first level is qualitative risk assessment in which the
analyst identifies major hazards of the system and possible accidents
that can occur, determines important aspects of system design or
operating procedure that pose an unusually high risk ('sore thumbs'),
and suggests possible remedies. On the next level of sophistication,
detailed analyses of 'sore thumbs' are performed and risk reduction
efforts are prioritized. The highest level of analysis involves a
fully quantitative assessment in which risks of the system are quantified, system risk is evaluated, and improvements are recommended on
the basis of risk reduction, if necessary. The appropriate level of
risk assessment is determined by the problem at hand.

We now turn to the concept of risk. The measurement of risk involves
the specification of the frequency and/or consequences of some
undesirable occurrence. In general, risk is calculated as the sum of
the consequences of a finite number of accident sequences weighted by
their probabilities. This defines a risk function, $r(c)$:

$$r(c) = \sum_{i} p_i f_i(c) \qquad (1)$$

where p_i is the probability that the 'ith' accident occurs and
$f_i(c)$ is the probability density function of the consequences given
that the accident has occurred. The sum is over all sequences being
analyzed. The density functions f_i represent stochastic variations
in the consequences. Equation (1) describes a density function for
the consequences of all accidents considered. Integration of the
curve yields measures of risk:

$$R(c) = \int_c^\infty \left\{ \sum_i p_i f_i(c') \right\} dc' \quad . \tag{2}$$

Here, $R(c)$ is a risk curve that represents the probability of
exceeding a particular consequence level, c. This curve is known as a
complementary cumulative distribution function. A typical example of
such a curve is shown in Figure 1.

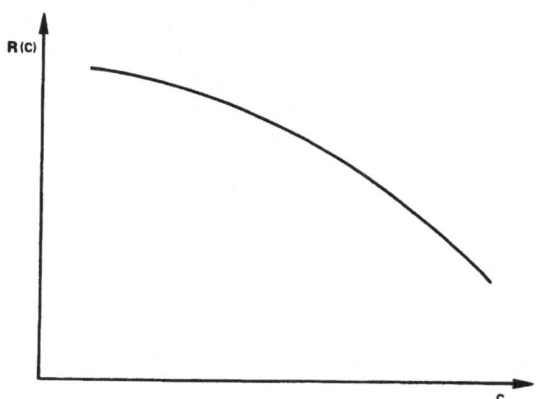

Figure 1. A Typical Risk Curve

The units of such <u>risk measures</u> are frequently of a particular type of
consequence. For example, the number of fatalities per year. Other
risk measures of the same <u>form</u> may also be of interest, for example,
the number of induced cancers per year. In some problems risk
measures of different forms may be more appropriate. In the case of
airplane crashes, often the focus of attention is on the frequency of
occurrence since the consequences are bounded and do not vary widely
from accident to accident. Traditional approaches to safety analysis
of engineered systems often focus on the consequences of maximum
credible accidents and do not specifically evaluate their probabilities. In these cases such risk measures as the monetary value of
destroyed equipment or the number of injuries are employed. Many
other risk measures and forms can be used. The choice depends on the
problem at hand. In some cases, it is desirable to employ more than

one risk measure in the same problem. For example, the total risk may
have several components such as health effects and property damage.
If, in such cases, a single measure of total risk is desired, the
difficult problem of converting to common units is encountered. The
controversial issue of converting fatalities or injuries into monetary
terms must then be faced. Conversion constants have been proposed and
employed but their value will depend on the particular circumstances
involved and will likely always remain controversial.

In some decision problems it may be desirable to employ several risk
measures to explore the effect of the choice on the decision taken.
Different decision makers may favor different risk measures and it can
then be important to explore the robustness of the decision taken to
the choice of risk measure.

In risk-based decision making it is important to account for public
perception of risk. Risk analysis provides an objective estimate of
risk to the extent practicable. This does not mean that subjectivity
is eliminated by risk analysis; indeed it is often an important element in developing data. However, risk analysis is intended to minimize subjectivity and to be as objective as possible. In addition,
attempts are made to display explicitly any subjectivity involved.
Often, calculated risks will differ significantly from the risks
perceived by the public. Where this occurs it is important to
familiarize the public with the calculated risks and their validity
since otherwise the management of risks in society will not proceed in
a balanced fashion. Sometimes irreconcilable differences exist
between calculated and perceived risk where the public displays marked
risk preferences. An example is provided by risk aversion. Most
individuals are less prepared to accept large-consequence events even
if the probabilities are proportionately lower than small-consequence
events. Such individuals are said to be risk averse. Some risk
criteria incorporate considerations of risk aversion as will be
described in the next section.

3. SAFETY GOALS AND RISK CRITERIA

In order to utilize risk estimates in decision-making it is necessary
that an acceptable level of risk be specified. The form of the
acceptability criterion depends on the risk measure that is to be
employed. Limits may be specified on event frequencies or event
consequences or both. For example, the Fatal Accident Frequency Rate
(FAFR) has often been employed as a risk measure in the chemical
industry. Limit lines for risk have also been proposed.[2,3] These
place limits on the probabilities of accident consequences, either for
individual accidents, as in the case of the Farmer limit line[2], or
for a spectrum of accidents, as in the case of a limit line in the
form of a complementary cumulative distribution function.[3]

An example of a limit line is shown in Figure 2. Obviously it is
desirable that the greater the consequence of an event, the lower
should be the associated probability and this situation is depicted in
Figure 2 where the probability of a given consequence is shown as a
function of the consequence. It can readily be seen that if the slope
of this line is -1, then all points on the line correspond to situations of equal risk since the product of probability and consequence
is constant. Thus, the probability of an undesirable event is
inversely proportional to its consequence. Clearly then, if an
acceptable value of risk can be determined, it can be used to con-
struct a graph such as that shown in Figure 2. All points falling on
or under the line would represent situations of acceptable risk, while
conversely, all points above the line would correspond to situations
of unacceptable risk.

Often it is desirable to reflect risk aversion in a limit line by
using a line of slope less than -1. This requires probabilities to
decrease more rapidly with increasing consequences than they would in
the 'constant risk' case described above. Such limit lines represent aversion to high consequence events by requiring acceptable
probabilities for these events to be lower than would be required by a
relation of inverse proportionality between probability and
consequence.

In the practical application of limit lines, it is convenient to make
two modifications to the line shown in Figure 2. These changes are
depicted in Figure 3. Here the low consequence part of the curve is
drawn with a constant probability value (horizontal line). This has
the desirable effect of avoiding the nuisance of high probability/low
consequence events. For the high-consequence part of the limit line,
a cutoff is shown. This could be used if it were decided that there
are some high-consequence events for which the only acceptable
probability is vanishingly small.

A limit line, such as that shown in Figure 3, defines an envelope of
acceptable risk. Events which fall within the envelope are acceptable; those which do not are unacceptable. The form and position of
the line need to be tailored to the particular application. This can
be done by evaluating the risk preferences of the risk takers (plant
operators, company management, shareholders, plant workers, members of
the public) and an analysis of the historical record. The former can
be a complex undertaking since consensus on acceptable risk can be an
elusive goal. The latter is more readily achievable and represents to
a certain degree the actual levels of risk that have been implicitly
accepted.

It is important to consider not only risk from single events but also
aggregate risk. While it may be acceptable for several events to lie
close to the limit line at any one time, it would probably not be
acceptable for the majority of events to lie close to the line.
Complementary cumulative distribution functions can be used to avoid

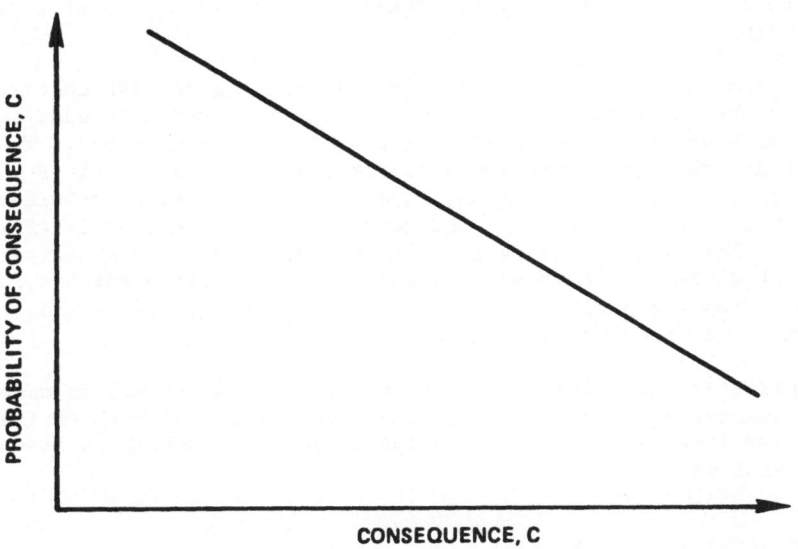

Figure 2.   Simple Limit Line

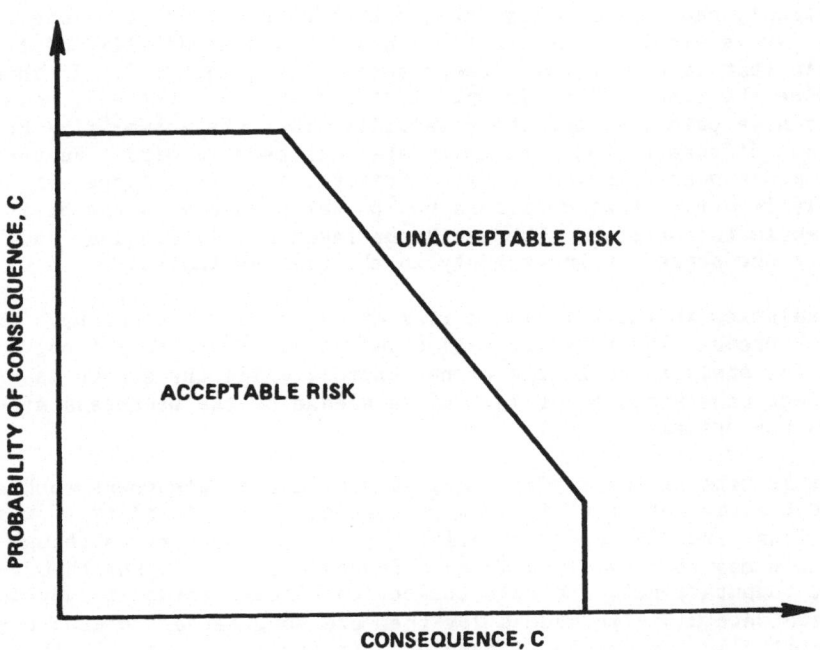

Figure 3.   Limit Line for Risk

this problem. However, they do require a full evaluation of risk to be performed.

It has been observed earlier that decision-making on risk must reflect not only the risks faced but also the costs involved. In addition, potential benefits of accepting the risk must be considered. Thus, an appropriate decision criterion should reflect both the risk (P x C), the costs of corrective actions, K, and the benefits, B, provided by the risk-producing activity. One possible such criterion is (P x C)/(K x B). The larger this ratio, the more desirable are remedial actions because of high risk, low costs for corrective actions, or low benefits from the activity. Conversely, the smaller the ratio, the less the need for remedial actions.

The application of a limit line in decision-making should be made with a full awareness of any uncertainties that may attend risk estimation. The treatment of uncertainties in decision-making is discussed in the next section.

4. THE TREATMENT OF UNCERTAINTIES

Although risk assessment entails the use of the best engineering and scientific methods available, in practice, owing to gaps in present engineering and scientific knowledge as well as data variability, some uncertainty can, and usually does, attach to the final estimate of risk. It is vital to the practical utility and credibility of the results that such uncertainties be adequately quantified. If this is not done the risk analyst is open to the criticism that only point values have been used and the possibility that the actual risk may be somewhat different than that calculated has been ignored. Furthermore, since uncertainties in risk estimates are often large, it is clearly important that decisions not be taken solely on the basis of best estimates of risk. The decisions taken should consider and reflect the degree of uncertainty in the risk estimates.

Uncertainties in the calculated risk of engineered systems have two basic sources. The first if real physical or random variation, over which the analyst has little or no control, while the second is imperfect understanding or lack of knowledge of the processes at work within the system.

The first type of uncertainty manifests itself in phenomena such as the stochastic nature of failure processes, the variability of weather conditions, and the unpredictability of human behavior. Although such phenomena may be deterministic on a fundamental level, the theoretical, computational, and data collection efforts needed to develop accurate deterministic models for them are so great as to make the task virtually impossible. Therefore, it appears inevitable that these phenomena will be modeled stochastically for the foreseeable future. Although the inclusion of such sources of variability in the calculation of risk is essential, since nothing can be done to reduce this uncertainty its consideration is not important in the development of strategies for improving confidence in risk estimates by reducing the associated uncertainties.

The second basic source of uncertainty, lack of knowledge, does have some controllable aspects. It manifests itself in imperfect modelling of physical processes that occur in accidents, such as the dispersion behavior of heavy gases, in uncertain values for physical constants and material parameters, and in doubts about the completeness of risk assessments. Continuing analytical and experimental programs can lead to improved models and more accurate physical and material data, with a consequent reduction in uncertainty. Problems related to completeness of the analysis are more difficult. It is always possible for an analyst to overlook some important accident sequences or to underestimate their impact. Such errors can be minimized through the use of formal techniques such as event and fault tree analysis. However, the validity of a risk assessment depends on appropriate quality control and the use of well-trained, skilled analysts.

In Section 2, various measures of risk were discussed. The simplest measures are the frequency and/or consequence of particular undesirable occurrences such as fires, explosions, and hazardous material releases. A more complete description of risk is provided by a risk curve which provides a frequency for exceeding each level of accident consequences.

Uncertainty estimates can be provided, in principle, for all these various risk measures. The form of the uncertainty estimate can vary from a simple bound or range about a best estimate of risk to complete information on variability in the risk in the form of a distribution. Simple bounds or ranges are often used when a low-level risk analysis is performed. The bounds are then usually determined by the application of engineering and scientific judgment. This can entail the use of group choice techniques in which several experts are polled and their estimates synthesized to provide a more reliable estimate than any of the individual estimates. An example of such bounds for a risk curve is shown in Figure 4. Uncertainty bounds can be related formally to the full distribution of risk by viewing them as percentiles of the distribution such as the 5th and 95th percentiles. Thus, considering the risk function definition of Equation (1), a new risk function that considers uncertainties in probabilities due to lack of knowledge can be defined as

$$r'(c) = \sum_i (P_i + n_i)f_i(c) \qquad n > 0 \qquad (3)$$

where σ_i is the standard deviation of the distribution due to lack of knowledge of probabilities for accident i. Uncertainties in consequences due to lack of knowledge can also be incorporated in the risk curve.

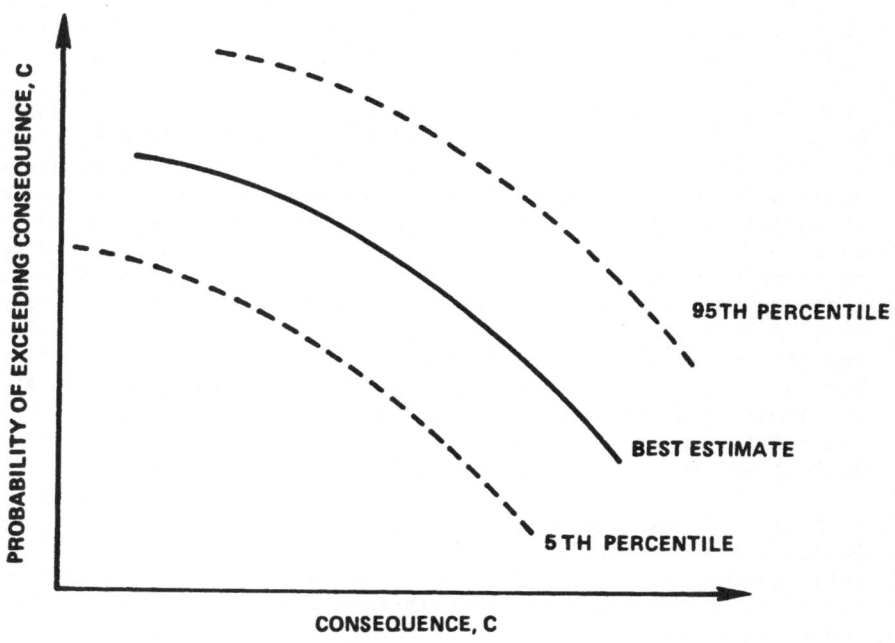

Figure 4. Display of Uncertainties in Risk Curves

Full distributional information on uncertainties is usually sought only when the highest level and most detailed risk analyses are performed. Specialized techniques are than used to propagate uncertainties in input data and models through the analyses to provide estimates of the uncertainties in the output. Uncertainty analyses can be carried out using a variety of methods including analytic techniques, response surface methods, Monte Carlo procedures, and differential sensitivity theory. A discussion and comparison of these methods has been published recently.[4]

Once uncertainties have been quantified, the problem becomes one of how to consider them in the decision-making process. The basic problem entails comparing risk estimates and their associated distributional information with a criterion for acceptable risk. In the case of a simple point estimate measure of risk this can be done in a relatively straightforward manner as shown in Figure 5. In this figure, $\gamma(\rho)$ is the distribution function for calculated risk, ρ, (a point measure of risk such as the fatal accident frequency rate), and ρ_A is an acceptable value of this risk. A statement can then be made about the degree of confidence with which it is believed that the calculated risk does not exceed ρ_A. The probability that the risk does not exceed ρ_A is given by

$$1 - \int_{\rho_A}^{\infty} \gamma(\rho)d\rho \quad . \tag{4}$$

A conservative decision posture would require a high value for this probability (say 0.95) and would seek the minimization of the shaded area in Figure 5.

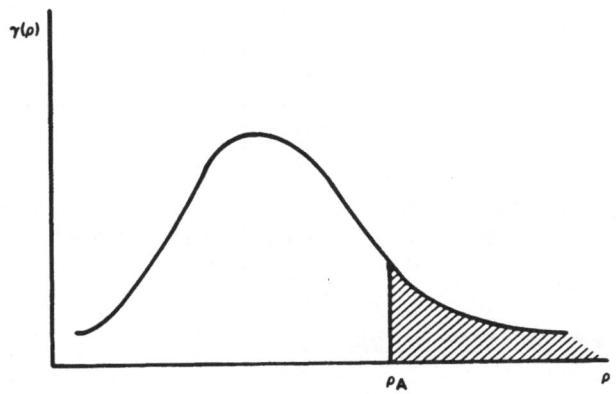

Figure 5.   Treatment of Uncertainty in
Point Estimate Risk Measures

In the case of risk curves, uncertainties can be displayed in the way shown in Figure 4. The comparison of risk curves with a suitable criterion of risk acceptability is not necessarily straightforward. Several cases are shown in Figure 6. Curve A is uniformly lower than the limit curve L and is thus acceptable. However, curves B and C cross the limit curve and thus are likely unacceptable. For curve B the probability of low consequences is higher than for the limit curve but the probability of high consequences is lower. Thus, for some people, curve B may represent acceptable risk even though for some levels of consequences it exceeds the limit curve. The reverse is true for curve C and thus it is likely unacceptable to all.

There is no straightforward manner in which the uncertainties in the risk curves can be incorporated into this comparison. One simple approach is to use some high percentile of the distribution about the best estimate risk curve, say the 95th percentile, in the comparison with the limit line. The more conservative the decision posture taken, the higher the percentile used.

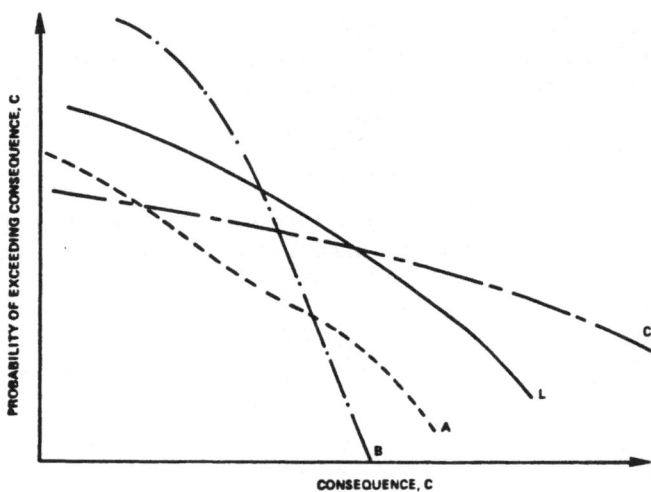

Figure 6.   Comparison of Calculated Risk
Curves with Limit Line

5.   PROCEDURE FOR DECISION MAKING

Decision making on risk is a multistep process that requires many
steps other than risk evaluation[5].   Indeed a number of steps must
be completed before risk calculation can begin.   The steps in one
approach to risk-based decision-making are shown and discussed below.

(1)   Define the problem

(2)   Identify decision alternatives

(3)   Select risk (or decision) attributes and their importance
        measures

(4)   Select risk measures for attributes

(5)   Determine safety goals and/or risk criteria

(6)   Determine appropriate level of risk analysis

(7)   Perform risk analysis and evaluate uncertainties in risk
        estimates

(8)   Formulate additional alternatives for risk control

(9)   Choose decision posture

(10)   Select decision technique

(11)   Formulate decision rules

(12)   Apply rules and evaluate alternatives

(13)   Formulate decisions.

The first step entails defining the problem. Although perhaps obvious, this should not detract from the importance of the step. The reasons for performing the analysis and the ultimate objectives must be stated clearly. This is followed by the identification of decision alternatives, i.e., the possible choices that are open to the decision-maker. Often these choices will be different system designs. The alternatives may be formulated by the decision-maker, but frequently they are provided to him, and it is his job to recommend a preferred alternative.

In the next step risk (or decision) attributes and their importance measures are selected. These are various aspects of risk that can be quantified and are of interest in the problem at hand, for example, human health effects and property damage. Importance measures such as simple weights and utility functions must also be developed for each attribute. This step is followed by the choice of risk measures for these attributes. Some possible risk measures were discussed in Section 2. They can range from the frequency of occurrence of events with consequences of particular magnitudes to simple statements of event frequencies or consequences. The choice of risk measure can impact the choice of decision analysis method and the way the chosen method is used. The risk measure should be selected with consideration of the form taken by any safety goals or risk criteria with which the risk estimates may ultimately be compared. The appropriate level of risk analysis must be determined. This depends on the problem being analyzed, the objectives of the analysis, resources available, and the importance of the problem (which partly depends on the expected magnitude of the risk). When this has been determined, the risk analysis is performed and uncertainties in the risk estimates are evaluated. The results of the risk analysis are then used to identify additional alternatives for the control or reduction of risk. This is achieved by using the risk analysis to identify root causes of risk and then formulating a strategy for their elimination or the reduction of their contribution to risk.

The next step is to choose a decision posture. This reflects the environment and perspective from which the decision is to be taken. For example, the decision maker may be required to assume a conservative position on risk and employ upper bounds rather than best estimates in the analysis. The decision maker may also be required to show high sensitivity to costs. Often, decision recommendations are made for several different postures. It is important that the posture be identified, understood by the decision analyst, and reflected in the analysis performed.

When the decision posture has been chosen, the decision technique must
be selected. This can range from a very simple approach such as the
matrix method to a more complex method such as multi-attribute utility
theory. The choice of method depends on the nature of the decision
problem and other factors. This is followed by the formulation of
decision rules which employ importance-weighted risk estimates for the
risk attributes. These rules are used to discriminate between the
decision alternatives. It is important that they account for uncertainty in the risk estimates since this can often influence the
alternative chosen. A decision is then formulated and an alternative
selected.

6. AN EXAMPLE OF THE USE OF RISK ANALYSIS IN DECISION MAKING

To illustrate the use of risk analysis in decision making, an example
has been constructed from the experience of the author. In this
synthesized example the problem of choosing between three alternative
designs for a liquefied natural gas (LNG) tank farm is considered.
The three designs were quite distinct and the construction costs were
significantly different. The company contemplating the construction
of the tank farm was concerned about the possibility of a gross rupture of the tanks and a vapor cloud explosion that could occur. There
was little information available on the relative integrity of the
tanks although the company suspected that design C was the safest.
However, it was not clear that the margin of safety over design A or B
was sufficient to warrant the additional cost. An independent consultant was called in by the company to perform a risk analysis and
make a recommendation on the design that should be chosen.

In discussions between the consultant and company officials it became
apparent that although they were concerned about the possibility of
the destruction of the tank farm in an accident, they were more
concerned about the possibility of injuries and fatalities to the
public in the vicinity of the tank farm site which was located in an
area of high population density. Most operations on the tank farm
were performed remotely and risk to tank farm personnel was not an
issue. The company had experienced a major failure in a tank at
another facility. Although a significant amount of LNG had escaped to
the atmosphere, weather conditions were such that the LNG dispersed
rapidly and no fire or explosion occurred. However, the company was
anxious to avoid a repeat of the incident. The company asked the
consultant to concentrate initial efforts on evaluating the risk to
the people living near the facility. Consequently, public fatalities
and public injuries were selected as the risk or decision attributes.
Fatalities were weighted by a factor of 1,000 over injuries. The
probabilities of fatalities and injuries occurring were calculated.

When queried on a suitable value for acceptable risk the company could
not make any recommendation but rather asked the consultant to conduct
a survey of accidents in storage facilities for LNG and other

hazardous materials in order to develop some insights into the levels
of risk that had been accepted in the past. This was done and the
data collected were analyzed and cast into the form of a limit line.

In discussions between the company and the consultant it was agreed
that the probability of tank failure by several mechanisms would be
calculated. Included were the growth of cracks in the storage
vessels, handling errors during transfer of LNG to and from the tanks,
and impacts by aircraft and other projectiles (tornado missiles in
particular). The dispersion of LNG in the environment was modeled as
were the effects of an explosion and the ensuing fire on the general
public. This corresponds to an intermediate level risk analysis.
Probabilities and consequences of several different accidents were
calculated and the uncertainties in these probabilities and consequences were estimated. During the risk analysis, it became apparent
that the use of a dike surrounding the tank farm could serve to
mitigate significantly the consequences of a spill. This was added as
an option to each of the three designs.

Owing to the concern and sensitivity of the company a conservative
decision posture was adopted. A simple decision technique and
decision rule was employed. The 95th percentiles of the calculated
risk curves were compared directly with the limit line that had been
developed. This comparison is shown in Figure 7. Design D consists
of design B with the dike. Curves for designs A and C augmented by a
dike are not shown in order to improve the clarity of the figure. It

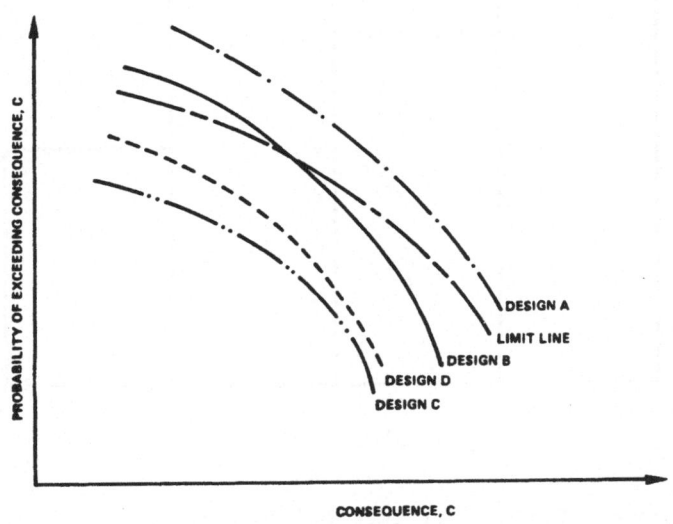

Figure 7. Comparison of Risks for LNG Tank Farm Design
           Options with Limit Line

can be seen from the figure that the risk of design A is greater than the limit line risk. The curve for design A augmented by a dike was also uniformly higher than the limit line. Hence, both designs are unacceptable. Design B is greater than limit line risk for low consequence accidents but crosses the limit line and has lower risks for high consequence accidents. The addition of the dike to produce design D reduces the risk of design B below the limit line for all values of consequences. The risk curve for design C is similar to that for design D and the curve for design C augmented by a dike is naturally lower than the design C curve. On the basis of risk alone design alternative C augmented by a dike would be selected. However, the cost of these design alternatives is also an important factor. In order to select among the design alternatives the expected values of the risk curves of Figure 7 were calculated and multiplied by the costs of each design alternative. The results are shown in Figure 8. The comparison shows that although design C has lower risk than design B, the expected risk/cost product is higher. Although the risk of design D is greater than design C the expected risk/cost product is lower. On this basis, the recommendation was made that the company adopt design D, that is design B augmented by a dike.

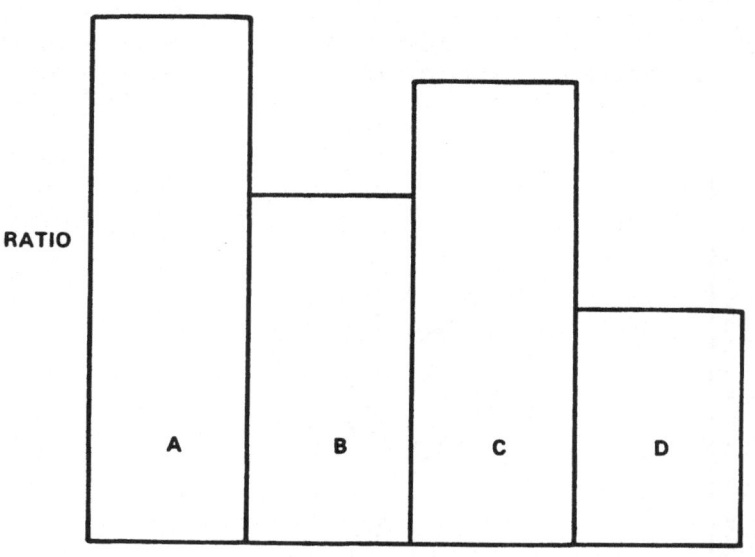

Figure 8. Products of Expected Risk Values and Costs for LNG Tank Farm Design Options

7. CONCLUSIONS

Risk analysis is a powerful technique for the control of risks
associated with the operation of industrial facilities. The results
of a risk assessment can be used by a decision-maker to help judge the
acceptability of risk and to aid in choosing between potential risk-
reduction measures. The effective use of risk assessment depends upon
the proper utilization of risk estimates in a decision-making process.
Especially important is the consideration of uncertainties in the risk
estimates. Other factors, such as risk aversion, must be taken into
account. Risk-based decision-making should also balance the costs of
risk reduction with the risks faced.

8. REFERENCES

(1) "Canvey: An Investigation of Potential Hazards from Operations
     in the Canvey Island/Thurrock Area", Health and Safety
     Commission, London: H. M. Stationery Office (1978).

(2) Farmer, F. R., "Reactor Safety and Siting: A Proposed Risk
     Criterion", Nuclear Safety, 8 539 (1967).

(3) Cox, D. C. and P. Baybutt, "Limit Lines for Risk", Nuclear
     Technology, 57 320 (1982).

(4) Cox, D. C. and P. Baybutt, "Methods for Uncertainty Analysis: A
     Comparative Study", Risk Analysis, 1 251 (1981).

(5) Baybutt, P., Cox, D. C., Denning, R. S., Kurth, R. E., Fraley,
     D. W., and Heaberlin, S. W., "The Treatment of Uncertainties in
     Risk for Regulatory Decision Making", International ANS/ENS
     Topical Meeting on Probabilistic Risk Assessment, September
     20-24, 1981, Port Chester, NY.

A COMPARISON OF CONSEQUENCES OF DIFFERENT TYPES OF UF$_6$ ACCIDENTS

W. Heudorfer, Battelle Institut e.V., Frankfurt am Main;
S. Hartwig, Universität Wuppertal and Battelle Institut e.V.,
Frankfurt am Main

1. INTRODUCTION

The objective of the investigations reported here is to compare the chemical toxic hazard associated with the road transportation of UF$_6$ with the handling hazard of UF$_6$ on enrichment or conversion plant sites. The reason for such a comparison is to get a better understanding of the risks of different phases of UF$_6$ treatment. UF$_6$ has a somewhat exceptional position within the nuclear fuel cycle because HF may be generated; HF is highly toxic and may occur as a heavy gas. This paper investigates only the UF$_6$ releases from the transport container. In principle the performance of a hazard analysis for dangerous materials can be subdivided into the following steps:
- definition of the system,
- investigation of initiating events,
- fault tree analysis,
- release analysis,
- dispersion calculation
- consequence analysis.

For the two cases mentioned above we will analyse the system according to the above mentioned steps.

2. THE INVESTIGATION OF THE TRANSPORTATION SYSTEM

2.1 Definition of the Transportation System

This system consists of:
- uraniumhexafluoride,
- the inner container (called cylinder),
- the outer container (called overpack),
- the van
- the environment.

At room temperature the transported UF$_6$ is a white, volatile solid. At a temperature of 64 $^{\circ}$C and a pressure of 1.52 bar (triple point), the UF$_6$ melts to form a colourless liquid of high density. The most

279

S. Hartwig (ed.), Heavy Gas and Risk Assessment - II, 279–292.
Copyright © 1983 by Battelle-Institut e.V., Frankfurt am Main, Germany.

important properties are listed in Table 1 (PNL-1978, Römpp-1981):
Table 1: Physical and chemical properties of UF_6

Sublimation point (1 bar)	56.6	^{o}C
Density, solid (20 ^{o}C)	5.09	g/ cm^3
Density, liquid (64 ^{o}C)	3.65	g/ cm^3
Density, liquid (121 ^{o}C)	3.26	g/ cm^3
Heat of sublimation (64 ^{o}C)	0.135	J/g
Heat of vaporisation (64 ^{o}C)	0.081	J/g
Critical pressure	46.1	bar
Critical temperature	230.2	^{o}C

Uraniumhexafluoride does not react with oxygen, nitrogen or dry air,
but is highly chemically reactive to water, ether and alcohol, forming
soluble reaction products. Gaseous uraniumhexafluoride , when released
into the atmosphere, reacts quickly with the atmospheric moisture to
form HF gas and particulate UO_2F_2, which tends to settle on surfaces:

$$UF_6 + 2H_2O \longrightarrow UO_2F_2 + 4 HF$$

The resulting UO_2F_2 is easy to recognize as a white cloud. A concentration of 1 mg of UO_2F_2 per cubic metre of air is visible. The clouds
resulting from major releases may obscure vision.
Dry HF gas associates to $(HF)_n$ molecules in the air. These molecules
are extended or else they may have cyclic forms. The quantity of the
associated molecules depends on the surrounding conditions. A typical
number is n=6. Therefore $(HF)_6$ can be 4.1 times as heavy as air. HF has
a considerable inhalation toxicity. The lowest lethal inhaled amount of
HF published is 73 mg. For the LD_{50} dose no values are known (BCL-1980).
Therefore in the calculations we use an arbitrary LD_{50}-value of 150 mg:
the container type considered is the 48 "Y. The general data of the
cylinder are listed in the following Table 2 they are quoted from the
ORO - Report (ORO- 651 - 1972).
Table 2: General data of the UF_6 cylinder model 48 "Y:

Nominal diameter	122 cm
Nominal length	381 cm
Wall thickness	1.59 cm
Nominal tare weight	2359 kg
Nominal net weight	12.501 kg

The cylinder is a steel cask, which is packed in a protective overpack.
The overpack is a double-walled steel cask; the thickness of the inner
wall is 0.48 cm and 0.32 cm for the outer wall. The space between these
walls is foamed.
The van is usually a trailer truck, which weighs,without cargo,10-15 tons.
Although UF_6 is transported on different types of roads,in this survey
we consider only highway transportation.

2.2 Initiating Events of the Transportation System

Road traffic accident research in the Federal Republic of Germany is mainly concerned with injuries of persons. To a far slighter extent, damage to load or vehicle is considered. As a consequence, little information is available on the severity distribution of accidents.
 In an investigation concerned with 3304 truck accidents, mainly on the autobahn, the causes for these accidents and their severity were evaluated (Hartwig 1981). The results are as follows:
Table 3: Accident frequencies with different kinds of vehicles

Type of vehicles	Accidents per driven km
Single trucks	$1.53 \cdot 10^{-6}$
Trucks with trailers	$1.90 \cdot 10^{-6}$
Trailer trucks	$1.39 \cdot 10^{-6}$
All	$1.65 \cdot 10^{-6}$

Classification according to the following causes of damage:
- fire,
- impact/ crush,
- puncture,
- faulty packaging

yields the following important information:
- Fire occurs in 1.5 % of all accidents in which trucks are involved, 20 % of these fires last longer than 0.5 h.
- In the absence of fire, impact and crush are the main causes of destruction in almost all accidents. Since the original accident reports do not give sufficient detailed information, it is not possible to distinguish between these two processes. They are therefore regarded as a single process. For medium-severity and high-severity accidents the location of damage on the trailer is given in Table 4.

Table 4: Frequency of damage at different locations of the trailer in medium- and high-severity accidents (in %)

Front	4.7
Side	16.7
Front + side	1.6
Back	6.2
Both sides	17.8
No damage on the trailer	48.0
All	100.0

Medium-severity impact accidents account for 8.5 % and high-severity impact accidents 2.5 % of all accidents. Puncture occurs in 6.5 % of all accidents. The accident reports provide no information about accidents caused by packaging errors that could be used for UF$_6$ transports.

We discussed American statistical data about UF_6 packaging
errors which show that packaging errors play a minor role in   the
release of UF_6 during the transport; therefore they are neglected.

2.3 Fault Tree Analysis for the Transportation System

As   stated previously, UF_6 at room temperature is a solid with a sub-
limation point of 56.6 ^0C. A major release of UF_6 is only
possible in accidents with fire; other accidents lead to a negligible
release only. Accidents involving fire can be accompanied by various
degrees of mechanical damage to overpack and cylinder. It is also
possible that these components are not damaged mechanically
Thus, in considering events with major consequences, it is only
necessary to discuss accidents involving fire.

   The top and   branches of the fault tree leading to major
consequences are shown in figure 1. In the head line of that figure
the severity of the mechanical damage by impact increases from the
left to the right branches.

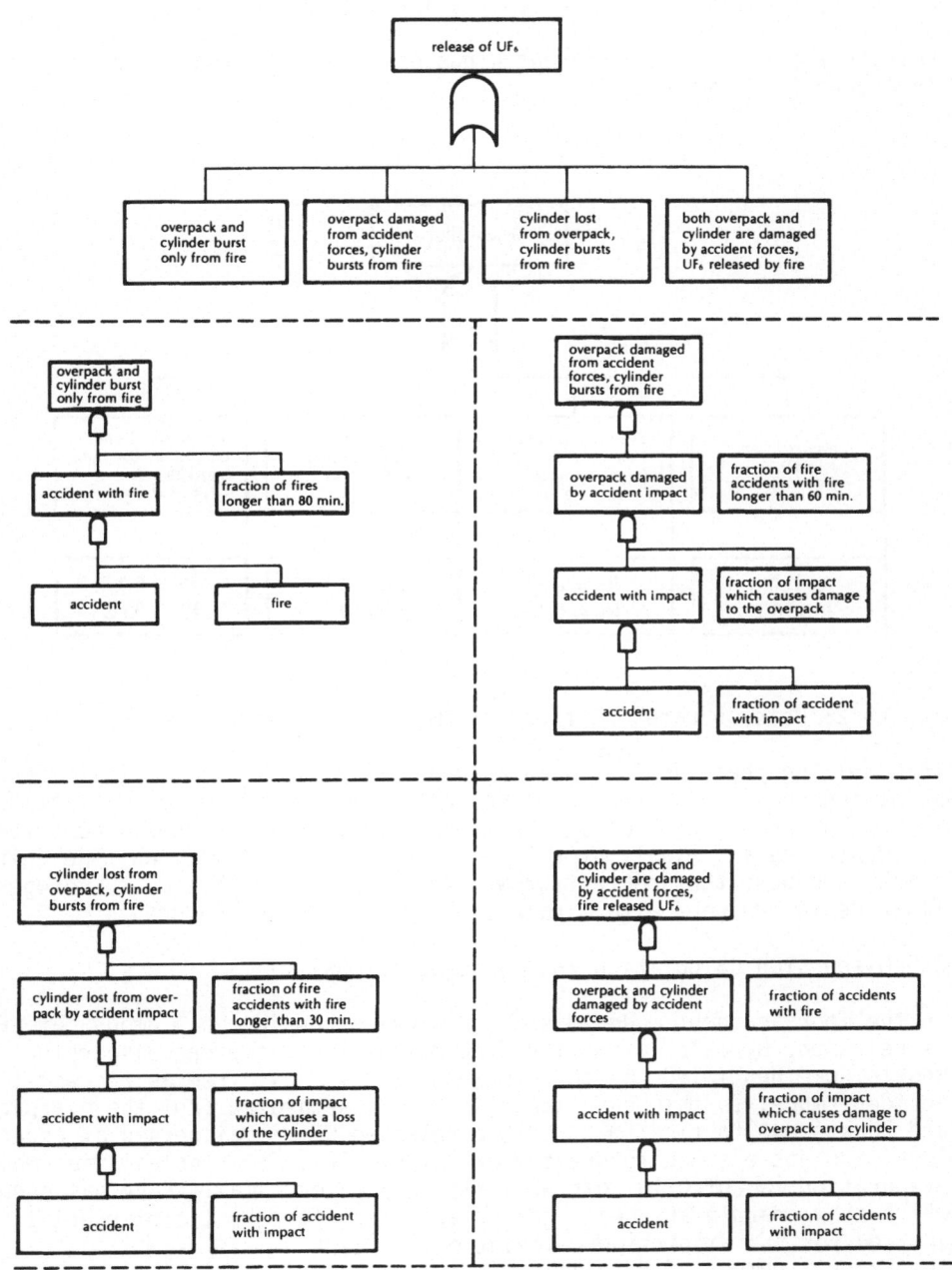

Figure 1: Top and Branches of the Fault Tree for the Transport System of UF$_6$

2.4 Release Rates for the Transportation System

Release rates   for   different kinds of releases are given below
in figure 2 (Hartwig 1981).

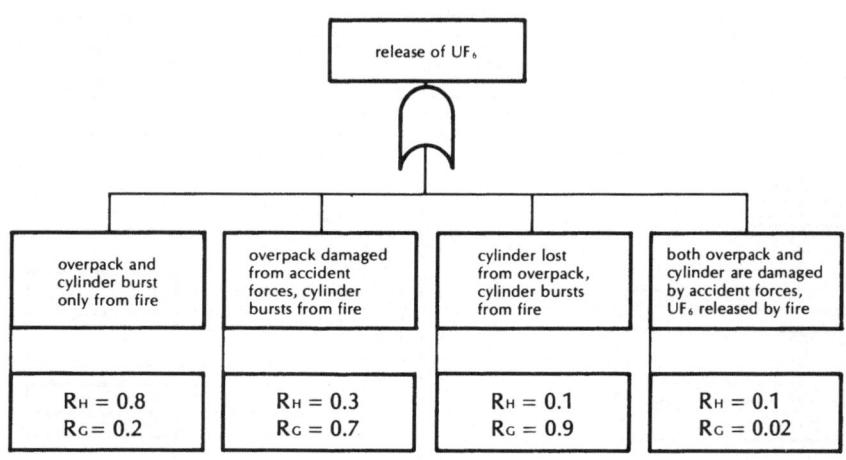

Figure 2: Release rates of the most important accident modes

In modelling these accidents we assume - with WASH - 1535 - that fraction
R_H of the UF_6 is released and rises with the fire plume to a height of
230 m. The remaining fraction R_G is scattered on the ground around the
accident site and slowly hydrolizes over the 4- hr time span. The third
branch produces the greatest ground fraction K_G. In this case hydrogenium-
fluoride is produced with a rate of 0.18 K_G/s near the ground.

2.5 Dispersion Calculation and Consequence Analysis

In the four accident modes discussed above a considerable amount of UF_6
is released. By calculating the dispersion with different dispersion
models, we identified the third branch of the fault tree as the worst
accident mode. In this case the cylinder is separated from the overpack
and bursts due to fire; 90 % of the released UF_6 is dispersed at ground
level and not ejected to greater height by fire. Nevertheless, the small
evaporation rate of 0.18 Kg/s HF forms only a very small heavy gas cloud
which show hardly any heavy gas effect. Therefore, a tracer model is
applied for the dispersion calculation.
The largest number of fatalities is to be expected during low and medium
wind speed and dispersion conditions of stability category I.
Figure 3   shows concentration values in 1.5 m above ground for the
worst case quoted above. To show the influence of the wind-speed concentration lines at wind speed of 2m/s and 10 m/s are drawn.

The solid lines were calculated with a Gaussian type model (Klug 1969).
The dashed lines were calculated with the numerical Battelle model
TRANSLOC. A comparison of the results of calculations with a Gaussian
model and TRANSLOC with experimental data has shown (Schnatz, Hartwig
1980) that in contrast to TRANSLOC Gaussian models tend to overestimate
the concentration near the source and underestimate the concentration
at greater distances. Therefore the TRANSLOC results are more realistic.

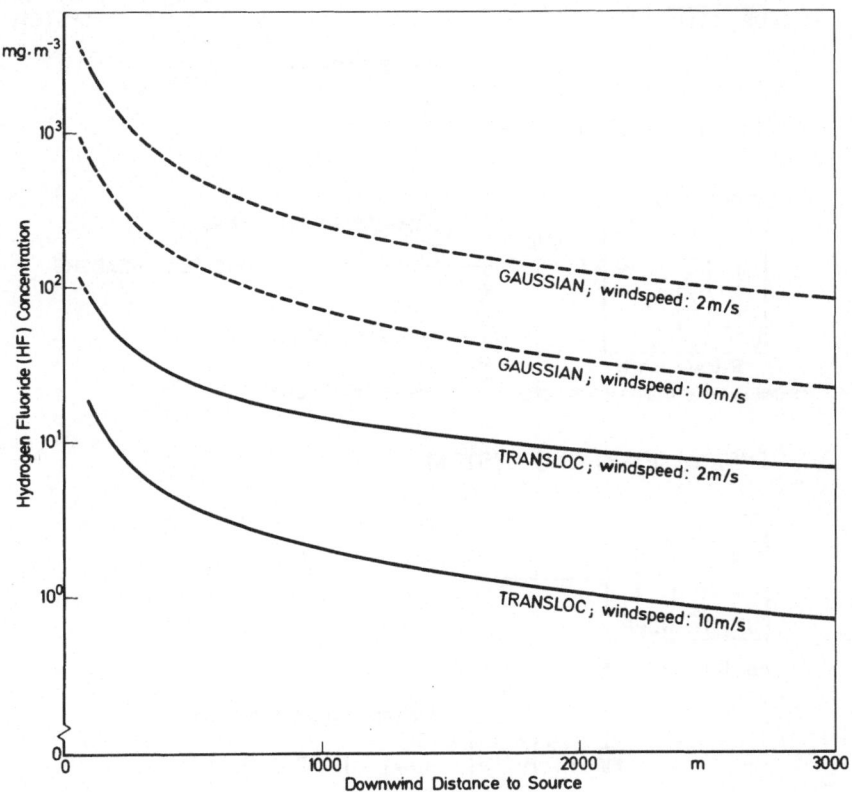

Figure 3: HF- Concentrations for the Worst Case Release in Trans-
          portation Accidents

According to TRANSLOC calculations, the greatest distance with lethal
values occurs at 360 m in a downwind direction with a wind velocity
of 2.0 m/s. With a wind velocity of 10 m the greatest distance is 50 m.
These calculations are based on the assumption that UF$_6$ evaporates
slowly from the solid fragments.

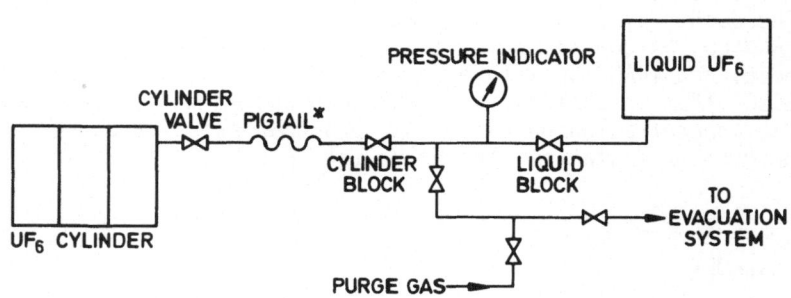

TYPICAL FILLING SYSTEM

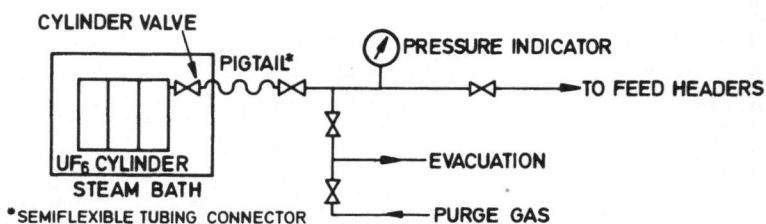

TYPICAL EMPTYING SYSTEM

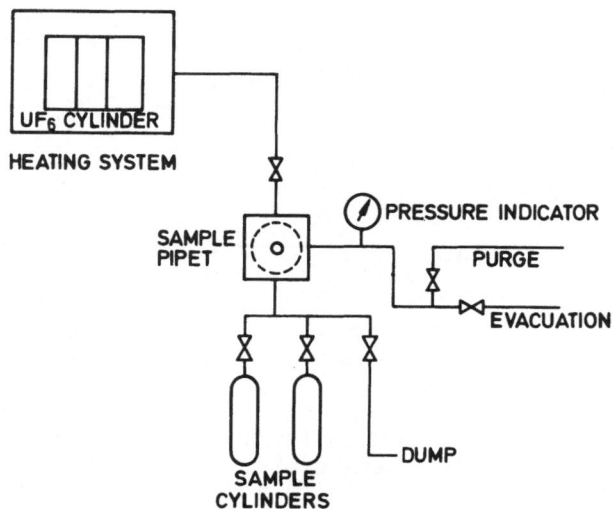

TYPICAL SAMPLING SYSTEM

Figure 4: Typical UF$_6$ Cylinder Filling, Emptying and Sampling Systems

3. INVESTIGATION OF THE HANDLING SYSTEM AT THE PLANT

3.1 Definition of the System

As mentioned earlier, this paper does not deal with all cases of release
which may occur with uraniumhexafluoride at enrichment and conversion
plants. It is restricted to releases that might   occur during filling,
emptying and sampling procedures of transport containers. We do not
discuss here problems that might arise with UF$_6$- storage.
   During filling, emptying and sampling procedures the system consists
of:
- uraniumhexafluoride,
- transporting cylinder,
- filling, emptying and sampling device,
- technical environment.

The properties of UF$_6$ and the cylinder are described in section 2.1.
The system is shown schematically in Figure 4.

The filling, emptying and sampling system consists of:
- tubes,
- valves,
- control devices
- and a heating device in the case of the emptying and
  sampling system.

During the filling prodcedure, liquid UF$_6$ is drained by gravity through
heated lines and valves into the cylinder. The liquid is kept at a temperature ranging from 79 °C to 121 °C. After the UF$_6$ has been solidified,
the cylinder is vented, if necessary, to a low pressure system in order
to remove contaminants that are volatile at room temperature.
During the emptying procedure the UF$_6$ usually is vaporized with either
steam heat or electrically heated air. The heat is controlled in order
to prevent temperature in excess of 121 °C.   Passing a network of
pipes that consists of a "pigtail" a"pressure indicator" and a valve,
the vaporous UF$_6$ is led to the "feed-heaters".
The sampling procedure is similar to the emptying procedure. The cylinder
and the entire system are heated up to 93 °C- 113 °C. The maximum
pressure is 5.4 bar. The sample cylinders are filled by a system of pipes.

3.2 Initiating Events

A qualitative analysis of the handling system leads to the following
basic events:
- damage of the tubes, valves, control devices or the cylinder
  caused by incorrect mechanical movements with the pigtail,
  the heating device, the container, the van, the truck or
  with elevators,
- overheating of the cylinder caused by a defective control
  device,
- erroneous opening of a valve when the cylinder is under
  pressure,
- material failure (material fatigue) of tubes or valves.

3.3 Fault Tree Analysis for the Filling, Emptying and Sampling Procedures

A UF$_6$ release is possible when there is a failure in the integrity of
the handling system. This failure can occur in the tubes, in the valves
or in the control system. (A failure of the integrity of the cylinder
is unlikely). From the basic events a fault tree of the handling system
is built.   Figure 5 shows the top of such a fault tree.

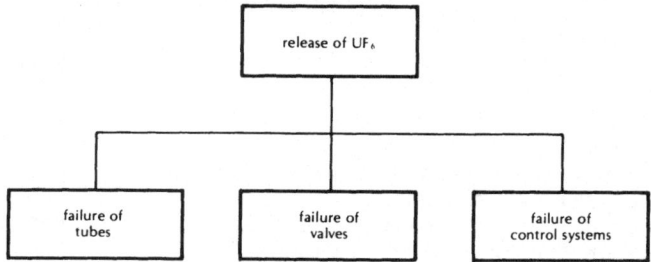

Figure 5: Top of the Fault Tree for a Handling System

3.4 Release Analysis

To assess the consequences it is necessary (similar   to the UF$_6$ transport) to perform dispersion calculations for the most probable and severe
event with a well defined release rate. To our knowledge sufficient data
of failure rates does not exist.   Therefore, the information of an
accidental release was used. The accident happened at the COMURHEX -
facility, in Pierrelatte ,   France on July, 1, 1977. We believe that
this accident is similar to the worst conceivable release scenario of
the system under consideration, except for the fact that the container
was not completely filled and the atmospheric conditions were favour-
able. The Accident:
During a sampling procedure at the COMHURHEX plant a sample of uraniumhexafluoride was drawn from a 48 "Y container holding 8827 kg of hexafluoride in the liquid phase. For this procedure the operator heated
the container with the aid of a mobile steam heater. The top cover of
the steam-oven was removed by a fork elevator. This handling operation
was the cause for the rupture of the container valve at the threads.
At that moment the following situation existed:
- the heating of the oven had been discontinued,
- the container was at a temperature of 90 $^{\circ}$C - 95 $^{\circ}$C,
- it was under a pressure of 3.7 bar,
- the cylinder valve was in the six o'clock position,
- the valve was closed, but still connected with the sampling
line.
Through the ruptured valve the uraniumhexafluoride in the container
immediately started to spill onto the ground. A heavy fog was formed.
About 23 % of the spilled UF$_6$ was vaporous by flash evaporation. The
duration of the spill was estimated 10-15 min. Once the level of the
liquid hexafluoride had fallen to that of the valve opening, the hexa-

fluoride continued to escape as a gas only. At this time 1727 kg UF_6 remained in a container and 7101 kg UF_6 were released. This corresponds to a release rate of 80, 4 %. Later it became obvious that an area of about 1000 m^2 was covered with liquid UF_6 during the accident. However, UF_6 and its reaction products were observed at a greater distance. Traces of uranium were observed in concentrations of 10 mg/ m^2 at a 600 m distance downwind from the source of emission. Hydrogen fluoride concentration of 2.4 mg/m^3 was detected up to a distance of 1200 m.
Calculations of enrichment plant scientists showed that 300 kg of uranium and 1600 kg of HF were released into the atmosphere. Consideration of the HF value leads to the conclusion that all the UF_6 released was changed into UO_2F_2 and HF.
In this manner 100 % of the ensuing HF and 6.2 % of the uranium in form of UO_2F_2, were airborne.
In our opinion, probably more than the above mentioned 300 kg of the uranium was emitted into the surrounding air at the beginning. But as the UO_2F_2 forms bigger or smaller aerosol particles, a larger portion of UO_2F_2 settled immediately on the ground.
We assume that a release of 80,4 % UF_6 corresponds to the worst case of release from a container during the filling, emptying and sampling procedure.
In order to model the COMURHEX accident a release rate of 1,0 kg HF/s over a time span of half an hour is used.
In case of a completely filled 48 "Y cylinder the release would be 1.6 kg/s in 30 minutes.

3.5 Dispersion Calculation

We performed model calculations for the COMHUREX accident with the numerical tracer model TRANSLOC. Sensivity tests indicated that the gravity spreading phase at the beginning of the dispersion plays a minor role. This holds especially for downwind distances greater than 200 m. The TRANSLOC calculations were made with atmospheric boundary conditions of July, 1, 1977   in Pierrelatte, at 14.30 h. These conditions are: Klug I, windspeed 5 m/s.
The concentration in downwind direction is shown in Figure 6_3
The measured concentration value which "exceeds" 2,4 mg HF/m^3 in a 1200 m source distance is reasonably well reproduced by the model. Lethal effects occur up to 60 m.
To estimate the worst consequences we have to consider a completely filled container and a spill occurring at low wind speed. To show the influence of the windspeed the concentrations in Figure 7 are calculated for a windspeed of 2 m/s and 10 m/s. Lethal effects occur up to 280 m.

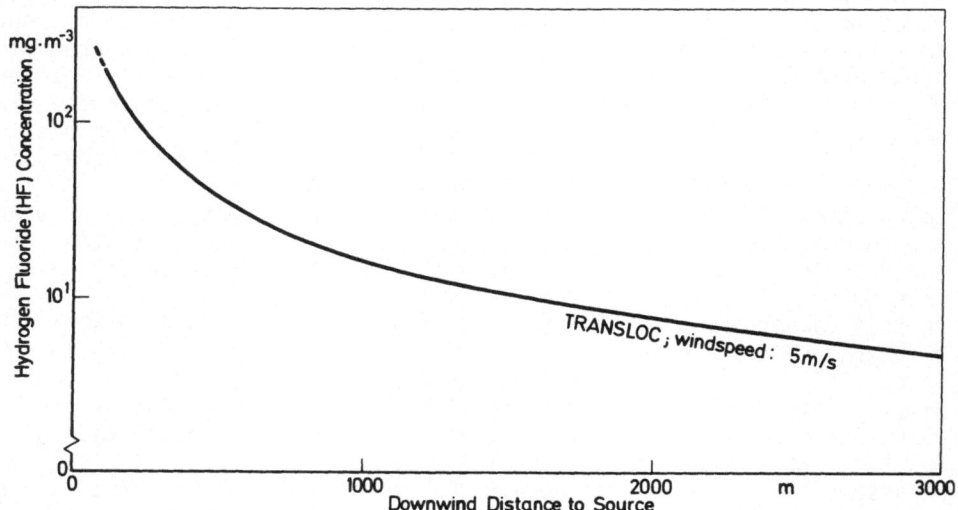

Figure 6 : Concentration of HF for the COMURHEX accident

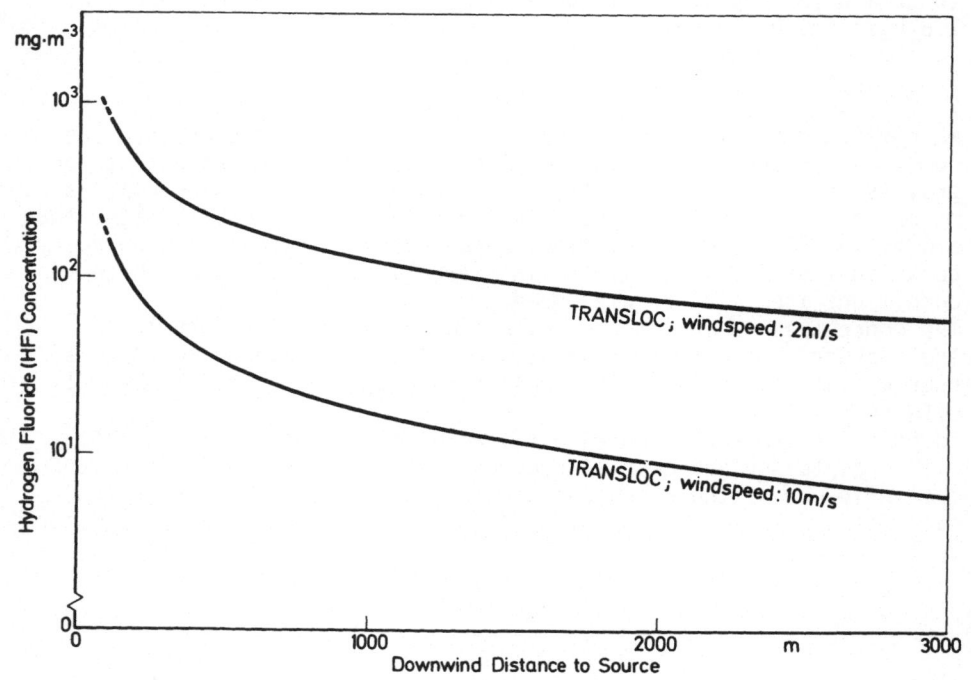

Figure 7: Concentration of HF for different windspeeds

4. COMPARISON

As mentioned previously the purpose of this analysis is to
better understand possible weak points in the UF$_6$- treatment system; in
other words, to learn whether phases in the UF$_6$- treatment of exceptional
high risk exist.
In table 4 the essential data for the two systems are given
to allow judgment of the different consequences.
Table 5: Important data of the two systems

Comparison	Transportation	Handling
release rate HF	0.18 kg/s	1.6 kg/s
duration	4.0 h	0.5 h
concentration at 500 m source distance (wind: 2 m/s)	24 mg /m^3	210 mg/m^3
maximum distance for lethal effects	360 m	280 m
population density	could be locally high	low
evacuation measures	difficult	–

It is obvious from the preceding discussions and table 4 that remarkable
differences in the value of the important data exist. Because of the
different heat content of the two systems (one system is already heated,
the other gets the heat from an external fire), the release rate and
the duration of release are very different. As a consequence the actual
downwind concentration values differ by one order of magnitude. Nevertheless,the maximum distance for lethal effects is very similar because
dose values are time integrated concentration numbers. Thus,the com-
bination of different release rates and different durations of release
results in very similar consequences. The values should be
discussed with some care because, firstly, no exact LD$_{50}$- value exists
(see paragraph 2.1), secondly the dose-effect function is nonlinear,
(Guide 1978) thirdly, because of different duration of the releases,
different mitigation measures (evacuation) could be effective.
With this reservation in mind, we conclude that the consequences for
the two discussed accident scenarios are similar. Because probability
data are scarce, we assume that the risk may be equal for both cases
or slightly higher for the transportation phase.

REFERENCES

/1/   BCL-1980: "Evaluation of Potential Health from
           Accidental UF_6 Releases: Chemical Hazards". Zaneto,
           M.A.; Warling, J.C.; Battelle Columbus Laboratories
           (1980)

/2/   DRS-1981; Deutsche Risikostudie Kernkraftwerke. Fach-
           band 8, Verlag TÜV Rheinland, S. 115

/3/   Ducouret-1978: An Experience of Accidental Release of
           UF_6. Meeting of Safety Problems Associated with the
           Handling and Storage of UF_6. Boekelo, Netherlands.
           June 27-29, 1978

/4/   Guide-1978 "Guide to Safe Practice in the Use and
           Handling of Hydrogen Fluoride:Chemical Industry Safe-
           ty and Health Council of the Chemical Industries
           Association". Alembic House, 93 Albert Embarkment.
           London SE1- 7 TU.

/5/   Hartwig-1981:" Schwachstellen und Risikoabschätzung
           beim Transport radioaktiver Materialien". BMI-SR58.
           Battelle-Institut e.V., Frankfurt am Main (1981)

/6/   Klug-1969:" Ein Verfahren zur Bestimmung der Ausbrei-
           tungsbedingungen synoptischer Beobachtungen". Staub,
           Reinhaltung der Luft 4. S. 193.

/7/   ORO- 651- 1972: "Uranium Hexafluoride: Handling
           Procedure and Container Criteria". (Rev.4). U.S.
           AEC, Oak Ridge Operations Office, (1972)

/8/   PNL- 1978:"An Assessment of the Risk of Transporting
           Uranium Hexafluoride by Truck and Train". Battelle-
           Northwest- Laboratory. Prepared for the US-DOE.
           PNL- 2211. UC-71 (1978)

/9/   Römpp- 1981: Römpps Chemie-Lexikon: Franckh'sche
           Verlagsbuchhandlung Stuttgart, 8. Auflage, Bd.2

/10/   Schnatz, Hartwig-1980:" TRANSLOC- ein numerisches
           Modell zur Simulation von Dispersionsvorgängen in
           der Atmosphäre und seine Anwendung bei einem Reak-
           torstörfall". Proceedings European Seminar on
           Radioactive Releases, Risc (1980)

/11/   WASH -1535: "The Environmental Impact of Trans-
           portation of Nuclear Materials in the LMFBR Pro-
           gram". Washington (1975)

SOME PROBLEMS IN RISK ASSESSMENT INVOLVING GAS DISPERSION

N C Harris
Imperial Chemical Industries PLC, Runcorn UK

1 ABSTRACT

Although quantified risk assessments which incorporate dispersion of
gases have now been conducted for over 10 years, there are still some
very fundamental problems to be resolved which can have a major effect
on the estimates produced. During the last 10 years there have been
tremendous improvements in our ability to predict the spread of gas
clouds and their dispersion, but this ability is largely restricted to
ideal conditions, for example over uniform flat ground.

Some of the various problems which are still inadequately resolved are
examined to show their current status, whether or not they are very
important or critical to the results to be obtained, and how it might be
possible to improve our understanding. Consideration is given not only
to specific aspects of gas dispersion, but to some primary source
problems and consequential effects which are just as important to the
overall assessment.

2 THE ESSENTIALS OF GAS DISPERSION

The basic and more detailed aspects of gas dispersion have been presented
at symposia for many years, and a very large number of mathematical
models have been described. They all have to incorporate a large number
of variables to characterise the atmospheric conditions, but they tend
to employ a variety of mechanisms for describing entrainment and dispersio
and it is in this area that much of the recent work has been directed.
They always require to make assumptions about the type of emission since
this has a major influence over the initial dispersion process, and
assumptions have also to be made about the influence of both up-wind and
down-wind topographical features. Both these aspects can have a major
bearing to the predicted dispersion of a gas release, and in very many
cases the model or models available are inadequate to deal with these
problems, so more simplifying assumptions have to be made which may or
may not be realistic. Some of the features of the typical mathematical
models will now be briefly described.

S. Hartwig (ed.), Heavy Gas and Risk Assessment - II, 293–303.
Copyright © 1983 by Battelle-Institut e.V., Frankfurt am Main, Germany.

2.1   Atmospheric Parameters

Air Stability

This is usually categorised according to one of the well known
classifications, eg Pasquill Categories A through F. Typically
six categories are used, occasionally more, but often less when
some simplification is attempted. In this case over conservative
results are usually obtained due to the method of simplification
adopted.

Difficulties also exist in equating some of these classification
systems to each other, but although this could be important if
weather probability data were used corresponding to another system,
it is not critical in any individual example where the conditions
in the atmosphere can be defined by each classification system.

Dispersion Coefficients

Corresponding to each of the air stability categories can be used
a set of coefficients to describe the rate of dispersion of the
pollutant into the wind field, based on the three co-ordinates, x
down-wind, y cross-wind and z vertical. Although there is some
debate on these values, a large amount of experimental observation
has provided reasonable values for these coefficients, normally
referred to as σ_x, σ_y and σ_z. Care is required in
selecting the data to use, since recent full-scale test data have
indicated different values from those to be found in past literature.

Wind Speeds

Again it has been normal practice to use a range of wind speeds,
a common system being the use of the Beaufort Scale. Although
this rises to over 10 levels, normally up to 6 levels only are
used. The higher levels are not only very low probability cases,
but are usually expected to exhibit reducing risk due to the
rapid advection of any gas cloud.

In some cases where simplification has been introduced, only one
wind speed is used for each air stability category, representing
the typical wind speed for that category. Although there may be
some loss of accuracy, it does not follow that this is necessarily
a conservative assumption.

Meteorological Roughness

This parameter, usually expressed as z_o, is related to the drag
effect of the ground on the lowest layers of the atmospheric wind
field. For instance drag is very low for mown lawns, is greater
for meadow grass or fields of corn, and greater still for forests
and buildings. Such effects are described by Pasquill and others
who ascribe notional meteorological roughness lengths to these
typical features.

Typically mathematical models for gas dispersion relate the wind velocity U_h at height h to the 10m velocity U and the roughness z_o such that the drag effect of the boundary layer can be properly incorporated. This has a significant effect on the results obtained overall, but there are several problems in application which will be discussed later.

2.2   Other Parameters

Source Terms

This term is of very great complexity, yet in order to be able to provide a mathematical model at all, only a very limited number of source terms can in fact be utilised, based on certain simple assumptions. Most models are capable of dealing with only a very few of these, as indicated in a recent review (Blackmore, 1982)

Examination of many of the major accidents which have been reported in sufficient detail to enable them to be closely studied, discloses that there is no such feature as a simple source term. Indeed in many cases the additional features which occur, such as air entrainment, would be expected to exert considerable influence over the early stages of the dispersion process. This aspect is also discussed in more detail later.

Terrain and Topography

There are three basic aspects which can conveniently be considered together at this stage. They are slope, ground level obstructions and uniformity. Both slope and obstructions in themselves cause major problems in modelling gas dispersion, especially when dense gases hugging the ground are concerned, but overall uniformity is a very serious problem, and its lack can often lead to inadequate or even incorrect modelling. A quick glance at a set of major accident scenarios will quickly indicate that there is usually a lack of uniformity in the local environment, making judgement essential. Yet experience to date has not provided much evidence on which to make sound judgements, and some are probably seriously in error. These aspects will also be discussed later.

Peak to Mean Ratio

Much of the atmospheric dispersion modelling, particularly for atmospheric pollution over longer distances, predicts mean concentrations measured over a minimum period of say 3 or 10 minutes. Although data for longer periods may also exist, this is less critical for accident scenarios and it also assumes that the mean wind direction remains constant, which may not be the case. Nevertheless it is well known that over short periods, measured more in seconds or fractions thereof, the instantaneous

concentration in a cloud from a steady source varies considerably.
In part this is caused by the variation in angular lateral spread
which is greatest in unstable conditions, and least in stable
conditions as the following table (Clarke, 1979) shows.

Stability	Variation in Angular Lateral Spread In Degrees	
	at 100 metres distance	at 100 kilo metres distance
A	60	(20)
B	45	(20)
C	30	10
D	20	10
E	(15)	(5)
F	(10)	(5)

It would therefore be expected that the peak to mean concentration
ratio for a point down-wind of the source would be greatest in
unstable conditions and least in stable conditions. Little data
had been available until recently, but the tests at Maplin Sands
and China Lake have now provided some which can materially assist
in solving this problem. This is of greater importance in
establishing the maximum distance to the LFL under peak conditions,
when ignition might occur and cause a flash-back should an
ignition source be present. A word of caution lest it be assumed
even with such data that the probability of cloud ignition can
then be calculated. It cannot with any certainty, since there
would appear to be other factors involved when one considers the
great difficulty in obtaining ignition or burn back conditions in
the two recent test series.

2.3   Consequential Effects

Combustible Clouds

In addition to the requirement for ignition calculations to be
able to assess the peak to mean concentration ratio, one may have
a requirement to assess the combustible content of the cloud.
Assuming that the total mass release were known, one in theory
needs to know the mass of fuel between the UFL and LFL, and also
the mass above the UFL. But there is another aspect which is
often overlooked and which has a controlling influence over the
procedure of events. It is the available oxygen. At the LFL
there is usually adequate oxygen available to complete combustion,
but this is rarely the case at the UFL where there is usually
inadequate oxygen. For instance, if the cloud contained say 15%
v/v of fuel, then there is only about 17% v/v oxygen present -
for most fuels this is inadequate.

However the turbulence of combustion would serve to lift the
cloud into a fireball entraining air, and continuing the combustion.
But it is unlikely to be complete, and recent spectral examinations
of flames of gas escapes which have been ignited clearly indicate
incomplete combustion. In addition, few clouds are in fact
homogeneous. Those which have been subject to air entrainment
initially may exhibit a degree of uniform concentration, but as
the atmospheric diffusion processes continue, so the cloud
becomes more graded in composition, as in a gaussian distribution.
It is an area of great complexity and difficulty where further
quantified research data will be of great help. In the meantime
one must guard against over-conservatism due to failure to
identify all the constraints which exist in practice.

Acute Toxic Gases

The application of toxicity data for toxic gases to the gas
concentrations calculated by mathematical modelling has resulted
in a very wide range of predictions, some of which are hard to
justify in the face of historical data. Whilst there are now
known to be many intervening factors playing a part, usually in
mitigating the effects, there are nevertheless some major problems
requiring explanation.

One of these has affected the concentration of toxic gas as
predicted by a mathematical model. In recent years the models
used have been rapidly developed to include a better appreciation
of the contributing mechanisms and in particular the inclusion of
density. This term is now known to lead to a marginally lower
(centre line) concentration at a particular distance than would
have occurred in otherwise similar circumstances with a gas of
atmospheric density. With the latest models available it is now
desirable to know their range of accuracy in predicting concentration,
since this can have an even more dramatic effect on accuracy in
predicting fatalities in certain cases.

Secondly the problems of toxicological data are very far from
being resolved. It is most important that the limited extent of
this data is properly understood, since all too often accuracy is
claimed in such risk assessments when it is not warranted as a
result of this data.

Most of the human data available is sparse and old. Some of it
is not very precise, for instance in its interpretation of the
time factor. Only a small proportion of this human data refers
to lethal concentrations, so interpretation of probability of
fatality in humans is virtually impossible, yet it is used in
risk assessments.

The way it is introduced is by looking at test data on small
animals, from which a large mass of lethal toxicity data can be
obtained. Even so there is a fairly wide variation, perhaps of
up to an order of magnitude, in the results. The exponent of the
concentration term in the probit equation can be deduced from the
animal data perhaps better than the constant. The latter can in
fact only be fixed by improved human toxicity data.

To illustrate the significance of the probit form of toxicity
data and its relationship to the accuracy of gas dispersion
predictions, the following table shows the range in the value for
the probability of fatality if the concentration were to be twice
or half as big, ie a factor of 4 overall in concentration.

Revised Chlorine Probit Equation $Pr = 0.82 \ln (C^{2.75}t) - 11.4$

Assume a uniform exposure time of t = 10 minutes

Consider an error factor of 2 in estimation of concentration C in
gas dispersion model (Ratio high/low of 4).

Concentration C ppm	Probability of Fatality for			Ratio of Probability for 2C:0.5C
	0.5C	C	2.0C	
100	6×10^{-9}	2×10^{-5}	5×10^{-3}	9×10^{5}
300	7×10^{-4}	.05	.464	700
1000	.309	.858	.996	3

This illustrates quite clearly that even having improved the
accuracy of the gas dispersion modelling, there is a critical
magnifying effect when the probit is added to the overall
assessment, and this, together with residual uncertainty in the
probit equation itself due to inadequate data, makes any sound
assessment of risk to life from toxic gas emission impossible.
The example shows that at 100 ppm the probabilities are small
anyway, so the error is academic, but at 300 ppm the error
becomes very significant indeed.

It is therefore imperative that the probit equations are derived
from better data. The one used differs from that used by USCG
(1975) in the constants deduced from the same toxicological data.

Escape Factors

Marshall (1977) has presented a series of papers in which he
develops a mortality index M for the number of people killed per
ton of toxic liquefied gas released (gross quantity) for several
groups of gases and vessels. The mean value indicated, and even
the maximum value, within the Group are below many of the indices
which one could calculate from assessments of risk which have
been published in the literature. This is a clear indication,
even for the same scale of release, that there are some deficiencies
in the overall risk assessment models used, such as the omission
of an escape factor. Thus the extension of heavy gas dispersion
to risk assessment for toxic gases is a far more complex problem
than most risk assessors seem to realise.

3   DEALING WITH THE PROBLEM AREAS.

Several key problem areas have been identified. Even though there
may be major difficulties in dealing with them when using gas
dispersion modelling, there are several possibilities. Quite a
lot of information is available to those conducting risk assessments
provided they scan the world technical literature thoroughly, and
this can be vital in supporting the judgements or interpretations
which will need to be made. The use of this background data
sensibly is far better than some of the wild guesses which risk
assessors have sometimes resorted to, but there is clearly a need
to develop the understanding in these areas quickly. Indeed in
relative terms it is now fast becoming the crucial issue, and
will replace the basic development of heavy gas dispersion models
before very long. This is because in many of these areas the law
of diminishing returns applies, and it becomes more and more
costly (and less and less necessary) to refine the detail of
particular aspects of a model any further. The stage is about to
be reached where effort must be turned to those factors which
have potentially a major effect on predictions and about which
little is so far known quantitatively. On the international
scene there is growing awareness of this situation and future
research programmes are now considering these revised needs. For
the present however it is necessary to examine more closely what
treatment can be given in our modelling of these problem areas.

3.1   Meteorological Roughness

This parameter, as described earlier, is used to provide the
relationship between wind speed and height for the boundary
layers. The problem in its use arises when the cloud height is
no greater than say twice the height of the obstructions on the
ground causing the roughness. Thus over open fields, the use of
z_o does not really present a problem. However when the cloud

is about the same height as the ground obstructions, as it may be
when passing through an area with buildings or plant structures,
then meteorological roughness ceases to be a true parameter in
assessing the local dispersion of the cloud. On the other hand
z_o remains relevant to the atmospheric layers above the ground
obstructions, and this is why it still has significance when
dealing with long range pollution problems. It is advisable in
cases where z_o is large, ie of the order of 1m (corresponding
to say an urban area), to consider first what will be actually
happening to the air flow in and around the ground obstructions.

A second problem affecting the use of z_o is that of uniformity.
As will be seen from the study of the topography down-wind of
many actual accidents, there is often total lack of uniformity.
For example in a gas release from a chemical plant it may first
pass through plant structures, then around major buildings like
offices, workshops or laboratories before passing on over open
areas surrounding the site. Many mathematical models are unable
to accommodate changes in the value of z_o with distance, but
some of the more sophisticated models may have this ability. It
is certainly not possible with most simple gaussian models which
require a simple (averaged?) value to be used.

3.2   Ground Level Obstructions

There are many observations of low level clouds of gas or smoke
blowing towards buildings and other large ground level obstructions,
and of the effects these have on the movement of the cloud. In
general one finds that at very low wind velocities the cloud
tends to move (at higher velocity) through the open spaces
between these obstacles, but as the wind velocity rises some of
the cloud may get drawn over the structure as well. Down-wind of
these obstructions down wash, down draught and other eddies
appear, such that there is increased turbulence, and increased
turbulence improves the efficiency of the dispersion process. It
is therefore not surprising that the impact of such large ground
level obstructions on the cloud is to enhance its dispersion
compared with the open field situation. A limited amount of data
to support this is available, and more recently experimental work
in wind tunnels has clearly demonstrated this effect. In one
example the dispersion of a standard source of emission both over
open ground in one direction, and through a chemical plant at
approximately 90° is directly compared, with all other variables
remaining constant. Thus the problems associated with scale up
of wind tunnels are not significant, and the comparative results
of the two modes is of real relevance. The distance reached by a
particular concentration isopleth is significantly reduced in the
plant structure case compared with the open field case.

Consideration is now being given to ways of utilising wind tunnels to provide an estimate of the degree of difference which could be expected from different structural layouts. This is believed to be more directly useful than attempting to model each and every situation in a wind tunnel, since the individual differences may be smaller than those identified in the more general case.

The preliminary results so far do not appear to accord with notional changes in z_o, presumably since building density also has an effect as well as building size in establishing a value for z_o . The more buildings there are in a certain area, the greater the expected turbulence.

It is recommended that proper consideration be given to this problem rather than simple manipulation of z_o to "equate" to tabular reference values.

3.3   Ground Slope

The experiments at Porton (Picknett, 1978) included a few where the 40 m^3 heavy gas cloud was released over ground having a small uniform slope. Under low wind speed conditions, a heavy gas cloud was observed to travel down a slope against the wind, and at the higher wind speeds such a cloud was blown up a slope with the wind. There must be some balance between cloud density, wind vector and ground slope which indicates the equilibrium condition above or below which the cloud will advect.

The data available so far does not yet provide this limiting set of parameters, and further work is desirable in order to obtain it. A limited amount of data may be available from the analysis of certain historical acidents, and these should also be examined.

In the real situation, terrain often has a dominating influence on the dispersion of a gas cloud and hence this apsect is one of great importance, justifying more attention to procurement of the basic information.

3.4   Source Term

Recently a lot more detailed consideration has been given to the mechanism of release of heavy gases from vessels and pipework which have failed in various ways. It is now apparent to those who have to conduct risk assessments that some wider degree of interpretation of these mechanisms is required instead of one notional type (often one of the worst cases) which originally was all that would have been considered. One must ask for instance whether in past accidents every tank of liquefied gases has

ruptured and released <u>all</u> its contents quasi-instantaneously. The
answer of course is that they have not. There is thus a need to
assess a wider range of scenarios, applying appropriate probabilities
to each type together with a more representative description of
the release mechanism. Several recent papers (for example,
Blanken, 1980) have indicated how this can be approached, but from
a gas dispersion viewpoint, the major problem comes with attempting
to define the various source terms, especially those where significant
air entrainment takes place at this stage. It must be remembered
that any high pressure gas discharge, or high pressure flashing
liquid discharge causes a turbulent jet to be formed. There are
several methods available for calculating the air entrainment into
these jets. Until the velocity of the gases has fallen close to
that of the wind, most of this type of release cannot be used,
without adjustment, as the source term, since air entrainment
occurs into such jets <u>before</u> the release comes under the influence
of atmospheric dispersion processes.

4   SUMMARY AND CONCLUSIONS

4.1   The extension of research into the fundamental dispersion
        processes is reaching the point where the return in further
        effort is rapidly decreasing to the point where alternative
        research is more desirable.

4.2   Some further work is desirable over sloping ground to
        demonstrate the basic relationships which exist between
        ground slope, wind vector and gas density.

4.3   There is an urgent need for basic research into the
        effect of ground obstructions, their type, size, disposition,
        etc, on the dispersion of a gas cloud. This should be
        tackled in a more general way in order to provide basic
        guidelines and data of reasonable confidence. It should
        not provide excessive detail for each and every site since
        this effort is not justified, and such additional improvement
        in level of confidence is probably very small.

4.4   The use of existing toxicological data for risk assessment
        of toxic gas releases is currently very unsatisfactory
        and has an unacceptable level of confidence due to the
        uncertainties in the probit equations and the data from
        which they are derived. Further research into providing
        this data must precede any further work on such risk
        assessments.

4.5 Improvements are essential in the correct application of the source term to the gas dispersion models in order that they truly reflect the wider range of releases which are possible, and also include in the source term the effect of air entrainment into turbulent jets.

4.6 The extention of heavy gas dispersion into risk assessments for both flammable and toxic gases is far more complex than is usually realised, and requires greater emphasis on a more thoroughly reasoned model to interpret the effects in a more realistic manner.

REFERENCES

Blackmore, D.R., Herman, M N., Woodward, J.L., 1982, J Haz. Mat, 6. "Heavy Gas Dispersion Models".

Blanken, J.M., 1980, Ammonia Plant Safety 22 pp 25-34, "Behaviour of Ammonia in the event of a spillage".

Clarke, R.J., 1979, NRPB-R91, "A Model for Short and Medium Range Dispersion of Radionuclides Released to the Atmosphere".

Marshall, V.C., 1977, Chemical Engineer, Aug, pp 573-577 "How Lethal are Explosions and Toxic Escapes?"

Picknett, R.G., 1978, CDE Report Ptn IL 1154/78 "Field Experiments on the Behaviour of Dense Clouds"

USCG, 1975, "Vulnerability Model" AD/A-015245

EXAMPLES OF ANALYSES OF GAS CLOUD EXPLOSION HAZARDS

H.J. Nikodem, Battelle-Institut e.V., Frankfurt am Main

1. INTRODUCTION

In normal risk assessments of rather large engineered systems significant simplifications are made. In particular, for dispersion
calculations the ground surface is mostly assumed to be a flat plane
with a certain roughness. Specific topographical features are neglected because the models for rapid calculations of dispersion which include topography are not readily available. In addition, it is difficult to estimate the extent of the errors that are introduced by
simplifications without making detailed calculations.
In the case of risk analysis of a large technical system, some errors
from the dispersion calculations are acceptable in view of the uncertainties of the final results.
There are numerous examples in which it is difficult to gain any
confidence in the sequence of events which can be expected to happen
in a specific accident situation. Models give inaccurate results
since they do not take into account the variety of real events.

Two cases are presented in which attempts were made to predict
accident consequences. Dispersion of LPG vapor clouds were involved
in both cases. Maximum consequences were identified and described for
given vapor sources. The effects of specific topographies on the
dispersion were also considered. In one case it was not possible to
describe the expected outcome with confidence; in the second case,
this should be achieved.
Because of the specific topographies, it would have been desirable
to use a 3-dimensional model for the calculation of the dispersion
of the heavy gas. At the present time, however, there are no such
models available which are sufficiently verified. We think the use
of such models to describe complex phenomena is not advantageous
compared to more straightforward approaches unless all relevant
physical phenomena are properly incorporated and the model can be
verified with experiments covering a wide range of cloud sizes.

S. Hartwig (ed.), Heavy Gas and Risk Assessment - II, 305–310.
Copyright © 1983 by Battelle-Institut e.V., Frankfurt am Main, Germany.

Therefore, heavy gas dispersion was modelled by modified versions of
a simple model /1/ tailored for the investigation of specific situations. The model describes top entrainment of air into the cloud,
 edge velocity , and the drift of the total cloud in the windfield.
The vertical concentration profile is assumed to be Gaussian. The
windprofile was assumed to be logarithmic.

Case 1

For a given vapor source and a specific topography it should be
determined if it is possible that a gas-air-mixture with flammable
concentration can reach air intakes within a target area at a significant elevation above ground surface.
The sponsor of the study was mainly interested in a solution to his
problem regardless of the scientific approach. The worst vapor source
case which had to be considered was a rather rapid release of many
tens of tons of vapor. The main feature of topography showed a
significant step in the ground surface close to the source area
(fig. 1). The vapor had to flow over some obstacles (low buildings)
before it could reach the larger building containing the air intakes.

In modelling the dispersion, the following serious problems had to
be considered:
- if the vapor cloud is pushed over the step at high wind velocity,
  what is the cloud's height after the step (at low windspeed the
  vapor is released at a low rate from the buffer which is formed
  by the step)?
- what is the overall effect of large obstacles on the vapor cloud?
- what is the local behaviour of the vapour cloud in a turbulent
  windfield around an obstacle?

A continuous source version of the model /1/ was used. In modelling
the orography at a step it was assumed that a certain fraction of the
kinetic energy of the turbulence generated by the step is transferred into potential energy of the heavy gas cloud after passing
the step. In this way the height of the cloud is increased by a
certain amount. The obstacles following the step slow down the drift
velocity of the cloud. This causes an additional rise in cloud height
by increased turbulence and gravitational spreading is more enhanced
to the crosswind directions.

By using various arguments (from boundary layer physics and energy
conservation) and parametric studies (wind velocity, roughness
length, source configuration) with the modified models, a rough
estimate was made for the height of the free-field vapor cloud at
the location of the target area obstacle. The resulting Gaussian
vapor concentration profile is shown in fig. 2. It indicates that the
height of the flammable cloud would not reach the elevated target area.
However, it was not possible to rule out with certainty that quantities
of flammable heavy gas mixtures can be forced up to air intakes by
a statistical occurrences of strong turbulence since there is no model
to describe a turbulent flow of gas with inhomogenious density around

an obstacle.

We reported to the sponsor that because of the limited state of our
knowledge his question could not be answered. We also cannot conclude
with complete confidence that the flammable gas mixture may reach the
air intakes.

Case 2

For a given LPG release scenario the vapor cloud with the largest
explosive hazard should be described. It was assumed that more than
10 tons of LPG vaporize rapidly in a lane between two large engineered
structures (fig. 3).

The model /1/ was used to describe top entrainment and the one dimensional front velocity of the heavy vapor cloud as it spreads within a
lane towards the lane ends in both directions. Wind drift was super-
imposed for the case in which the wind blows along the lane direction.
Taking into account the back pressure of the cloud, we assume that
the outflowing vapor disappears after the cloud reaches an end of the
lane and creates a standing cloud front.

This is acceptable because it seems reasonable to assume that the
heavy gas outside the lane ends disperses rapidly. The velocity of
the outflowing gas was taken as the sum of the wind drift velocity of
the cloud and a gravitational spreading velocity. The influence of
turbulence on the dispersion generated by the large structures within
the lane was neglected.

The model predicts that the largest volume of gas mixture in the concentration range between the flammability limits, LFL and UFL, occurs
for high wind velocities and wind direction that is perpendicular to
the lane direction. This is reasonable considering the rapid air entrainment into the cloud through the top at high wind velocities when
the cloud is not pushed out of the lane by the wind.

The resulting vertical concentration profiles of the vapor cloud for
such a case are shown in fig. 4 for various points in time. The largest
height of the flammable vapor cloud is 5 m.

It was found that the flammable vapor cloud is highest when more than
half of the vapor has flown out of the lane driven only by gravitational forces.

In this case we noticed that the height of the flammable cloud and
the corresponding explosive hazard potential was extremely insensitive
to the amount of released vapor.

CONCLUSION

In one case, it was not possible to describe with confidence what can
be expected to happen; in the second case this could be achieved.

The first case underlines the need for a model that can handle density
effects and orography. We hope it will be developed within the near
future.

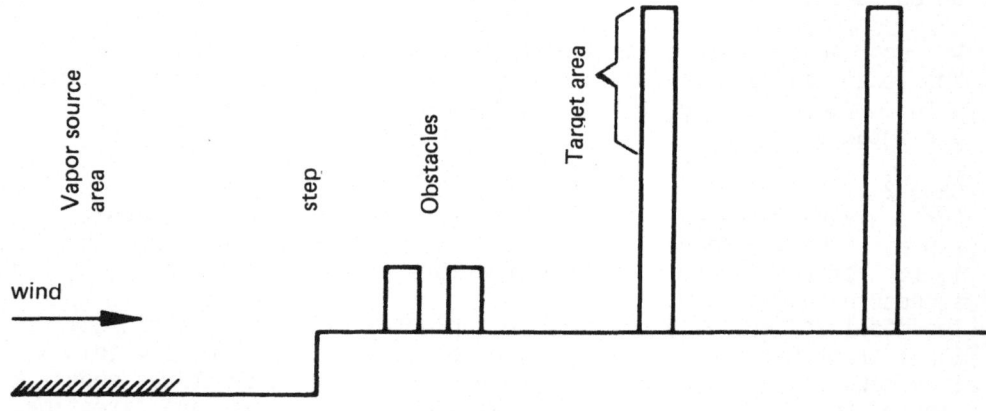

<u>Fig. 1:</u> Case 1 Topography, elevations are exaggerated

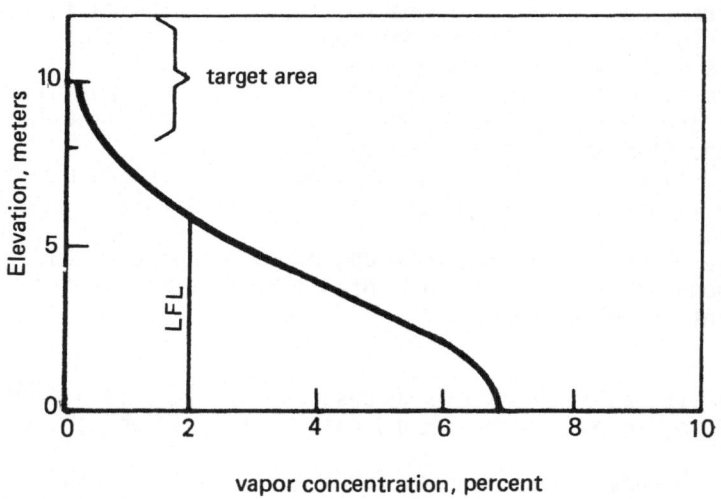

<u>Fig. 2:</u> Case 1, percent vapor concentration

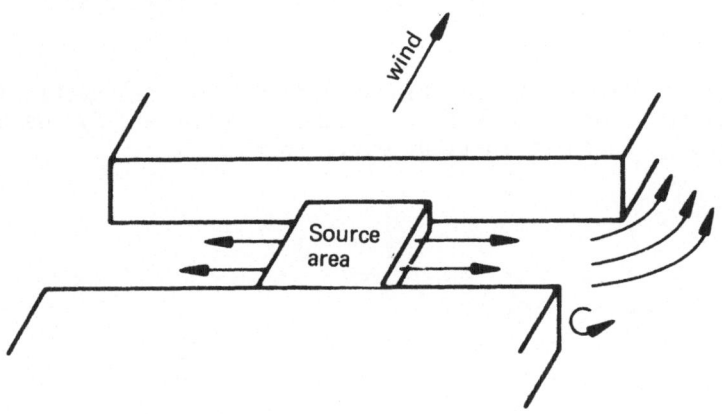

Fig. 3: Case 2 Topography

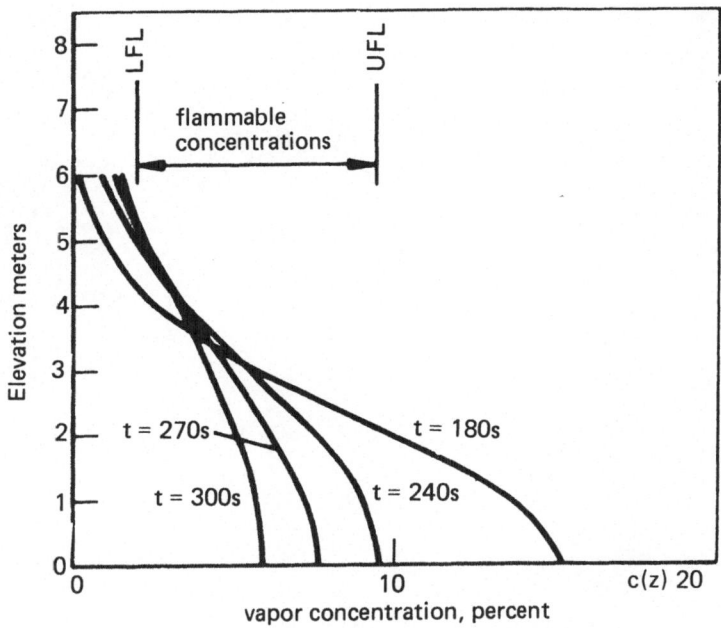

Fig. 4: Case 2, percent vapor concentrations in lane
between obstacles 180, 240, 270 and 300 seconds
after release of vapor

REFERENCES

/1/ Flothmann, D.; Nikodem, H.J.; "Ein Schwergasmodell mit stetigem
Übergang zwischen Gravitations- und Dispersionsphase"; Symposium
"Schwere Gase", Battelle-Institut e.V., 1979

LIST OF PARTICIPANTS

Abeln, P.P.J., Ing., Aramco Overseas Company,
        The Hague, Zestienhovensecade 278,
        3043 Rotterdam, NL

Aptroot, R., Ing., N.V. Nederlandse Gasunie,
        Laan Corpus den Hoorn 102, Groningen, NL

Atallah, S., Director, Gas Research Institute,
        8600 West Bryn Mawr Avenue, Chicako,
        IL 60631, USA

Bagerman, J.E., Ing., Provinciale Waterstaat,
        Eendrachtskade z.z. 2, 9726 Groningen CW, NL

Baybutt, P., Dr., Battelle Columbus Laboratories
        505 King Avenue, Columbus Ohio 43201, USA

Behr, H., Dipl.-Chem., Münchner Rückversicherungs-
        Gesellschaft, Königinstr. 107, 8000 München 40

Blackmore, D., Dr., Shell Research Ltd.,
        Thornton Research Centre, P.O. Box 1,
        Chester, CH1 35H, United Kingdom

Bol, J., Battelle-Institut, Sicherheitstechnologien
        Am Römerhof 35, 6000 Frankfurt/M. 90

Braig, A., Dr., Battelle-Institut, Angewandte
        Informatik, Am Römerhof 35, 6000 Frankfurt/M. 90

Campbell, P.G., Safety Manager, BP Chemicals Ltd.,
        Belgrave House, 76 Buckingham Palace Road,
        London SW1W OSU, GB

Carpenter, R.J., British Gas Corporation, Research
        and Development Div., Wharf Lane, Solihull,
        West Midlands B91 2 JW

Colénbrander, G.W., Koninklijke/Shell Laboratorium
        Badhuisweg 3, 1031 Amsterdam - Noord, NL

Dauwe, R.M., Dr., Dow Chemical (Nederland) BV
        P.O. Box 48, 4530 AA Terneuzen, NL

Dawson, R.P.M., BP International Ltd., BP Research
        Centre, Chertsey Road, Sunburry-on-Thames,
        Middlesex, TW16 7 LN, GB

Deaves, D.M., Dr., Atkins Research & Development,
  Woodcote Grove, Ashley Road, Epsom,
  Surrey, GB

Dewerdt, M., Gas de France, 33/35 Rue d Alsace,
  92531 Levallois-Perret, France

Diepold, W., Dr., Battelle-Institut, Sicherheitstechnologien, Am Römerhof 35, 6000 Frankfurt/M. 90

Drenckhahn, W., Dr., Kraftwerk Union AG, Abt. VRS 13,
  Hammerbacherstr   12+14, 8520 Erlangen

Duerlod, J., Ing., NSM
  Industrieweg 10, 4541 Swiskill, NL

Dunst, M., Dr., Meteorologisches Institut der
  Universität Hamburg, Bundesstraße 55,
  2000 Hamburg 13

Elsner, W., Mobil Oil AG, Steinstraße 5, 2000 Hamburg 1

Emblem, K., Sintef Div. 15, Aero-and Gas Section,
  N-7034 NTH-Trondheim, Norway

Englisch, W., Dr., Battelle-Institut, Laser u. Optik,
  Am Römerhof 35, 6000 Frankfurt/M. 90

Eyre, I.A., Dr., Shell Research Ltd., Thornton Research
  Centre, P.O. Box 1, Chester CHI 3 SH,
  Chesire, GB

Filipsen, M., Ministry of Social Affairs,
  P.O. Box 69, Voorburg 2270 MA, NL

Fischbach, D., Dr., Staatliches Gewerbeaufsichtsamt,
  Holzstraße 11 b, 6200 Wiesbaden

Fischbach, G., NUKEM GmbH., Postfach 110080,
  6550 Hanau 11

Fischer, F., Dipl.-Ing., BEB Gewerkschaften
  Brigitta und Elwerath, Betriebsführungs-GmbH.,
  Riethorst 12, 3000 Hannover 51

Fischer, I., Dipl.-Met., Meteorologisches Institut
  der Universität Hamburg, Bundesstraße 55,
  2000 Hamburg 13

Flothmann, D., Dr. Battelle-Institut, Sicherheitstechnologien, Am Römerhof 35, 6000 Frankfurt/M. 90

Förster, H., Dr., Physikalisch-Technische Bundesanstalt,
        Bundesallee 100, 3300 Braunschweig

Franke, G., Dipl.-Phys., Regierungspräsident in
        Darmstadt, Postfach 110740, 6100 Darmstadt

Fransen, H., Ajax Fire Protection
        Cruquiusweg 118 Postbus 4105, 1009 Amsterdam, NL

Freund, H.U., Dr., Battelle-Institut, Sicherheits-
        technologien, Am Römerhof 35, 6000 Frankfurt/M. 90

Friedel, L., Dr.-Ing., Hoechst AG, Techn. Prüfung,
        Postfach 80 03 20, 6230 Frankfurt/M. 80

Geiger, W., Dr., Battelle-Institut, Sicherheits-
        technologien, Am Römerhof 35, 6000 Frankfurt/M. 90

Giesbrecht, H., Dr., BASF AG, Abt. D-DET/ES
        Carl Bosch Str., 6700 Ludwigshafen/Rhein

Golz, C., Dipl.-Ing., Union Rheinische Braunkohlen
        Kraftstoff AG, Postfach 8, 5047 Wesseling

Gundelach, V., Dipl.-Phys., DECHEMA,
        Postfach 970146, 6000 Frankfurt/M.

Haan de, F.J., Ing., Netherlands Railway Ltd.,
        P.O. Box 2025, 3500 HA Utrecht, NL

Haeske, H., Dr., Geschäftsf. Vorstandsmitglied,
        Battelle-Institut, Am Römerhof 35,
        6000 Frankfurt/M. 90

Harris, N.C., I.C.I. Plc.,
        P.O. Box 13, Runcorn, Cheshire, WA7 4QfF, GB

Hartwig, S., Prof. Dr., Battelle-Institut, Sicherheits-
        technologien, Am Römerhof 35, 6000 Frankfurt/M. 90
        Universität Wuppertal, Gaußstr. 20,
        5600 Wuppertal-Elberfeld

Havens, J., Prof. Dr., University of Arkansas,
        227 Engineering Building, Fayetteville,
        Arkansas 72701, USA

Harst van der, L., Vice President, CER Corporation
        P.O. Box 15090 Las Vegas, Nevada 89114, USA

Heemst van, M.V., Ing., Akzo Engineering bv,
        P.O. Box 209, 6800 LV Arnhem, NL

Heike, Th., Dipl.-Ing., DVGW, Forschungsstelle
        Engler Bunte Institut,
        Richard Willstätter Allee 5, 7500 Karlsruhe 1

Heinrich, M., Dr., Technischer Überwachungsverein Nord-
        deutschland, Große Bahnstraße 31, 2000 Hamburg 54

Hempel, P., Deutsche Shell AG, Hohe Schaar Str. 36,
        2102 Hamburg 93

Herberg, G., Hoechst AG,
        Postfach 80 03 20, 6230 Frankfurt/M. 80

Hertel, F.C.J.K., Ing., Deo Boer B.V./Ajax Brandbeveiliging,
        Postbus 4105, 1009 AC Amsterdam, NL

Heudorfer, W., Dr., Battelle-Institut, Sicherheits-
        technologien, Am Römerhof 35, 6000 Frankfurt/M. 90

Hirst, W.J.S., Dr., Shell Research Ltd., Thornton Research
        Centre, P.O. Box 1 Chester CHI 3SH, Cheshire, GB

Hofmann, J., Dr., Battelle-Institut, Sicherheits-
        technologien, Am Römerhof 35, 6000 Frankfurt/M. 90

Hogan, W.J., Dr., Lawrence Livermore National Laboratory,
        P.O. Box 808, Livermore, California 94550, USA

Hogh, M.S., Dr., The British Petroleum Company Ltd.,
        BP International Ltd., Britannic House,
        Moore Lane, London EC2Y 9BU, GB

Jaeschke, M., Dr., Ruhrgas AG, Huttroppstr. 60,
        4300 Essen 1

Jagger, S.F., United Kingdom Atomic Energy Authority,
        Wigshaw Lane, Warrington, GB

Johann, W., Dipl.-Ing., Ruhrgas LNG, Flüssigerdgas Service
        GmbH., Huttroppstr. 60, 4300 Essen 1

Kirsch, J., Dr., Battelle-Institut, Sicherheits-
        technologien, Am Römerhof 35, 6000 Frankfurt/M. 90

Köppner, M., Dr., Chemische Werke Hüls,
        Postfach 1320, 4370 Marl

Köster, H., Dr., Rheinische Olefinwerke GmbH.,
        Postfach 31, 5047 Wesseling

Korjuslommi, E., Ing., Neste Oy,
        Porvoo Works, 06850 Kulloo, Finland

Krux, P., Dipl.-Ing., UHDE GmbH.,
        Deggingstraße 10-12, 4600 Dortmund 1

Kvaal, E., Det norske Veritas,
        P.O. Box 115A, Oslo 2, Norway

Lämmerzahl, D., Dipl.-Ing., Salzgitter AG
        Postfach 41 11 29, 3320 Salzgitter 41

Lautkaski, R., Technical Research Centre of Finland,
        Nuclear Engineering Laboratory,
        P.O. Box 169, SF-00181 Helsinki, Finland

Leuckel, W., Prof. Dr.-Ing., Lehrstuhl für Feuerungs-
        technik der Universität Karlsruhe,
        Richard Willstätter Allee 5, 7500 Karlsruhe 1

Lohmeyer, A., Dr.-Ing., Universität Karlsruhe,
        Kaiserstraße 12, 7500 Karlsruhe

Maeder, R., Dipl.-Ing., Bayer AG, Ing.-Wiss.Abt.,
        5090 Leverkusen

Mameren van, A.C., Ing., Bureau of Industrial Safety TNO,
        Lange Kleiweg 117, 2280 AA Ryswyk, NL

Mass, J., Dipl.-Ing., Drägerwerk AG,
        Moislinger Allee 53/55, 2400 Lübeck

Masznyik, Dipl.-Ing., Deutsche Flüssigerdgas Terminal
        GmbH., Frankenstr. 336, 4300 Essen 1

Menashe, J., Dr., Insurance Technical Bureau,
        Terminal House, 52 Grosvenor Gardens,
        London, SW1W OAU, GB

Medrow, W., Dipl.-Met., Rheinisch-Westf. Techn. Über-
        wachungsverein e.V., Postfach 103261, 4300 Essen 1

Morlock, G., Dipl,-Phys., Gesellschaft f. Reaktorsicher-
        heit, Schwertnergasse 1, 5000 Köln 1

Müller, Dr.-Ing., Deutsche Shell AG, Raffinerie Godorf,
        Godorfer Hauptstr., 5000 Köln 50

Munday, G., Dr., The Insurance Technical Bureau,
        Terminal House, 52 Grosvenor Gardens,
        London SW1W OAU, GB

Nauta, T., Ing., Aramco Overseas Company,
        Laan van Meerdervoort 55, 2517 AG The Hague, NL

Neuhoff, S., Dipl.-Ing., Berufsfeuerwehr Köln,
        Mommsenstraße 10, 5000 Köln 41

Nikodem, H.J., Dr., Battelle-Institut, Sicherheits-
        technologien, Am Römerhof 35, 6000 Frankfurt/M. 90

Nyren, K., Försvarets Forsknings Anst., Stockholm,
        Schweden

Noha, K., Dipl.-Ing., Hoechst AG, Sicherheitsüberwachung
        Postfach 80 03 20, 6230 Frankfurt/M. 80

Opschoor, G., Ing., Prins Maurits Lab., TNO,
        Postbus 45, 2280 AA Rijswijk, NL

Pankrath, J., Dr., Umweltbundesamt,
        Bismarckplatz 1, 1000 Berlin 33

Peiter, K.H., Dipl.-Ing., Lurgi Kohle- und Mineralöl-
        technik GmbH., Bockenheimer Landstr. 42,
        6000 Frankfurt/M.

Pikaar, M.J., Dr., Shell Internationale Petroleum,
        P.O. Box 162, The Hague, NL

Pilz, V., Dr., Bayer AG, Ing.-Wiss. Abt.,
        5090 Leverkusen

Puttock, J.S., Dr., Shell Research Ltd., Thornton Research
        Centre, P.O. Box 1, Chester, CHI 3SH, GB

Rhoads, R.E., Battelle Pacific Northwest Laboratory,
        P.O. Box 999, Richland, WA 99352, USA

Riethmuller, M.L., Prof. Dr., von Karman Institute for
        Fluid Dynamics, Chausseé der Waterloo 72,
        1640 Rhode-St-Genese, Belgium

Rinnan, A., Ing., Norsk Hydro,
        P.O. Box 2594, Oslo, Norwegen

Rippen, G., Dr., Battelle-Institut, Chemische Analytik,
        Am Römerhof 35, 6000 Frankfurt/M. 90

Roulet, Elf France, Tour Gan, Cedex 13 Courbevoie -
        La Defense

Rozsondai, Z., Dr.-Ing., Rheinisch-Westfälischer
          Technischer Überwachungsverein e.V.,
          Postfach 103261, 4300 Essen 1

Rulkens, P.F.M., DSM Central Laboratory
          P.O. Box 18, 6160 MD Geleen, NL

Sandstede, G., Dr., Battelle-Institut, Direktor Forschung
          und Technik, Am Römerhof 35, 6000 Frankfurt/M. 90

Schecker, H.-G., Prof. Dr., Universität Dortmund,
          Postfach 500500, 4600 Dortmund 50

Schenk, H., Dipl.-Ing., Technischer Überwachungsverein
          Rheinland, TÜV,
          Postfach 101750, 5000 Köln 1

Schildknecht, M., Dr., Battelle-Institut, Sicherheits-
          technologien, Am Römerhof 35, 6000 Frankfurt/M. 90

Schimmel, G., Dr., Direktor Auftragswesen,
          Battelle-Institut, Am Römerhof 35,
          6000 Frankfurt/M. 90

Schlüter, M., Dipl.-Ing., BEB Gewerkschaften Brigitta und
          Elwerath Betriebsführungsgesellschaft mbH,
          Riethorst 12, 3000 Hannover 51

Schnatz, G., Dipl.-Met., Battelle-Institut, Sicherheits-
          technologien, Am Römerhof 35, 6000 Frankfurt/M. 90

Schneider, W., Battelle-Institut, Techn. Wirt. System-
          planung, Am Römerhof 35, 6000 Frankfurt/M. 90

Schreurs, P., Ing., Kul - Cit
          De Croylaan 2, B-3030 Leuven, Belgium

Schrödter, W., Dr.-Ing., Bundesanstalt für Materialprüfung,
          Unter den Eichen 87, 1000 Berlin 45

Schulz, N., Dr.-Ing., Bayer AG, 5090 Leverkusen

Seifert, H., Dr., BASF AG, Abt. D-DET/ES,
          6700 Ludwigshafen/Rhein

Singh, S., Dr., National Maritime Institute,
          Teddington, Middlesex, TW11 OLW, GB

Sluman, T., Ing., Shell Nederland, Raffinaderij B.V.,
          P.O. Box 7000, 3000 HA Rotterdam, NL

Smiet, L.G., Ing., Shell Nederland Raffinaderij B.V.,
        P.O. Box 7000, 3000 HA Rotterdam, NL

Sonnberg, M., Dipl.-Ing., ICI,
        Steinstraße 26, 2940 Wilhelmshaven

Staginnus, B., Dr., Bundesamt für Wehrtechnik und
        Beschaffung, BWB, Postfach 7360, 5400 Koblenz

Stammler, M., Dr., Battelle-Institut, Sicherheits-
        technologien, Am Römerhof 35, 6000 Frankfurt/M. 90

Steen, H., Dr., Physikalisch Technische Bundesanstalt,
        Bundesallee 100, 3300 Braunschweig

Stock, M., Dr., Battelle-Institut, Sicherheits-
        technologien, Am Römerhof 35, 6000 Frankfurt/M. 90

Süssenberger, J., Dipl.-Phys., Gesellschaft für Reaktor-
        sicherheit, Schwertnergasse 1, 5000 Köln 1

Tauschek, H., Ing., Gelsenberg AG,
        Remstedterstraße 48, 2000 Hamburg

Thevenard, P., Ing., GAZ DE FRANCE,
        33/35 Rue d Alsace, 92531 Levallois-Perret, France

Thon, H., Dipl.-Phys., Technischer Überwachungsverein
        Rheinland, TÜV, Postfach 101750, 5000 Köln 1

Treissl, K., Ing., LGA Gastechnik GmbH.,
        Postfach 604, Bonnerstraße 10, 5480 Remagen 6

Valckenaers, M., Ing., Office of Industrial Promotion,
        Square de Meeus 26, 1040 Bruxelles, Belgien

Willicks, W., Dr., Battelle-Institut, Prod. u. Verf. Entw.,
        Am Römerhof 35, 6000 Frankfurt/M. 90

Waite, P.J., Cremer & Warner,
        Buckingham Palace Road, London SWIW 9SQ, GB

Wiekema, B.J., Department Industrial Safety TNO,
        P.O. Box 342, 7300 AH Apeldoorn, NL

Wild, G., Ing., Statens Sprengstoffonspeksjon,
        P.O. Box 355, N-3101 Tønsberg Norway

INDEX